CHEMICAL ENGINEERING METHODS AND TECHNOLOGY

RECENT PROGRESS IN CHEMISTRY AND CHEMICAL ENGINEERING RESEARCH

CHEMICAL ENGINEERING METHODS AND TECHNOLOGY

Additional books in this series can be found on Nova's website under the Series tab.

Additional E-books in this series can be found on Nova's website under the E-books tab.

CHEMICAL ENGINEERING METHODS AND TECHNOLOGY

RECENT PROGRESS IN CHEMISTRY AND CHEMICAL ENGINEERING RESEARCH

A.K. HAGHI
EDITOR

Nova Science Publishers, Inc.
New York

For permission to use material from this book please contact us:
Telephone 631-231-7269; Fax 631-231-8175
Web Site: http://www.novapublishers.com

LIBRARY OF CONGRESS CATALOGING-IN-PUBLICATION DATA

Recent progress in chemistry and chemical engineering research / editor, A.K. Haghi.
p. cm.
Includes index.
ISBN 978-1-61668-501-0 (hardcover)
1. Chemical engineering. I. Haghi, A. K.
TP155.R374 2009
660--dc22

2010036442

Published by Nova Science Publishers, Inc. ✦ New York

CONTENTS

PREFACE

As chemistry becomes more and more amenable to mathematically rigorous study, it is likely that chemistry will also become an alert and demanding consumer of new mathematical results. From theoretical chemistry and quantum chemistry, to applied fields such as molecular modeling, drug design, molecular engineering, and the development of supramolecular structures, mathematical chemistry is an important discipline providing both explanations and predictions. This book has an important role in advancing chemistry to an era of detailed understanding of molecules and reactions.

Chapter 1 - Clothing manufacture is the last link in a long and complicated process by which two-dimensional fabric is transforms into three-dimensional garments by sewing the fabric pieces together. However, whatever the quality of the fabric, if the sewing operation is not performed appropriately, the resultant garment will be second-rate. Hence the sewing operation is most important in clothing manufacture and the sewing threads play a very important role in this operation. Generally, sewing threads are manufactured from either natural or manmade fibres in either staple or filament form. Sewing threads are primarily used in garments, upholstery, air-supported fabric structures and geotextiles to join different components by forming a seam. Different kinds of sewing threads are manufactured depending upon their specific field of application. A number of factors should be considered when choosing the sewing thread: to achieve good sewing performance, sewing thread must be chosen according to the fabric characteristics and end-use of the material. In this paper a glimpse on processing and properties of sewing threads is given.

Chapter 2 - Various methods have been developed to solve transient flow in pipes. These ranges have been formed from approximate equations to numerical solutions of the nonlinear Navier-Stokes equations. Hydraulic transient flow is also known as unsteady fluid flow. During a transient analysis, the fluid and system boundaries can be either elastic or inelastic: a) Elastic theory, describes unsteady flow of a compressible liquid in an elastic system (e.g., where pipes can expand and contract). b) Rigid-column theory, describes unsteady flow of an incompressible liquid in a rigid system. It is only applicable to slower transient phenomena. Both branches of transient theory stem from the same governing equations. In this work a computational method was used for prediction of water transmission failure.

Chapter 3 - In this work pure Oxygen was introduced in wastewater by a 40 *lit* pure *O2* container. The main objective of this research was to decrease *BOD5* amount equal to 80-90 percent. This paper presents the computational performance of numerical methods for modeling of pure oxygen diffusion in wastewater transient flow.

Chapter 4 - Three kinds of polypropylene resins with different melt flow index (MFI) values were melt-spun at different spinning temperatures and take-up speeds. The extrusion rate was set to result in fibers with similar linear density. Thus the effects of fiber diameter on the mechanical properties were omitted. The effects of different spinning conditions on the physical (birefringence, density and shrinkage) and mechanical properties (tenacity, modulus and elongation at break) of the fibers were investigated. It was shown that increasing spinning temperature caused an increase or a decrease in the properties of the fibers due to different effects of temperature on the melt viscosity, mobility of polymer chains, chains orientation and nucleation and growth rate of crystallites. Moreover, it was found that the mechanical properties of the fiber is mainly controlled by the characteristics of the starting polymer and variation of the spinning conditions could only adjust these properties in the limits set by the initial polymer properties.

Chapter 5 - In this study a non-conventional electrospinning technique was designed for production of high bulky polyacrylonitrile (PAN) nanofiber mats. Optimum nanofiber mats are achieved with 15 wt% solution of PAN in DMF. Such mats result in a bulk porosity which is as high as 99.9 and density as low as $0.84x10^{-3}$ g/cm^2. Effect of nanofiber mats porosity on the air permeability and moisture transfer was investigated. Based on the results, high bulky nanofiber mats possess high moisture transfer. Experimental data reveals that with a slightly decrease in the bulk porosity, air permeability decreased noticeably, while moisture transfer changed slightly.

Chapter 6 - Polypropylene (PP)/Montmorillonite (MMT) nanoclay based composite was prepared by melt compounding with maleic anhydride grafted polypropylene (MA-g-PP) in a twin screw extruder and an internal mixer. Test specimens were prepared by compression molding. Mechanical properties such as tensile modulus, tensile strength and maximum percent strain were measured. The clay dispersion was investigated using wide angle X-ray diffraction and scanning electron microscopy. The XRD patterns and SEM images show that introducing the proper compatibilizing agent ratio and suitable processing condition like a twin screw extruder decrease the immiscible clay component in the matrix. However, the increase of tensile strength is not achieved.

Chapter 7 - Polypropylene nanofibres containing silver nanoparticles were successfully prepared via melt-electrospinning process. The electrospun nanofibres were characterized using SEM and XRD. AATCC 100 test, a quantitative method, was employed to evaluate the antibacterial properties of samples. The results of the intensity pattern of X-ray diffraction indicated that the crystallinity of the samples containing silver nanoparticles was slightly less than those of pure polypropylene nanofibres. In order to determine fibres diameter and diameter of the silver nanoparticles, two sets of SEM micrographs along and perpendicular to fiber axis were taken. It was found that while fibre diameter lies in the range of 110 to 120 nm. The diameter of the imbedded silver nanoparticles in the nanofibres was 40 nm. It was also noticed that some aggregation of the silver nanoparticles had occurred. The results of antibacterial evaluation showed that the nanofibres, which contained silver nanoparticles exhibited excellent antibacterial properties.

Chapter 8 - One limitation of electrospun scaffolds fabricated with synthetic polymers such as PCL is low cell affinity to them due to low hydrophilicity and lack of surface cell recognition sites. To overcome this drawback of the synthetic materials, natural polymers have been widely used to modify the synthetic material to improve the cell adhesion properties.

In this study polycaprolactone (PCL) was blended with gelatin for improvement the cell adhesion and biocompatibility of PCL nanofibrous scaffolds. Fiber diameter and hydrophilicity of PCL and PCL/gelatin nanofibrous scaffolds were evaluated. *In vitro* cell culture study was carried out using nerve stem cells (NSCs). Proliferation and morphology of NSCs on PCL and PCL/gelatin nanofibrous scaffolds were investigated.

Fiber diameter was found to decrease and hydrophilicity was found to improve with incorporation of gelatin in electrospinning solution. The results of MTS assay and SEM micrographs showed that PCL/gelatin nanofibrous scaffolds serve as better substrate for NSCs compared to PCL nanofibrous scaffolds.

Chapter 9 - Polymers that exhibit high electrical conductivity have successfully been synthesized in the last few decades. Polypyrrole (PPy) has been one of the most extensively studied conductive polymers. Polypyrrol was deposited on the Poly vinyl alcohol (PVA) nanofibers during electrospinning process. Polymerization of the PPy was performed in the vapor phase at various conditions such as oxidant type and oxidant content in the electrospinning solution. For this purpose, a chamber was designed and electrospinning was done in this chamber. Possibility of the chemical deposition of PPy on the surface of the electrospun PVA nanofibers using ammonium persulfate as oxidant was shown. Fiber formation and morphology of the coated nanofibers were studied using a scanning electron microscope (SEM) and conductivity of the nanowebs was determined. The electrical conductivity of the coated nanofiber mats depends on the coating conditions and increases at PVA/oxidant weight ratio of 3/1. Measurement of the electrical conductivity of the coated mats shows electrical conductivity in the range of 10^{-3} to 1 S/cm depending on the chemical coating conditions.

Chapter 10 - This chapter portrays the phenomenal development that has taken place in the field of catalyst science. It attempts to outline various catalyst preparation methods, characterization techniques, catalyst deactivation and regeneration. This chapter further delineates about the strategies for the kinetic testing of industrial catalysts (e.g. SSITKA (steady-state-isotopic-transient-kinetic-analysis and TAP technique). It explains the interaction of inorganic compounds with oxides and impact of drying of dying on supported catalyst. The salient feature of this chapter is the selection of best transition metal for the reaction using modern density functional theory (DFT) utilizing a d-band model to explain it, Electric field effect, BEP (Brønsted–Evans–Polanyi) relation and volcano curve in heterogeneous catalysis. The details regarding SMSI (strong metal support interaction) and NEMCA for the modification of catalytic activity and selectivity embodied with the future of catalysts. Within this framework, catalysis will still play a leading role in technical and economical development in the 21st century.

Chapter 11 - The most important challenge for chemists of 21st century is to combat against ever increasing environmental pollution. To have a neat, clean, healthy and green environment, there is an urgent desire to search such an approach, which may be applicable at room temperature, safe in handling economic, ecofriendly, and above all, the main requirement it itself is not harmful any more to environment in any manner. In the last two-three decades, photocatalysis has attracted the attention of scientific community all over the world due to its multifarious applications in various fields like energy conversion, environment, technology, waste water treatment, self cleaning process, decomposition of offensive atmospheric odors as well as toxins, the fixation of CO_2 and the decomposition of NO_X and chlorofluorocarbons on global scale.

Chapter 12 - The demand for fresh or mildly preserved fish products is on rise globally. Chilling is the best method to preserve the fish in fresh condition. However the limited shelf life of chilled products has prompted the industry to search for new and more advanced preservation methods. Modified atmosphere packaging is one such advanced preservation technique which considerably extends the shelf life of fish products. This involves the replacement of the air surrounding the product with desired gases mainly carbon dioxide, oxygen and nitrogen depending on the type of fish products. By applying this technique shelf life of fish products can be doubled which helps in improved distribution of this nutrient rich perishable commodity to distant places.

Chapter 13 - Some vegetables and dairy products have nutritional properties, but its consumption is somewhat limited in part of the population due to some digestion problems or intolerances it provokes. This chapter focuses on foods containing high level of some oligosaccharides and disaccharides, and the importance of α-galactosidases and β-galactosidases to improve health and quality of these products. Several studies have shown the production and the application of pure α-galactosidases to reduce undigested components naturally present in foods to minimize digestion problems. β-galactosidases are another group of galactosidases also important in food processing mainly in dairy-products processing to achieved lower lactose content to people who has lactose intolerance. Researches of β-galactosidases production and purification are being done to apply in new dairy-products. The development of new food products containing healthy benefits and which minimized digestive problems are world-wide consumption trends.

In: Recent Progress in Chemistry...
Editor: A. K. Haghi, pp. 1-21
ISBN: 978-1-61668-501-0

Chapter 1

DETAILS ON PROCESSING AND PROPERTIES OF SEWING THREADS

Akshay Kumar* and M. Subramanian Senthil Kannan
[1]Department of Textile Technology, Indian Institute of Technology, New Delhi -110016, India
[2]Department of Textile Technology, PSG College of Technology, Coimbatore,Tamilnadu, India

ABSTRACT

Clothing manufacture is the last link in a long and complicated process by which two-dimensional fabric is transforms into three-dimensional garments by sewing the fabric pieces together. However, whatever the quality of the fabric, if the sewing operation is not performed appropriately, the resultant garment will be second-rate. Hence the sewing operation is most important in clothing manufacture and the sewing threads play a very important role in this operation. Generally, sewing threads are manufactured from either natural or manmade fibres in either staple or filament form. Sewing threads are primarily used in garments, upholstery, air-supported fabric structures and geotextiles to join different components by forming a seam. Different kinds of sewing threads are manufactured depending upon their specific field of application. A number of factors should be considered when choosing the sewing thread: to achieve good sewing performance, sewing thread must be chosen according to the fabric characteristics and end-use of the material. In this paper a glimpse on processing and properties of sewing threads is given.

1. INTRODUCTION

According to the definition given by ASTM, sewing thread is a flexible small diameter yarn or strand usually treated with a surface coating, lubricant or both, intended to be used to stitch one or more pieces of material or an object to a material It may be defined as smooth,

* e-mail: akshay.sardana@gmail.com

evenly spun, hard-twisted ply yarn, treated by a special finishing process to make it resistant to stresses in its passage through the eye of a needle and through material involved in seaming and stitching operations [1].

Sewing threads are used in garments, upholstery, air-supported fabric structures and geotextiles to join different components by forming a seam. The primary function of a seam is to provide uniform stress transfer from one piece of fabric to another, thus preserving the overall integrity of the fabric assembly.

Seam can be formed by the following techniques:

- Mechanical: stapling, sewing.
- Physical: welding or heat-setting.
- Chemical: by means of resins.[2]

The formation of seams by physical and chemical methods is restricted to a few specialized applications, as these processes tend to alter certain properties of the textile material. Among mechanical sewing techniques, sewing maintains its prevailing position by virtue of its simplicity, sophisticated and economical production methods and the controllable elasticity of the seam produced.

2. Historical Background

Sewing thread is considered to be the first textile material excavated. Archaeologists concluded that caveman made the first sewing threads some 25,000 years ago. Caveman or jungle-man prepared them by rubbing and twisting together either fur fibres from skin of animals or bast fibres. These were single twisted yarns, which were unbalanced and snarly, making stitching difficult. The difficulty was overcome by plying single threads to balance torque. Out of the natural fibres, cotton and silk were extensively used for stitching purposes. Until the middle of 19th century, all the threads were used for hand sewing. However, Isaac Singer, who invented sewing machine in 1850, brought about a considerable change in the field. James and Patric Clark, who were engaged in manufacture of loom, produced Cotton thread in 1812 in Paisley. The technology of Clark was further developed to a great extent by James Coat and became leading sewing thread manufacturer as M/s. J and P Coats of Scotland.

In India, Calico Mills commissioned India's first sewing thread plant in 1922 and introduced 100% polyester sewing thread in 1968. During pre-independence period M/s. J and P Coats were manufacturing most of the sewing threads in India for a long period. [2] Then M/s Madura Coats (in collaboration with M/s. J and P Coats) and M/s Modi Threads were the leading sewing thread manufacturers in India. At present M/s. Madura and M/s. Vardhman are the two leading sewing thread manufacturers of India.

3. Different Types of Sewing Threads

Usually, sewing threads are manufactured from either natural or manmade fibres in either staple or filament form [3]. A broad classification of different types of sewing threads is given below:

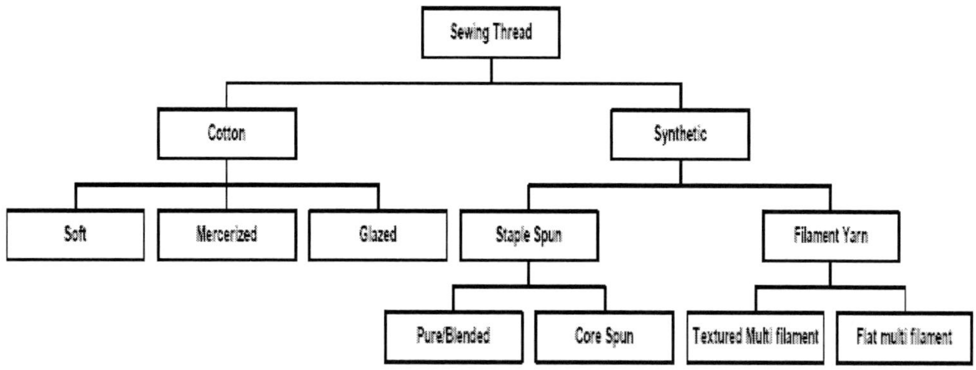

1. Cotton Threads: Cotton is still used for sewing thread. High quality fibres are used for the manufacture of sewing threads. The very good resistance to heat of cotton is responsible for its high duty performance. Cotton threads are however susceptible to acid damage, mildew and certain forms of bacterial attack, and have relatively low extensibility and abrasion resistance compared to synthetic threads.

The spinning process including combing provides yarns of highest strength and quality. These singles yarn generally have an S- twist direction and are next twisted together in the Z-direction, the twist being balanced to eliminate snarling. In case of cotton sewing threads, wet folding is usually used on ring system to lay-in the surface fibres and thereby obtain smoother, stronger and more compact yarn than could be obtained by dry processing; the single cotton yarns are passed are passed over the surface of rotating conditioning roller on folding machine before they are combined together with twist insertion. Cabled yarns may be used with various twist combinations. Frequently three two-fold yarns are cabled together with Z cabling twist following S spinning twist or vice-versa, with folding twist approximate 1.5 x spinning twist, cabling twist about 1.3 x spinning twist if a balanced yarn is required. Spinning twist has only a minor influence on the cabled yarn properties; the cabling twist has a major influence on cabled yarn strength, extension, torque and yarn length contraction during cabling. The proportion of cabled yarn used in sewing has reduced largely because of the higher cost of production and because strong folded synthetic yarns are now available; nevertheless cabled yarns may still be required for applications such as buttonholes and decorative topstitching [4].

A medium thickness (size 50) is available in a wide range of colors, and is used for sewing on light and medium weight cottons, rayons and linens. Cotton thread is usually mercerized a finishing process that makes it smooth and lustrous, also helping it to take dye better.

The highest quality cotton threads are mercerized to make them lustrous and to improve strength. The alkali treatment is done after twisting and before dyeing.

Some cotton yarns are supplied as glace or polished sewing threads; these are often used when severe condition is encountered during sewing and in use such as with heavy fabrics; and for decorative stitching. The yarn has a mixture of size and lubricants added to it before being polished by high speed rotating brushes while under tension. The surface finish is not permanent and is removed by subsequent wet treatments.

Process for 2, 3 and 4 cord sewing threads (ply yarn):

Ring Frame – steaming- cone winding- cheese winding- doubling (wet) – cheese winding – steaming – singing – reeling (if necessary) – wet processing (bleaching/ mercerizing/ dyeing) – rewinding (hank to cheese) – spooling (finish application) – labeling – packing.

Process for Embroidery Threads:

Final twisted yarn (grey yarn) – cone winding (never on hanks) – ball warping/ singing – warp mercerizing – quilling (operation in which the mercerized warp is separated into individual threads) – rewinding on perforated stainless steel cones – package bleaching – package dyeing (vat, reactive, naphthol, i.e. fast colours- never direct dyes) – rewinding on spools (sewing finish application, which gives a protective surface coating to the thread that minimizes thread or needle friction and hence reduces thread breakage during sewing operations. Generally, yarns here pass through two/three waxing units or silicone finish in liquid form) – labeling – packing.

2. Spun Synthetic Threads:

Modern high-speed sewing machines, imposing very high strains on the threads, require sewing thread of high toughness for satisfactory performance. Synthetic fibres, with very high strength and durability, have therefore become popular.

However, the thermoplastic nature of these fibres makes them susceptible to change in their properties when they heated either during sewing or in subsequent processing or use. Stabilization treatment at suitable temperature and stretch during manufacture is therefore essential. To improve the threads gliding and cooling properties, special finishes are also used.

The fibres used for these threads are normally polyester with a staple length similar to that of cotton fibres so they can be spun and twisted on the same machinery. Polyester yarn and multifilament threads have almost replaced cotton-sewing threads in the market [5]. The main reason for the increasing demand for polyester threads is their higher breaking force, low degree of shrinkage while washing, and good wearing properties in comparison to cotton threads.

Polyester sewing threads have one negative feature: at high sewing speeds and high friction with the metal parts of a sewing machine, these threads heat up and start melting at ~250°C. Due to their high strength, good chemical properties, acceptable elasticity and good dye fastness, polyester multifilament yarns are considered a perfect raw material for sewing threads.

Due to its even surface and circular cross-section, polyester is more lustrous than natural threads. Synthetic yarns are also produced with fibres of long staple length. Special machines are used for processing and generally higher strength is achieved.

The essential differences between cotton and synthetic threads are illustrated in Table 1.

Table 1.

Cotton threads	Synthetic threads
Good strength	High strength
Low extensibility	High, adjustable extensibility
Low laundry shrinkage	Very low laundry shrinkage
Low dry-heat shrinkage	Very low dry-heat shrinkage
Do not melt	Melt
Good abrasion resistance	High abrasion resistance

3. Core Spun Threads:

To achieve the optimum strength to fineness ratio of continuous filament threads together with the sewing performance and surface characteristics of spun fibre threads, core spun threads are produced in two main versions. In one version, polyester filament threads are covered and protected by cotton fibres; strength provided by the core and the heat resistance of the sheath permitting high sewing speed.

In the other, known as poly. / poly. polyester filament threads are covered by polyester fibres. The filaments are brought together with the fibres in the yarn spinning operation, these composite yarns being then twisted to form 2, 3 or 4 ply threads.

As with all synthetic threads core spun threads require a setting and stabilizing process and special dyeing technique is needed to obtain the same shade on both core and sheath where cotton is used. Because of their cotton cover, it is possible to produce glace core spun threads by polishing in a similar way to the 100% cotton threads. Mercerized core spun threads are also very popular.

4. Silk Threads:

Although they are not widely used, some sewing threads are made from spun silk. An important feature of such yarns is the high extent at break of about 20%, along with a high breaking load. Some continuous filament silk yarns are used for buttonholes and decorative stitching because of their attractive appearance and smoothness.

5. Monofilament Threads:

Monofilament is not a conventional thread, as it comprises only a single filament, suitably lubricated. Monofilaments are produced by melt spinning (as are continuous multifilament yarns), but the spinneret has only one aperture which is much larger than for conventional yarns.

6. Multifilament Threads:

Plied or corded threads produced from 100% synthetic (polyamide or polyester) filaments are used as continuous filament threads. Where flexible threads are required, the filaments are twisted, set, dyed and lubricated only.Where optimum sewing performance and abrasion resistance are required, the plied threads are treated with synthetic co-polymer, which bonds the individual filaments together. Plied continuous filament multilobal polyester threads are favored for embroidery applications on account of its superior lusture.

Very fine threads are produced from a single ply of multifilament polyester yarn, suitably twisted and then treated with a light bonding finish - just sufficient to consolidate the individual filaments without stiffening the final product.

7. Bonded Continuous Multifilament Threads:

The continuous multifilament threads are low twisted and uniformly coated with suitable bonding agent to form bonded filament threads. The bonding agent protects the thread from possible heat damage in sewing, prevents cut ends from fraying and also prevents run-back of twist. Bonding of threads makes them easier to sew and retain their tensile strength during the sewing operation. Additional UV inhibitors are often applied during the bonding process by incorporating them into the bonding resins. The bonding of nylon is comparatively easier than that of the polyester. Most nylon-bonded threads are hot-stretched to give high modulus together with low extension at break. Bonded nylon thread is an ideal choice for footwear, luggage, upholstery, wet suits, seat belts, leather goods, canvas, heavy-duty fabrics and similar materials.

8. Textured Threads:

Texturing is a general term to describe modifications to the appearance and surface characteristics of flat synthetic filament yarns, obtained by various means such as false twisting, air jet, stuffer box etc. depending on the application for which the thread is intended, the degree of crimp rigidity can be varied. Textured threads have a soft handle and are used primarily as under threads where a particularly soft seam is required. Texturing is the process where filaments are entangled by various methods, which impart softness and bulk to the product. Loops made in this manner may be fixed by thermo-setting depending on the origin of the raw material processed, i.e. if a thermoplastic polymer is under texturing. Non-thermoplastic polymers are not fixed by thermo-setting. For each polymer, an adequate thermo-setting temperature must be selected.

It is ideal for very extensible seams in knitwear, underwear, swimwear, foundation wear and tights. Textured sewing threads are most economical sewing threads available today. The textured threads can be used in most of the stitches except for the lock stitches. A German patent gives a method for producing low shrinkage bulky air jet textured yarns for sewing thread from synthetic POY. J and Protection Coats Ltd. reported a method of producing sewing thread that consists in passing two drawn continuous filament yarns, to an air jet-texturing device to form a single bulked yarn in which the filaments are intermingled and looped [5].

9. Direct Spun Sewing Threads:

Tow to yarn 'direct spun' sewing threads are made by feeding a continuous filament tow into a stretch breaking zone and inserting twist into the strand of fibres so produced, with delivery on to a ring spinning package. Compared with yarn spun from staple fibre, direct spun yarns are more uniform, compact, strong have a lower extension at break, a better abrasion resistance, and are free of defects such as neps. Unless high temperature setting is used on the direct spun thread, it will have a greater relaxation in subsequent hot wet processing than an equivalent spun yarn. Because of their characteristics, direct spun yarns may be used to advantage in some critical applications, but only when the higher cost can be justified.

The tow produced in this purpose is relatively thin compared with tow produced for staple fibre production and consequently it usually costs well over twice the cost of staple fibre; hence relatively expensive.

4. Essential Properties Required For Sewing Threads

Industrial sewing techniques make specific and often very exacting demands on the threads involved in the sewing process. The sewability of sewing threads is of major importance [6], having a very profound effect on seam quality and production costs. The sewing and the seam performance of a sewing thread are largely influenced by the material to be sewn, the sewing technique and the end-use for which the sewn material is intended. These requirements can be defined as:

- ❖ The ability of the sewing thread to meet the functional requirements of producing the desired seam effectively.
- ❖ The ability of the sewing thread to provide the desired aesthetics and serviceability in the seam.
- ❖ The cost of sewing thread and that associated with producing the desired seam.

The different important properties required by a sewing thread are discussed below:

1. Needle thread must pass freely through the small eye of the needle; consequently they must be uniform, knot-free, non-torque and fault free.
2. Tensile strength / breaking strength is one of the essential properties of the thread. It must be capable of withstanding several kinetic/ lateral movements during sewing. The strength of the sewing thread must be higher than that of the fabric so that the thread does not rupture during use. During sewing at high speeds, the needle thread is subjected to repeated tensile stresses at very high rates. The thread also comes under the influence of heat, bending, pressures, torsion and wearing. The value of these stresses depends on the sewing speed, machine settings and the thread used. The stresses created within the thread have a negative effect on the processing and functional characteristics of the thread, and there is significant reduction in the thread strength after sewing.

This is a function of the dynamic and thermal loading of the thread and is influenced by the thread frictional properties, thread tensioning during sewing, needle size, stitch length and number of fabric layers in the seam. The thread should therefore possess adequate strength and elongation in order to perform satisfactorily during sewing and in seam [7].

3. For good performance in sewing machine moderate to low extension-at-break of the thread is usually preferred. Needle thread with different elongation-at-break has been found to behave quite differently during stitch formation. The determinants of success of sewing a thread with certain elongation percent without any problem are the machine setting and special properties of the sewing thread itself [6].
4. The elasticity of the sewing thread must be uniform along its length in order to enable equal length stitches to be formed, and it must closely match the elasticity of the fabric being sewn; otherwise either seam thread fracture, or tearing of the adjacent fabric may arise during garment use. Clearly the requirements of woven and knitted fabrics will be different.
5. The forces that are developed in the sewing thread are mostly due to the friction between the thread and machine parts, the most severe action taking place between:

- The thread and the needle.
- The thread and the fabric being sewn.

A controlled level of both static and dynamic friction is required; this must not be too high, which could cause lack of thread control. High static friction values are necessary to allow the stitches to lock and prevent "run-back" of seams. Spun threads are particularly good in this respect when compared with filament thread. The worst is the monofilament threads. The frictional properties are affected by lubrication. The factors that influence the frictional properties are:

✓ Uniform application of lubricating agents.
✓ Adhesion of the finishing agent on the thread.

The quantity and quality of finishes are very important. Special finishes like silicone compounds have been found to exhibit clear advantage over standard paraffin wax.

6. Good abrasion resistance is essential for good sewing performance. The thread is under tension condition, especially when the stitch is being set. The thread must be resilient enough to return to shape after the distortions, and then must maintain its physical properties to provide good performance in the seam after the sewing process is complete. Nylon and polyester offer the best resistance to abrasion.
7. Good resistance to heat is a very important requirement of a sewing thread. The temperature reached by the sewing needle during sewing very much depends on:

❖ the nature of the fabric to be sewn (density, thickness, finish)
❖ the speed of the sewing machine
❖ the type of needle used (size, shape, surface finish)
❖ size and finish of the sewing thread.

The needle temperature is especially critical for fabrics and sewing threads of thermoplastic fibres, where it may exceed their melting temperature.

Needle heating causes sewing thread breakage, cross-thread, skipped stitches, seam damage and physical damage to the needle.

Various studies show that the sewing thread influences the needle temperature significantly.

Its movement through the needle reduces the needle temperature by an average of 21-45%, the amount of reduction depends on the sewing condition and the structure, fineness and composition of sewing thread.

Lubrication of sewing thread with a mixture of wax, emulsions with synthetic resins, and silicon based products may minimize heat generation, and the fibres surface of spun yarns may be an advantage in that a thin layer of the surrounding air will move with the thread and promote needle cooling.

1. The hairiness of sewing thread also affects the appearance of the seam. Sewing thread for decorative seams are singed, squeezed and gloss-brushed.
2. The final direction of twist insertion may be important to enable the stitch forming mechanism of the sewing machine to perform correctly; most sewing machine require Z twist, but there are a few where performance is better with S twist.
3. Colour Fastness is a general requirement for sewing thread. It is important that the selected shade retain its colour throughout the life of the garment. Two aspects of fastness are important:

- the thread must not change colour.
- The thread must not stain any material adjacent to the seam.

4. Low shrinkage during washing and ironing is required. Shrinkage due to fibre swelling causes seams to pucker, especially if the fabric exhibits less shrinkage than threads. Synthetic threads suffer less from this problem than cotton threads owing to their much lower moisture absorbency; however they are liable to residual shrinkage problems if unsuitable manufacturing processes are employed. Synthetic threads can suffer from the problem of thermal shrinkage during ironing but this difficulty can be solved by the use of high temperature setting, which stabilizes the thread at temperature above those normally encountered during the ironing process.
5. The sewing threads should possess better evenness and should contain minimal number of knots, faults and neps etc. Thread should have very low level of imperfections and classimat faults.
6. Good luster in the thread improves appearance of the seam.
7. Threads must be uniformly dyed in a good match to the materials being sewn and also the dyed thread should have properties like colourfastness to washing, light, perspiration, and sublimation.
8. The ability of the thread to perform efficiently in the sewing machine is defined sewability. It can be assessed by the number of breaks that occur during the sewing of a certain number of stitches. However, owing to the generation of needle heat in high-speed sewing, the threads could be damaged without breaking. The long knot free evenner yarns in case of rotor and air-jet can give better sewability.
9. The characteristics of properly constructed seam are strength, elasticity, durability, stability and appearance. The relative importance of these qualities is determined by

the end-use of the sewn product. The factors that govern these properties are seam and stitch type, thread strength and elasticity, stitches per unit length of seam, thread tension, seam efficiency of the material. The hairiness of sewing thread is important to decide seam appearance. The shrinkage potential of the thread and hence the seam is also major importance for proper seam appearance. The serviceability of a garment depends not only on the quality of the fabric but also on that of the seam. The seam quality is measured by stitching parameters of the threads and seam parameters such as size, slippage and strength.

The failure of seam produced by traverse loading can generally be classified as: Type I: the failure due to thread breakage, Type II: the failure due to fabric breakage, Seam breakage: the failure due to the slippage of cloth yarns at right angle to the seam.

Seam slippage is the most probable cause for seam failure that leads to garment rejection in wear.

The durability of a seam depends largely on its strength and its relationship with elasticity of the material.

It is measured in terms of seam efficiency, where Seam Efficiency = (Seam tensile strength/fabric tensile strength) x 100, generally ranges between 85 to 90 %. The minimum loop strength correlate well with the stitch breaking strength. Further resistance to abrasion and wear of the seam during everyday use, including laundering is also essential for the longer seam.

Seam pucker can be defined as a differential shrinkage occurring along the line of a seam and is mainly caused due to seam instability, due to high tension imposed during sewing. Though currently available threads have a certain amount of controlled elasticity and elongation they get over-stretched when the sewing tensions are high. During relaxation the thread recovers its original length thus gathering up the seam. Threads for use in apparel are also required to have good stability to laundering, ironing and other treatments since differential shrinkage between the sewing thread and the fabric of a garment can cause puckering.

Further, Seam pucker can be determined by measuring the differences in fabric and seam thickness under a constant compressive load. The seam-thickness strain is calculated by using the formula:

Thickness strain (%) = (seam thickness – 2 x fabric thickness) x 100 / 2x fabric thickness {ref}

5. Process Flow Chart of Sewing Thread

The process flow chart for the manufacture of sewing threads is shown in the flowchart given below.

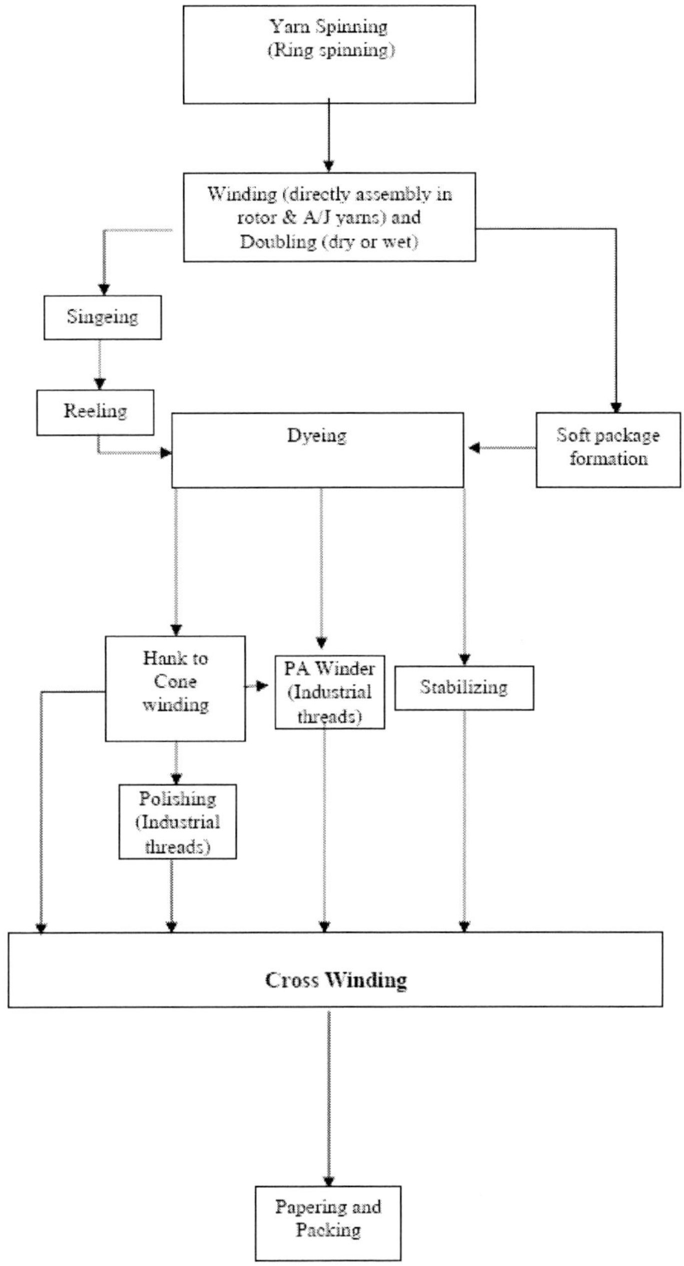

6. Function of Various Machines in Sewing Thread Manufacturing Process

6.1. Yarn Singeing

Sewing thread must be singed to ensure that the projecting fibers do not interfere with downstream processing. Percentage of singeing can be achieved varied by varying the yarn collection speed. Hair removal efficiency at singeing machine is normally 30-50%. Flame temperature is around 800°C. Singeing is mainly done in Cotton Sewing Threads.

Features: The heart of the machine - the burner, serves to singe reliably the projecting fibres of yarn running through at high speed, without inflicting burns on yarn itself. There is a choice between the gas burner and the electric burner. Gas burner are widely used. The gas burner consumes about 55 gm of natural, propane or butane gas an hour, depending on singeing rate and yarn type.

- Speed: 300-1200 m/min.

6.2. Hank to Cone winding

- Conversion of hank in to cone of suitable weight
- Waxing for reducing co-officient of friction in sewing thread .
- Features of new machines:
- Twin Input Rollers: At the front of the head are the twin-input rollers, set to a fixed speed but proportional to the winding speed. The main functions of this unit are to eliminate unwanted tensions prior to yarn entering waxing unit.
- The speed of machine ranging from 400-700 mpm, with possible traverse from150-200mm.

6.3. Polishing

Some threads for special end uses like leather industries, bag stitching, kite flying are treated with starch, softeners, whitener etc. on this machine. Cooked starch is mixed with other chemicals and different recipes are made for different qualities depending on the end uses.

Main objectives of polishing are:

- Extra ordinary smooth surface
- Thread becomes round.
- Stiffness increases.
- Increase in tensile strength (7-10%).

6.4. Cross Winding and Lubrication

- Winding in various types of sewing thread packages like cone, cop , tube, ball, vicone and spool.
- Threads are treated with special waxes for achieving best workability during sewing operation.
- Lick roller lubrication is applied on industrial sewing thread where thread has to run on high speed sewing machine; the basic ingredient of the most of the lubricants is

parafin wax. Although silicones are also used because of their stability to heat and various additives are also included to give some special properties.

7. Geometry of Packages

Threads are wound in many forms. Small length spools are employed in retail store distribution, whereas somewhat larger spools are used to a limited extent industrially. Some of the very fine soft threads are wound on cones, very coarse soft threads are in skein form, but the largest proportion by far is wound on the one headed tube with base or straight tubes. These packages in some instances are put on a weight basis; however, the larger percentage is marked on a length basis.

Ready-wound bobbins in a number of styles to accommodate the various sewing machines employed are also available.

Cross winding threads are generally laid in with traverse ratio 1: 6 (one double traverse = six spindle revolution) and 1: 4 for coarser counts.

Following types of packages are used in sewing threads (with commonly used dimensions).

7.1. Spool: Spools are small flanged plastic or wooden bobbins, they are both with tapered (so-called Diabolo spools) and straight flanges. Mainly parallel winding (because side unwinding is easy) is done. Cross-windings are also possible on the spools. They contain relatively short length of 100-500 meter thread. The length of traverse on spool is 2.9-3.8 cm.

End uses: Upholstery, footwear, leather goods manufacturing, and in hand sewing operations.

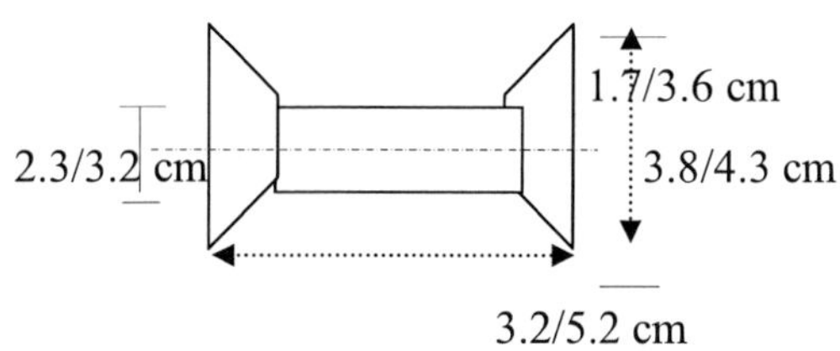

7.2. Cop: Cops are small cylindrical flangeless spools, with precision cross winding. They are mostly made of paper and plastic. They are of two types, small Cop (Tube) and medium Cop (Cop). The lack of flanges facilitates regular off winding on industrial sewing machines although their small diameter makes them less well suited to the faster thread take off machines. Smaller Cops are popular make-up in fashion trades, where a variety of shades are used and production runs for any one colour or style of garments. The length of thread wound ranges from 100-2000m on small cop and 400-4000m on cop. The length of traverse on small cop is 5 - 6.3 cm and on cop is10 cm.

End uses: Kite flying, Upholstery, ready-made garment, tailoring, hosiery, umbrella, and shoe stitching

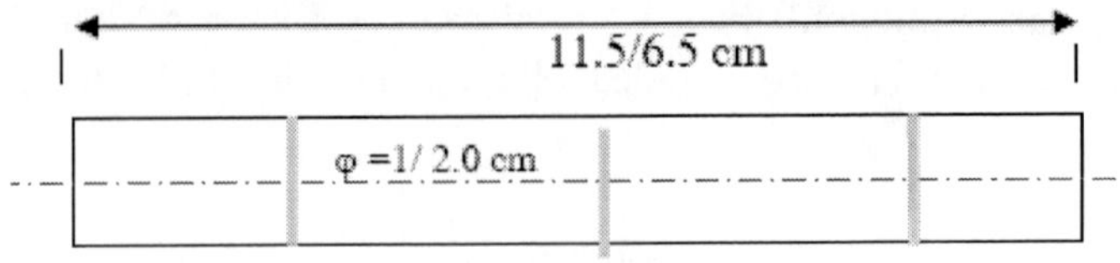

7.3. Cone: This is self-supporting, cross wound conical package. It is easier to withdraw yarn over end from a cone than from a cheese and because of this, cone is more widely used. They contain relatively long lengths 1000-25000m with length of traverse ranging from 10-15cm. They give trouble free thread unwinding at intermittent or continuous high speeds. Cones are the most economical packages for conventional sewing threads in situations where thread consumption is high and production runs are long.

End uses: ready made garment, tailoring, hosiery, leather stitching, Upholstery, shoe stitching, denim, embroidery, and kite flying.

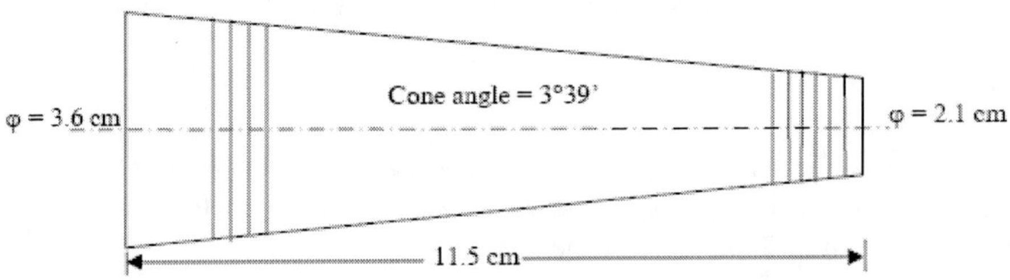

7.4. Vicone or King Spool: Vicones are parallel tubes or low angled cones with an additional base in the form of a raised flange, which may incorporate a small tip. The build of vicone depends on the exact conformity of the taper with the angle of vicone's base. Coarse yarns require a large traverse for the taper - fine yarn a small one. They contain lengths of 1000-5000m with length of traverse 6.5-9cm.

End uses: embroidery, core-spun, and filament threads.

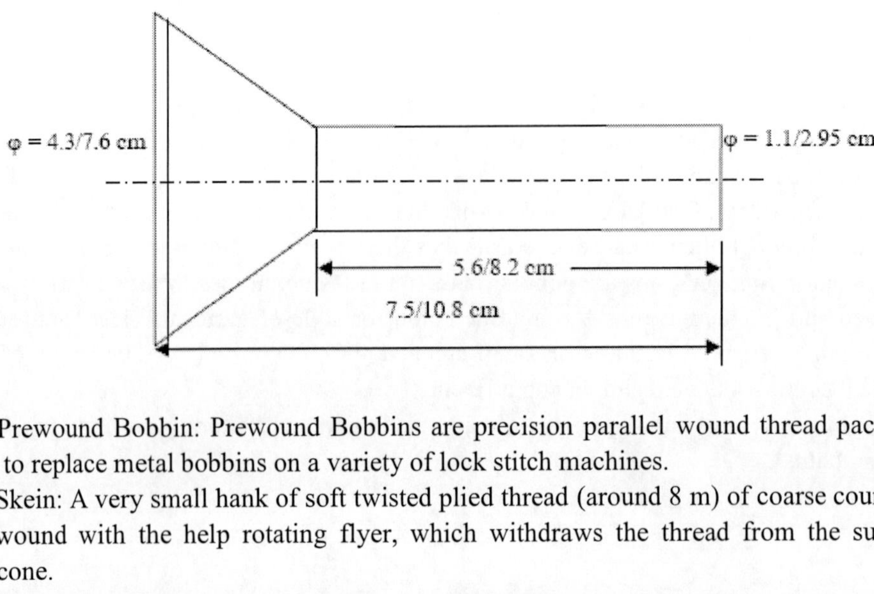

7.5. Prewound Bobbin: Prewound Bobbins are precision parallel wound thread package designed to replace metal bobbins on a variety of lock stitch machines.

7.6. Skein: A very small hank of soft twisted plied thread (around 8 m) of coarse count, is parallel wound with the help rotating flyer, which withdraws the thread from the supply package cone.

End uses: embroidery.

7.7. Ball: A typical ball like structure, wound with the help of four types of different winding. They are: Rough base winding, form winding, surface layer winding, and circumference winding. The initial winding provides firmness at the base. Next winding process makes space for placing the identification ticket. The third stage of winding provides firm gripping of the ticket. The last stage of winding makes a band over the ball, which retains its shape. An easy unwinding of thread is possible.

End uses: embroidery, fishing net and bag closing.

7.8. Cocoon: Cocoons are self-supporting i.e. center-less thread package specially designed for the insertion in the shuttle of multi needle quilting and some types of embroidery. Cocoons are used in the shuttle of multi needle quilting and some types of embroidery machines.

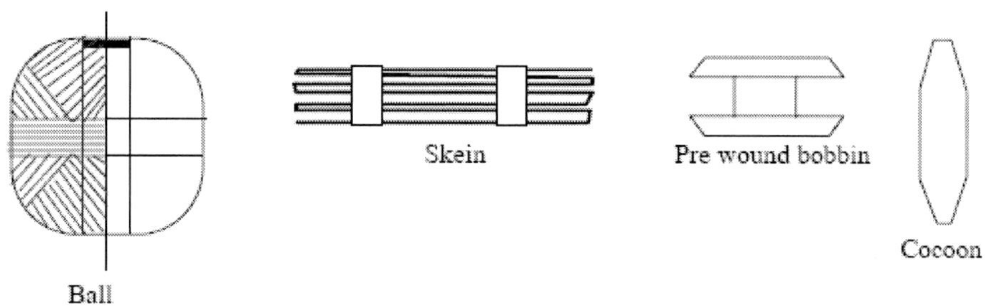

8. Different types of Winding System

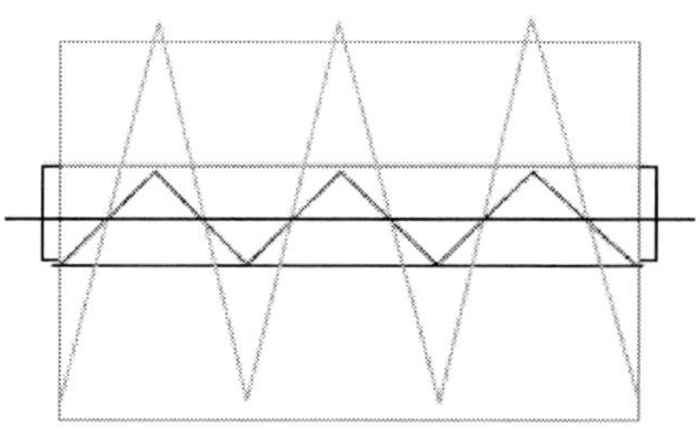

Precision winding:

- constant winding ratio
- winding angle reduces with increasing diameter
- No pattern areas
- Good off-winding characteristics
- High package density

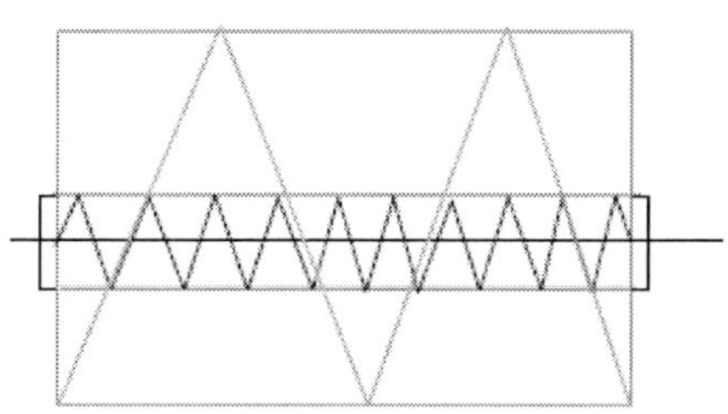

Step precision or Digicone winding:

- almost constant winding angle
- the wind ratio is reduced in steps
- combines the advantage of random and precision winding
- no pattern areas
- higher consistent package density
- perfect unwinding characteristics
- straight sided packages

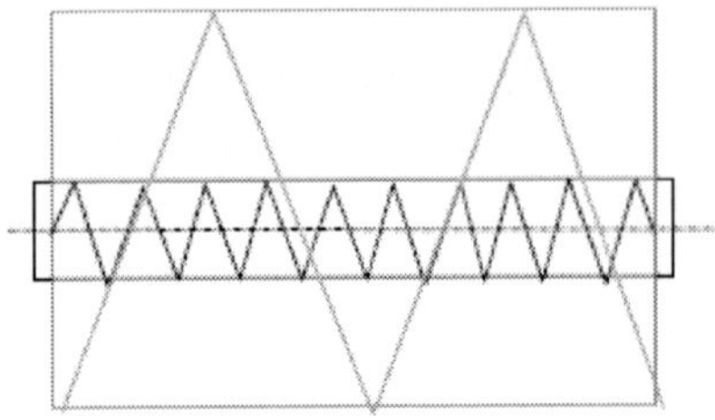

Random winding:

- winding angle is kept constant since the winding ratio reduces with increasing diameter
- stable packages
- even density

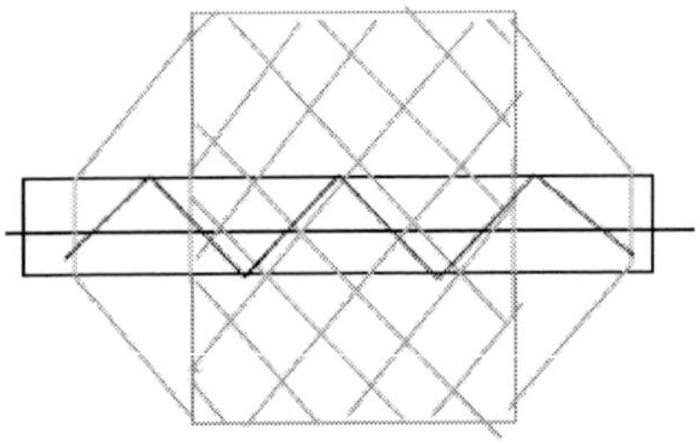

Pineapple winding:

- winding traverse reduces to produce packages with tapered edges
- required for filament winding operations
- all three types of winding applicable

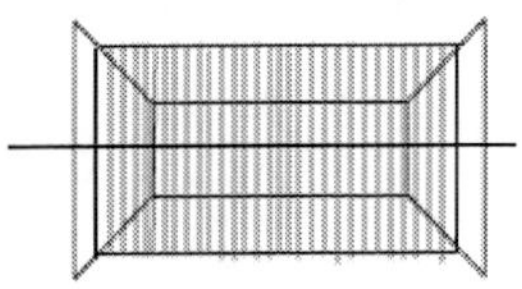

Parallel winding:

- very high package density
- thread vertical to package axis
- relatively short lengths of thread
- suitable for side unwinding
- no pattern areas[5,6]

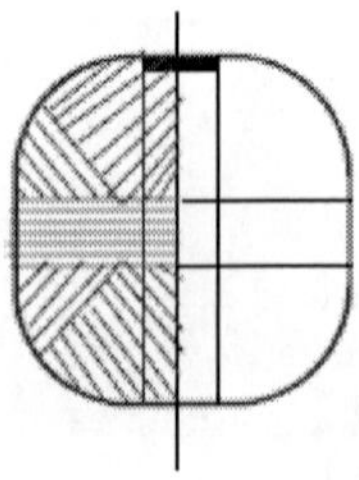

Ball winding:

- very easy unwinding
- winding takes place in 4 stages:

1. rough base winding
2. form winding
3. surface layer winding
4. circumferencial winding

Skein Winding:

- easy unwinding
- Very small parallel strand of soft twisted thread.

9. Proper Selection of Sewing Thread

A number of factors should be considered when choosing the sewing thread: to achieve good sewing performance, sewing thread must be chosen according to the fabric characteristics and end-use of the material. \

The choice of sewing threads according to fabric characteristics can be classified as in Table 2.

Table 2.

Fabric properties	Sewing thread
Fabrics and garments that are not tumble-dried	Sewing thread shrinkage must be less than 1%. To dry at the same time with the fabric, moisture absorption of the thread must be low.
Fabrics and garments that can be ironed	Sewing thread must be resistant to high temperature (ironing and pressing), and must not give its colour. If it is cellulosic, the thread must be little thicker.
Water resistant fabrics	Texturised or core-spun yarns must be preferred to fill the needle holes in the fabric.
Fabric that are dried under tension	The threads' extensibility must be equal to the fabric's.

9.1. Selection According to Seam Type

9.1.1. Lock stitch: If cotton yarn is to be used in this type of seam, 50% extensibility can be achieved and this ratio is enough for the comfort of the garments. If standard synthetic sewing thread is used, extensibility value reach 60% and no slacking occur when the seam is under tension. If plied nylon thread is used as bobbin thread, 20% more extensibility can be achieved. Cotton thread must be preferred for normal lock stitches. When fabric with 30% to 100% extensibility are to be sewn, synthetic threads or blended plied threads must be used.

9.1.2. Chain stitch: As this kind of seam is generally used on elastic fabrics, synthetic or blended threads are preferable to cotton. As looper thread, nylon threads or filled give better results.

9.2. Selecting the right size: The ticket number system (related to Ne, Nm and denier) is now generally used by most thread manufacturers and industrial thread consumers to describe approximately the threads of the finished product. The ticket number of cotton threads are generally based on cotton system Ne, these ticket numbers being approximately equal to three times the cotton count of the thread. The ticket number of synthetic threads are based on the metric count system Nm ; approximately three times the metric count of the thread.

Selecting the correct size of thread for a particular application is very important to the thread performance during sewing and afterwards in the seam. However the choice is not always easy. It depends on many factors, including seam strength, fabric weight and type, end use, stitch type, seam type, needle size and many others.

In general, the heavier the fabric, the thicker the thread is. The ultimate choice lies with garment designer and the garment technologist, taking account of aesthetic and technical considerations.

9.3. Colour selection: There are many hazards to be avoided when visually matching samples, and especially:

- Dichroism: An apparent change of colour as the direction of viewing is changed. This tends to occur more with fabrics than with threads, but can affect the judgement of thread colour.

- Metamerism: A change of shade which occurs when lighting conditions are changed. For colour matching this is a serious problem because while two samples may match in daylight, they may not change colour under artificial light to the same degree.

9.4. Colourfastness: When a suitable shade of thread is chosen that matches the fabric to be sewn, there remains a very important factor: it is important that the selected shade retain its colour fast through the life of the finished article. It must not stain any article adjacent to the seam. The colour fastness is affected not only by washing and light but by chlorinated water, rubbing/ crocking, dry cleaning, hot pressing, bleaching and sublimation. So before selection all these things are to be considered.

10. Applications of Sewing Threads

Approximately 80% of all sewing threads produced are used by the clothing industries. The main classes of clothing requiring threads are as shown in Table 3 and 4.

Table 3.

Class	Approximate percentage of all threads produced
Dresses and blouses	13
Shirts and nightwear	5
Suits, coats and other outwear	
Underwear	28
	16
Knitwear	
Work clothing	
Sportswear	
Footwear	
Other clothing	21
accessories	

Table 4. End uses of different types of sewing threads

Cotton	Spun Polyester	Core Spun	Textured Threads	Airjet Textured	Bonded Fil.(nylon)	Bonded Fil.(PET)
Garment	Shirts	All	Slacks	Jeans	Automotive	Rainwear
Embroidery	Blouses	apparel			Interiors	
		(lingerie	Home	Mattresses	Seat belts	Furniture
Selvage	Lingerie	to jeans)	furnishing	Slacks	Air-bags	Tents
Threads	Slacks	Furniture	Shirts	Mops	Footwear	Footwear
Tea bag	Knits	Shoes	Blouses	Flags	Leather	Mattresses
Strings	Jeans	Tents	Jeans	Caps	Canvass	Sporting goods
Bag closing	Mattresses		T-shirts		goods	Embroidery
	Embroidery		Brief		Luggage	
			Fleece		Sporting	
					goods	

Sewing threads for technical applications:

Different kinds of sewing threads are manufactured depending upon their specific field of application. Threads for use in apparel are required to have good stability to laundering and ironing, with better seam appearance. The threads are also required to perform satisfactorily during high speed sewing operations [8]. However, for technical applications sewing threads are engineered to meet certain critical requirements. Sewing thread characteristics are required to be different for their use in flame retardant garments, clothing for extreme weather conditions etc. for different application, the types of sewing thread and their characteristics vary widely.

High-temperature applications:

Threads for high-temperature applications are required to hold the seam and secure it in extreme temperature conditions (between 260 and 1100°C). Threads are usually made from quartz, glass, carbon, polytetrafluoroethylene, steel and aramide fibres. Military requirements and growing demands of safety legislation have built a viable market for high-temperature threads in spite of their higher cost.

Table 5. Sewing Threads for Technical Applications

Generic type	Availability	Applications
p-Aramide	Staple fibre	Heat protection clothing, personal protection clothing, welder's clothing etc.
PTFE	Monofilament	Awnings, sun protection shades, boat covers, filter bags etc.
Polyester(Normal tenacity)	Multifilament	High load seams, safety belts, tents, suitcases etc.
Polyester(High tenacity)	Multifilament	Safety belts, cargo slings, tarpaulins, air bags, geotextiles, convertible car hoods etc.
m- Aramides	Multifilament	High temperature fields
Polyester/cotton	Spun yarn	Tents, tarpaulins, flags and bunting, book binding, convertible car hoods etc.
Polypropylene	Multifilament	Filtration, binder twine for agriculture use etc.
Polyamides(4.6an d 6.6)	Multifilament	Safety belts and cargo slings, car seatings, tarpaulins, air bags, footwear, tents, geotextiles, sails, rucksacks, convertible car hoods etc.

The most commonly used and comparatively cheap threads are manufactured from glass fibre, Nomex and Kevlar. Threads from carbon and quartz fibres are quite expensive and are produced exclusively for military and aerospace applications. Nomex threads are widely used in the filtration of hot gases and in thermal protection clothing. Quartz fibre is also used in atomic plants since it does not contaminate the atmosphere and can withstand temperature as high as 1093 °C.

Geo-textile applications:

In many geotechnical and air-supported fabric structures, the seam constitutes one of the most critical components in forming a wider fabric from the available narrow fabrics. Sewing is generally performed because of its simplicity and its ability to maintain the elasticity of the fabric structure.

Polyester and polypropylene fibre threads are the most used threads, since these are the principal materials used for geotextiles. For demanding applications, high tenacity fibres such as Kevlar are used.

CONCLUSION

The sewing thread is of considerable importance and playing a major role in retaining the fabric appearance, look, and life of the garment in the long run, even though it usually represents much less than 1% by mass of a garment.

Now-a- days, a numerous variety of sewing threads are prevailing in the market due to diverse demands from the sewing industry, increasing use of different types of fibres in the garment industry and expanding application of textile materials in various fields like apparels, technical applications as well. Better understanding of the sewing process and its requirements as obtained through studies by modern instrumentation techniques has also greatly contributed to the development of new threads. It is also very much required and appreciable to have different types of sewing threads which can suit various applications, since various end-uses demand specific property requirements.

It's beyond anybody's doubt that the success of garment manufacturing process majorly depends upon the operation of sewing, though a very better quality of fabric is selected for the garment manufacturing process. Again, the sewing threads play a vital role in the success of sewing operation, since a wrong thread may collapse a very high quality fabric and even a best sewing machine used for the sewing, the whole process will be failed. It can add to waste of both time and money. Hence it is very much imperative to select a right type of sewing thread which can suit one's requirements exactly. This is possible by the correct understanding of the type of fibre used to manufacture, manufacturing processing sequence and properties of different types of sewing threads existing on earth, which was touched upon in this technical article to a greater extent.

REFERENCES

[1] J.O. Ukponmwan, A.Mukhopadhyay and K.N.Chatterjee, " Sewing Threads", Textile Progress, *The Textile institute*, Volume 30, No. 3 /4, Page 1.

[2] S. D. Pai, V. G. Munshi and A. V. Ukidue, "Cotton and synthetic sewing thread- I", *The Indian Textile Journal*, August 1984, page 63-75.

[3] Alan Crook (information updated by Ron Lioyd), "Sewing threads", Textiles, Issue No. 2, 1991, page 14-16.

[4] K. N. Krishna Swamy, "Prerequisites for manufacturing sewing thread- I", *The Indian Textile Journal,* March 1982, page 103- 111.

[5] H. U. Bhonde, "Sewing thread: types and some recent developments", MANTRA Bulletin, Manmade Textiles Research Association, Vol. 19/11, November 2001, page-5-10.

[6] Pradip V. Mehta, satish K. Bharadwaj, "Managing Quality in the Apparel Industry", National Institute of Fashion Technology, New delhi, India, New Age Publihers, 1998.

[7] K. R. Salhotra, P. K. Hari and G. Sundaresan, "Sewing thread properties", Textile Asia, September 1994, page 46-57.

[8] Walter Fung, Mike Hardcastle, " Textiles in Automotive Engineering" Woodhead Publishing, 2001.

In: Recent Progress in Chemistry…
Editor: A. K. Haghi, pp. 23-34
ISBN: 978-1-61668-501-0

Chapter 2

IMPROVED MODELING FOR PREDICTION OF WATER TRANSMISSION FAILURE

Kaveh Hariri Asli*[1], Faig Bakhman Ogli Naghiyev[2] and Soltan Ali Ogli Aliyev[3]

[1]Department of Mathematics and Mechanics,
National Academy of Science of Azerbaijan" AMEA", Baku, Azerbaijan,
[2]Azerbaijan State Oil Academy, Baku, Azerbaijan
[3]Department of Mathematics and Mechanics,
National Academy of Science of Azerbaijan" AMEA", Baku, Azerbaijan

ABSTRACT

Various methods have been developed to solve transient flow in pipes. These ranges have been formed from approximate equations to numerical solutions of the nonlinear Navier-Stokes equations. Hydraulic transient flow is also known as unsteady fluid flow. During a transient analysis, the fluid and system boundaries can be either elastic or inelastic: a) Elastic theory, describes unsteady flow of a compressible liquid in an elastic system (e.g., where pipes can expand and contract). b) Rigid-column theory, describes unsteady flow of an incompressible liquid in a rigid system. It is only applicable to slower transient phenomena. Both branches of transient theory stem from the same governing equations. In this work a computational method was used for prediction of water transmission failure.

NOMENCLATURES

λ=coefficient of combination I= moment of inertia (m^4)
w = weight (kg)
P =fluid power (pa)

* Corresponding author E-mail: HARIRI_K@YAHOO.COM

z = elevation at the centroid (m)
t=time(s) λ ꝭ= unit of length
p = pressure (N/m^2)
α= pipe cross section area (m^2)
V = velocity (m/s)
s=length (m) f = friction factor
τ=shear stress (pa) $H2$-$H1$=pressure difference (m-$H2O$)
C=surge wave velocity (m/s)
V=volume (m^3)
F= fluid force (N)
W=frequency
q =flow rate (m^3/s)
D = diameter of each pipe (m)
μ=fluid dynamic viscosity ($kg/m.s$) R=pipe radius (m), γ= specific weight (N/m^3)
J=junction point (m)
y= Surge tank and reservoir elevation difference (m)
K=Volumetric coefficient (GN/m^2)
T=period of motion
A = pipe cross-sectional area (m^2)
dp =is subjected to a static pressure rise(m)
h_p = head gain from a pump (m),
h_L = combined head loss (m)
Ev=bulk modulus of elasticity (pa) α =kinetic energy correction factor
P=surge pressure (m), ρ= density (kg/m^3)
C = velocity of surge wave (m/s), g=acceleration of gravity (m/s^2)
K = wave number
Tp = pipe thickness (m), Ep = pipe module of elasticity (kg/m^2)
Ew = module of elasticity of water (kg/m^2),
$C1$=pipe support coefficient
$Ymax$=Max. fluctuation
v =fluid dynamic viscosity ($kg/m.s$)

1. Introduction

Among the approaches proposed to solve the single-phase (pure liquid) water hammer equations are the Method of Characteristics (MOC), Finite Differences (FD), Wave Characteristic Method (WCM), Finite Elements (FE), and Finite Volume (FV). One difficulty that commonly arises relates to the selection of an appropriate level of time step to use for the analysis. The obvious trade-off is between computational speed and accuracy. In general the smaller the time step, the longer the run time but the greater the numerical accuracy. The challenge of selecting a time step is made difficult in pipeline systems by two conflicting constraints. First, to calculate many boundary conditions, such as obtaining the head and discharge at the junction of two or more pipes, it is necessary that the time step be common to all pipes. The second constraint arises from the nature of the MOC. If the adjective terms in the governing equations are neglected (as is almost always justified), the MOC requires that ratio of the distance Δx to the time step Δt be equal to the wave speed in each pipe. In other words, the Courant number should ideally be equal to one and must not exceed one by stability reasons.

For most pipeline systems, having as they do a variety of different pipes with a range of wave speeds and lengths, it is impossible to satisfy exactly the Courant requirement in all pipes with a reasonable (and common) value of Δt. Faced with this challenge, researchers have sought for ways of relaxing the numerical constraints. Two contrasting strategies present themselves.

The method of wave-speed adjustment changes one of the pipeline properties (usually the wave speed, though more rarely the pipe length is altered) so as to satisfy exactly the Courant condition [1].

Balancing the energy across two points in the system yields the energy or Bernoulli equation for steady-state flow: the components of the energy equation can be combined to express two useful quantities, the hydraulic grade and the energy grade:

$$\begin{aligned}&(P_1/\gamma)+Z_1(V_1^2/2g)+h_p= \\ &(P_2/\gamma)+Z_2+(V_2^2/2g)+h_L,\end{aligned} \tag{1.1}$$

Pressure Head: p/γ Elevation Head: z Velocity Head: $V^2/2g$

where: p-pressure (N/m^2), γ-specific weight (N/m^3), z-elevation (m), V-velocity (m/s), g-gravitational acceleration constant (m/s^2, ft/s^2).

2. Materials and Methods

The method of characteristics "MOC" is based on a finite difference technique where pressures are computed along the pipe for each time step [2]. Two cases are considered for modeling:

1- The combined elasticity of both the water and the pipe walls is characterized by the pressure wave speed (Arithmetic method -combination of Joukowski formula and Allievi formula):

$$H_2 - H_1 = (C/g)(v_2 - v_1) = \rho C(v_2 - v_1), \text{ (Joukowski formula)}, \tag{2.1}$$

$$c = 1/[\rho((1/k) + (d.C_1/E.e))]1/2, \text{ (Allievi formula)}, \tag{2.2}$$

With combination of Joukowski Formula and Allievi Formula:

$$\lambda\begin{bmatrix}(\partial v/\partial t)+(1/\rho)(\partial p/\partial s)+\\ g(dz/ds)+(f/2D)v|v|\end{bmatrix}+$$

$$C^2(\partial v/\partial s)+(1/p)(\partial p/\partial t)=0, \tag{2.3}$$

$$\lambda={}^{+}c\ \&\ \lambda={}^{-}c$$

Hence, water hammer pressure or surge pressure (*ΔH*) is a function of independent variables (*X*) such as: *ΔH*≈ *ρ, K, d, C1, Ee, V, g,* (2.4)

2-The method of characteristics (MOC) based on a finite difference technique where pressures are computed along the pipe for each time step,

$$H_P=1/2\begin{pmatrix}C/g(V_{Le}-V_{ri})+(H_{Le}+H_{ri})\\ -C/g(f\ \Delta t/2D)(V_{Le}\ |V_{Le}|-V_{ri}|V_{ri}|)\end{pmatrix}, \tag{2.5}$$

$$V_P=1/2\begin{pmatrix}(V_{Le}+V_{ri})+(g/c)(H_{Le}-H_{ri})\\ -(f\ \Delta t/2D)(V_{Le}\ |V_{Le}|+V_{ri}|V_{ri}|)\end{pmatrix}, \tag{2.6}$$

f=friction, *C*=slope (*deg.*), *V*=velocity (*m/s*), *t*=time(*s*), *H*=head (*m*)

Present work used the method of characteristics "MOC" to solve virtually any hydraulic transient problems. Dateline for field tests and lab. Model data collection was: at 12:00 a.m., 10/02/07until05/02/09.Locaton of work field tests and lab. Model was at Rasht city in the north of Iran. Pilot subject was:" Interpenetration of two fluids at parallel between plates and turbulent moving in pipe ".

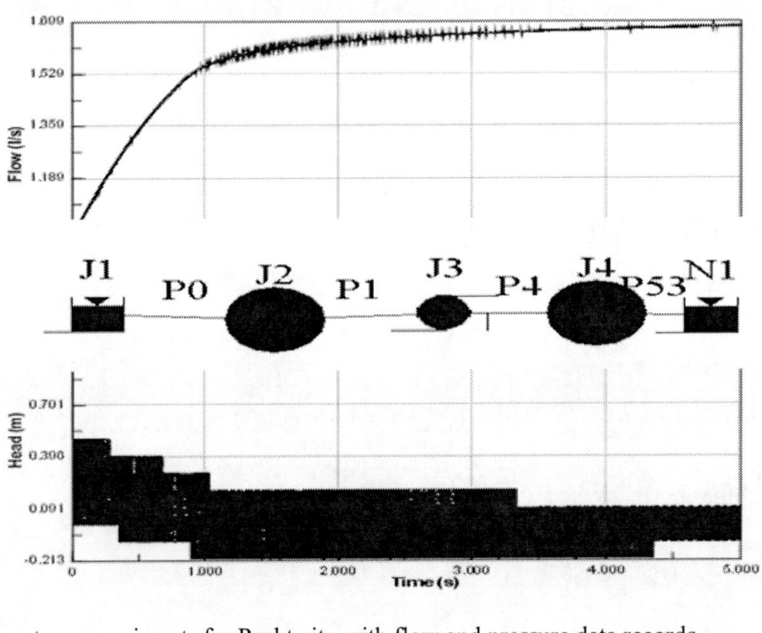

Figure 1. Laboratory experiments for Rasht city with flow and pressure data records.

For data collection process, Rasht city water main pipeline have been selected as a field tests model. The pipeline was included water treatment plant pump station (in the start of water transmission line), 3.595(*km*) of two lines of 1200(*mm*) diameter pre- stressed cement pipes and one 50000(m^3) water reservoir (in the end of water transmission line). All of these parts have been tied into existing water networks.

2.1. Regression

The curve estimation procedure allows quickly estimating regression statistics and producing related plots for different models. Curve estimation is most appropriate when the relationship between the dependent variable(s) and the independent variable is not necessarily linear. Linear regression is used to model the value of a dependent scale variable based on its linear relationship to one or more predictors. Nonlinear regression is appropriate when the relationship between the dependent and independent variables is not intrinsically linear. Binary logistic regression is most useful in modeling of the event probability for a categorical response variable with two outcomes. The Auto regression procedure is an extension of ordinary least-squares regression analysis specifically designed for time series. One of the assumptions underlying ordinary least-squares regression is the absence of autocorrelation in the model residuals.

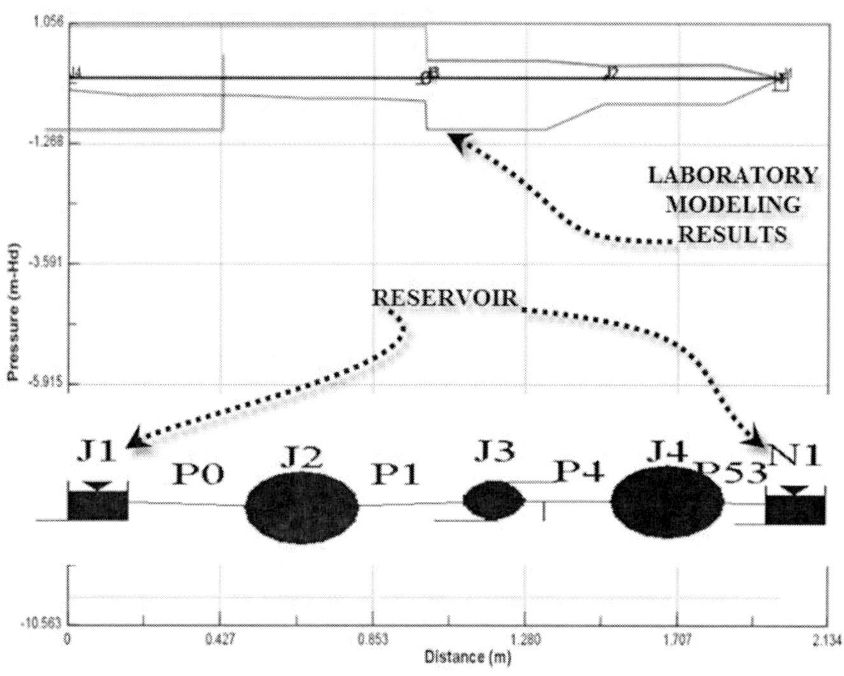

Figure 2. Laboratory experiments for Rasht city with flow and pressure data records.

Time series, however, often exhibit first-order autocorrelation of the residuals. In the presence of auto correlated residuals, the linear regression procedure gives inaccurate estimates of how much of the series variability is accounted for by the chosen predictors. This can adversely affect choice of predictors and hence the validity of the model. The auto regression procedure accounts for first-order auto correlated residuals. It provides reliable estimates of both goodness-of-fit measures and significance levels of chosen predictor variables. The Auto regression procedure by regression software "SPSS 10.0.5" has been selected for the curve estimation procedure in present work. Regression model (3.1-3.8) has been built based on Field tests data and in the final procedure it has been compared (Figure 5) with method of characteristics "MOC" numerical modeling and simulation results.

3. Results and Disscusion

Water hammer software -version 07.00.049.00 results have been compared with regression software "SPSS 10.0.5". Regression software "SPSS" has fitted the function curve and provided regression analysis. So the regression model has been found in the final procedure. By this model, Field Tests results have been compared by Lab. model results (Figure 1-2). This was the main practical aim of present work. In the final procedure a condition base maintenance (CM) method has been founded for all water transmission systems. The results are as flowing: At fast transients, down to 1 second, surge pressure and velocity of surge wave have been recorded. Curve estimation procedure has used these data which have been detected (Figure 3-4) on actual systems (Field tests). Also flow and pressure were collected by Lab. tests model. Those data have been compared by flow and pressure data which have been collected from actual systems (field tests). The model is calibrated using one set of data, without changing parameter values. It is used to match a different set of results [3]. Regression model (3.1-3.8) has been built based on field tests data. Curve estimation procedure (Figure 3, 4) was formed by estimating regression statistics and producing related plots for field tests model with two assumptions:

Assumption (1): p=f (V), Assumption (2): p= f (V, T, L)

Assumption (1): p=f (V), V - velocity (flow parameter) is the most important variable. Dependent variable: P - pressure (*bar*), for starting point of water hammer condition. The independent variable is Velocity ($m.s^{-1}$). Regression software "SPSS 10.0.5" performs multidimensional scaling of proximity data to find least-squares representation of the objects in a low-dimensional space.

$$\mathbf{Linear\ function \therefore pressure = 6.062 + .571 Flow,}$$
$$\mathbf{Quadratic function \therefore pressure = 6.216 - .365 Flow^4 + .468 Flow^3,}$$
$$\mathbf{Cubic function \therefore pressure = 6.239 - .057 Flow^2 + .174 Flow,}$$
$$\mathbf{Compound function \therefore pressure = 1.089(1 + Flow)^n, n = compoundirg\ period}$$
$$\mathbf{Growth\ function \therefore pressure = 1.804(.085)^{Flow/.05},}$$
$$\mathbf{Exponentid\ function \therefore pressure = 6.076 e^{FlowLn.085},}$$
$$\mathbf{Logitic function \therefore pressure = 1/(1 + e^{-Flow})\ or\ pressure = .165 + .918 Flow} \quad (3.1\text{-}3.7)$$

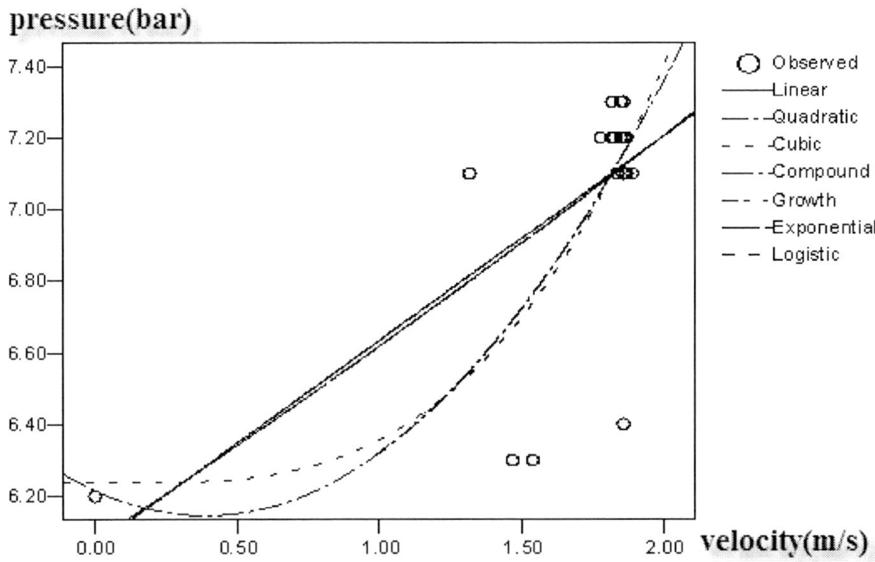

Figure 3. Scatter diagram for Rasht city water transmission lines (Start of water hammer condition).

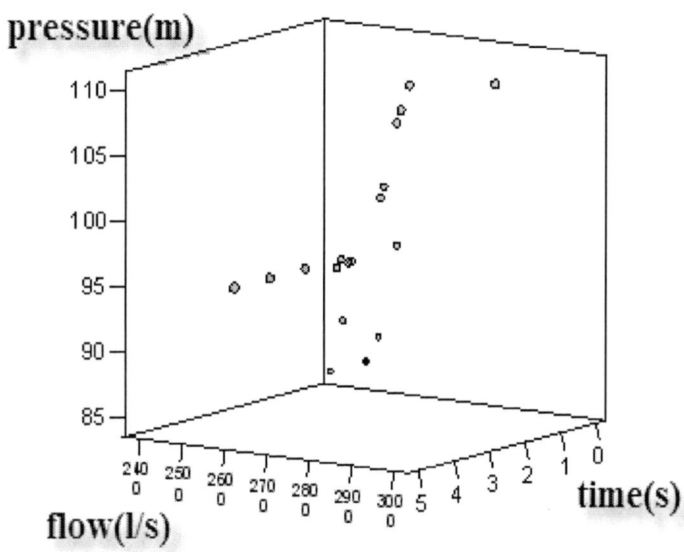

Figure 4. Scatter diagram for Lab. Tests (Rasht city research Field Tests Model).

Assumption (2): p= f (*V, T, L*), *V*-velocity (flow) and*T*- time and *L*-distance, are the most important variables. Input data are in relation with water hammer condition. Regression software "SPSS 10.0.5" fitted the function curve (Figure 3-4) with regression analysis for water hammer condition.

$$\textbf{pressure} = \textbf{28.762} + \textbf{.031Flow} - \textbf{.005Distance} + \textbf{.731Time}, \quad (3.8)$$

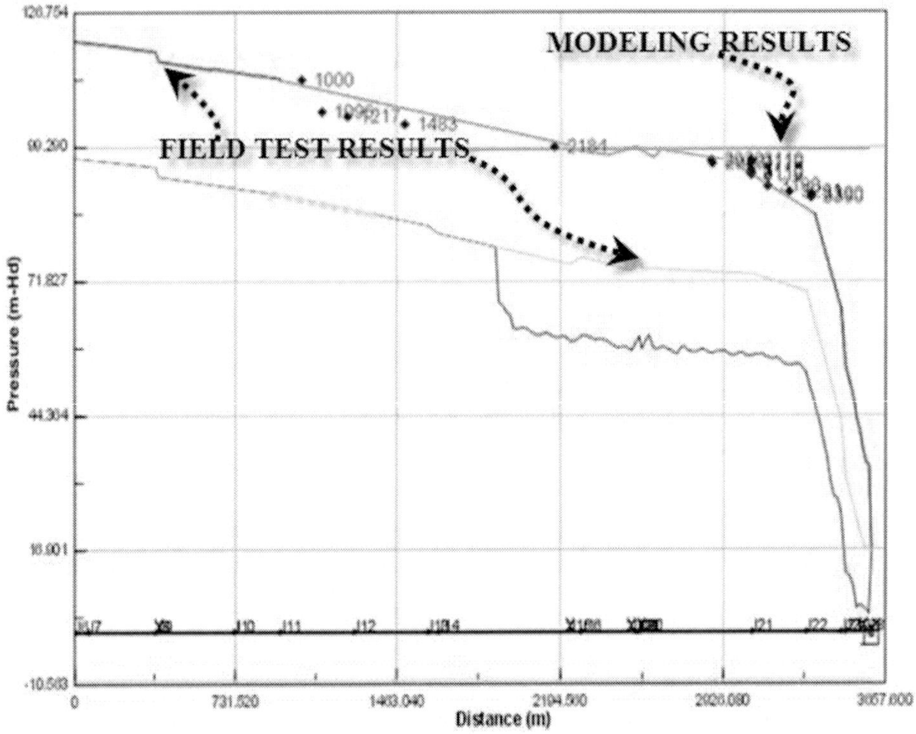

Figure 5. Analysis and comparison results for calculations of modeling results and Field Tests results Rasht city water pipeline pilot research.

3.1. Comparison of Present Work Results with Other Expert's Research

Comparison of present work (Figure 5) results (water hammer software modeling and SPSS modeling), with other expert's research results, shows similarity and advantages:

Apoloniusz Kodura and Katarzyna Weinerowska, 2005:

In present work water hammer has been run in pressurized pipeline with the local leak. The experiments and numerical analysis results in were presented. If the pipeline with local leak was considered, the water hammer phenomenon was influenced by some additional factors [5, 6]. Detailed conclusions were drawn on the basis of experiments and calculations for the pipeline with a local leak. Hence the most important effects that have been observed were as flowing: The influence of the ratio of discharge from local leak has been restricted. Total discharge effect in the pipeline on the values of period wave oscillations has been investigated. The outflow to the overpressure reservoir from the leak was effected the value of wave celerity. (Single straight pipe of the length *L,* extrinsic diameter *D* and the wall thickness *e* or the pipeline consisted of sections of varied parameters. The pipeline was equipped with the valve at the end of the main pipe, which was joined with the closure time register. The water hammer pressure characteristics were measured by extensometers, and recorded in computer's memory. The supply of the water to the system was realized with use of reservoir which enabled inlet pressure stabilization.

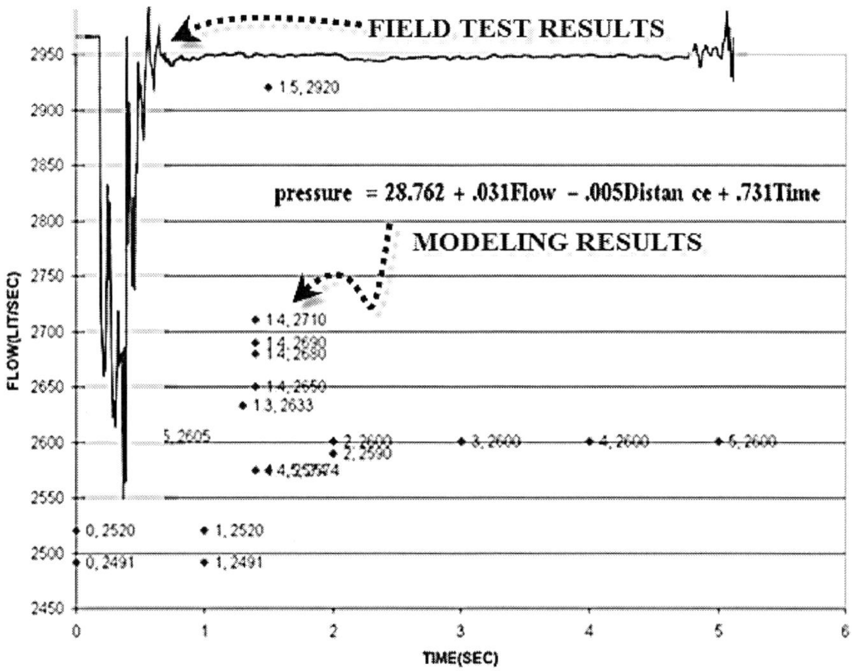

Figure 6. Rasht city water pipeline (Modeling results and Field Tests results, comparison).

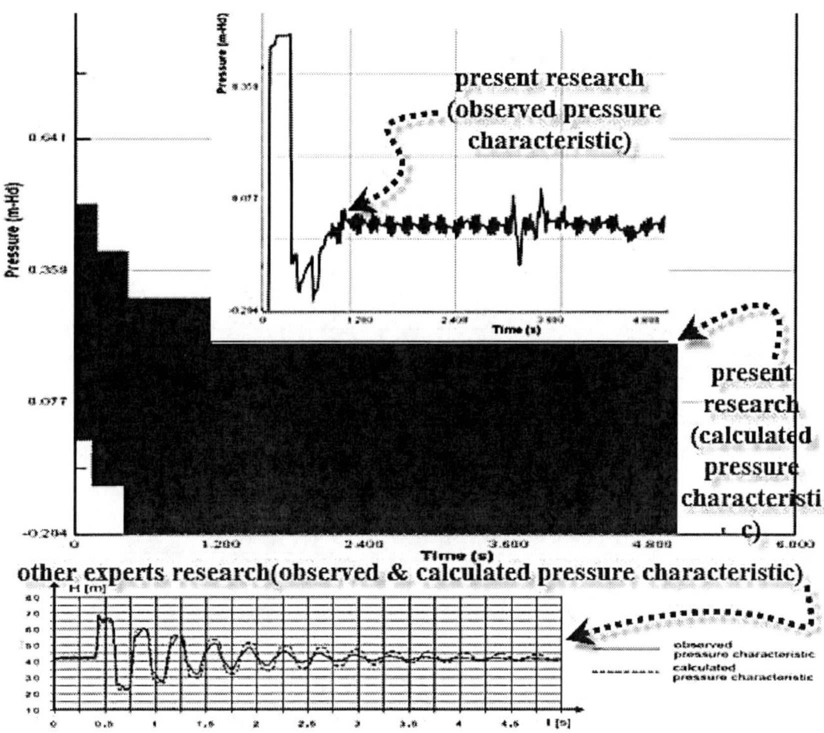

Figure 7. Laboratory Model Pressure Characteristic (observed and calculated) for present research and Kodura and Weinerowska research.

Local Leakage Rate Effect on the Maximal Value of Pressure

The reason for the high pressure drop in water transmission was related to the leakage effect (local leakage effect on high pressure deceasing in□ Water Pipeline). It was done to explain repeated pipe breaks. This work led to improved standards for precession designs and installation techniques in the field of Sub atmospheric transient pressures that can suck contaminants into the water system.

The leakage has been happened near the location of water treatment plant. So water flow was decreased from 3000(*l/s*) to 2500 (*l/s*). In this case pipeline was equipped with sure tank. So leakage happened on point P3:J7 (P3:45.00%) at elevation 36.2 (*m*) and Distance 140(*m*) far from water pump station , Max head at 155(*m*) dropped to 135.1 (*m*) .vapor pressure was equal to -10.0 (*m*). Initial head and Min head had not any changes (Table 1).

Table 1.Field tests model (water pipeline of Rasht city in the north of Iran -Software hammer version 07.00.049.00 - leakage location at extreme heads)

Point Distance Elev. Init Head Max Head Min Head Vap.Pr
+ P3:J7 .0 36.3 132.8 156.8 105.1 .000 -10.0
P3:5.00% 15.6 36.3 132.7 156.6 105.3 .000 -10.0
P3:10.00% 31.1 36.3 132.6 156.3 105.1 .000 -10.0
P3:15.00% 46.7 36.3 132.4 156.1 105.0 .000 -10.0
P3:20.00% 62.2 36.3 132.3 155.8 104.9 .000 -10.0
P3:25.00% 77.8 36.3 132.2 155.6 104.7 .000 -10.0
P3:30.00% 93.3 36.2 132.1 155.5 104.4 .000 -10.0
P3:35.00% 108.9 36.2 131.9 155.3 104.3 .000 -10.0
P3:40.00% 124.4 36.2 131.8 155.2 104.2 .000 -10.0
P3:45.00% 140.0 36.2 131.7 155.0 104.0 .000 -10.0
P3:50.00% 155.5 36.2 131.6 135.1 103.8 .000 -10.0
P3:55.00% 171.1 36.2 131.4 134.9 103.7 .000 -10.0
P3:60.00% 186.6 36.2 131.3 134.7 103.5 .000 -10.0
P3:65.00% 202.2 36.2 131.2 134.4 103.4 .000 -10.0
P3:70.00% 217.7 36.2 131.1 134.6 102.9 .000 -10.0
P3:75.00% 233.3 36.2 130.9 134.4 102.7 .000 -10.0
P3:80.00% 248.8 36.1 130.8 134.0 102.7 .000 -10.0
P3:85.00% 264.4 36.1 130.7 133.9 102.4 .000 -10.0
P3:90.00% 279.9 36.1 130.6 133.6 102.3 .000 -10.0
P3:95.00% 295.5 36.1 130.4 133.5 102.1 .000 -10.0

+ P3:J8

CONCLUSION

Comparison showed similarity in results between this work discussion and other expert works. Field Tests results showed in leakage condition at negative phase air was sucked into the Rasht city water pipeline. There was minus pressure in the zone of the near 50000(m^3)

water reservoir. So this volume of air must be removed from the system. It was showed that at 110M surge pressure in the near to pump station (start of transmission line). The leakage has been happened near the location of water treatment plant. So water flow was decreased from 3000(*L/S*) to 2500(*L/S*).This was unaccounted for water "UFW" alarm. This work has found the location and rate of unaccounted for water "UFW" in the pipeline. Results have showed location and rate of unaccounted for water "UFW" in the pipeline .This was the main practical aim of present work. In the final procedure a condition base maintenance (CM) method has been founded for all water transmission systems.

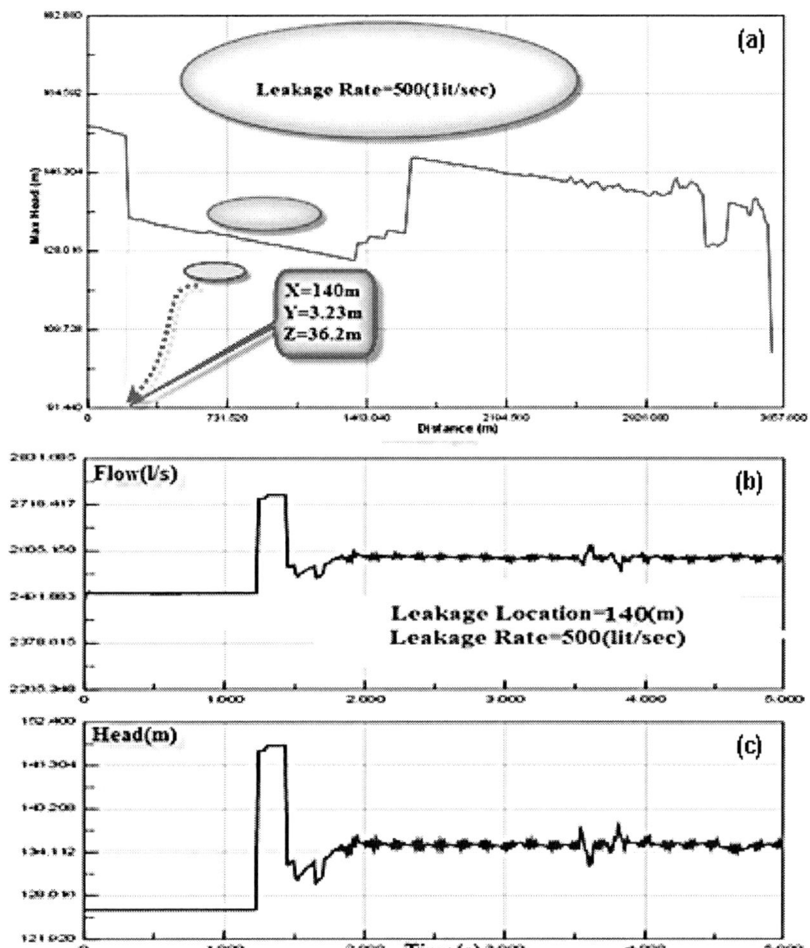

Figure 8. a) location and rate of unaccounted for water "UFW", b) flow variation, c) head variation related to "UFW" in Rasht city Water Pipeline.

REFERENCES

[1] Arturo Leon,S.: improved modeling of unsteady free surface ,pressurized and mixed flows in storm-sewer systems,Submitted in partial fulfillment of the requirements for

the degree of Doctor of Philosophy in Civil Engineering in the Graduate College of the University of Illinois at Urbana-Champaign, 57-58 (2007).

[2] Asli, K.H., Nagiyev, F.B., Hagi, A.K. Three-dimensional Conjugate Heat Transfer in Porous Media, International Journal of Journal of the Balkan Tribological Association, *Thomson Reuters Master Journal List*, ISSN: 1310-4772, Sofia, Bulgaria, 336-346(2009).

[3] Asli, K.H., Nagiyev, F.B., Hagi, A.K.: Physical modeling of fluid movement in pipelines, Nanomaterials Yearbook - , USA, 2009, (has been accepted for publication in 2010, 1st quarter), https://www.novapublishers.com/catalog/product_info.php?products_id=11587.

[4] Wylie, E.B., Streeter, V.L.: Fluid transients, Feb Press, Ann Arbor, MI, 1983, corrected copy: 166-171 (1982).

[5] Apoloniusz, Kodura, Katarzyna, Weinerowska. : Some aspects of physical and numerical modeling of Water Hammer in Pipelines, 125-133 (2005).

[6] Asli, K.H., Nagiyev, F.B., Hagi, A.K. "Some Aspects of Physical and Numerical Modeling of water hammer in pipelines", Nonlinear Dynamics An International Journal of Nonlinear Dynamics and Chaos in Engineering Systems, ISSN: 0924-090X (print version) ISSN: 1573-269X (electronic version) Journal no. 11071 Springer , USA,http://nody.edmgr.com/,2009, (has been accepted for publication in 2009).

In: Recent Progress in Chemistry…
Editor: A. K. Haghi, pp. 35-46
ISBN: 978-1-61668-501-0

Chapter 3

PURE OXYGEN PENETRATION IN WASTEWATER FLOW

Kaveh Hariri Asli**[*1]**, Faig Bakhman Ogli Naghiyev**[2] **and Soltan Ali Ogli Aliyev**[3]
[1]Department of Mathematics and Mechanics,
National Academy of Science of Azerbaijan" AMEA", Baku, Azerbaijan,
[2]Azerbaijan State Oil Academy, Baku, Azerbaijan
[3]Department of Mathematics and Mechanics,
National Academy of Science of Azerbaijan" AMEA", Baku, Azerbaijan

ABSTRACT

In this work pure Oxygen was introduced in wastewater by a 40 *lit* pure *O2* container. The main objective of this research was to decrease *BOD5* amount equal to 80-90 percent. This paper presents the computational performance of numerical methods for modeling of pure oxygen diffusion in wastewater transient flow.

NOMENCLATURE

f–friction,
C–slope (*deg.*),
V–velocity (*m/s*),
t–time(*s*),
H–head (*m*)
V – Aeration Tank volume (m^3)
LO – total influent pollution (*kg – BOD 5*)
B_V – Tank volume rate ($kg - BOD\ 5/m^3.day$)

* Corresponding author E-mail: HARIRI_K@YAHOO.COM

B_V – Volume rate (*kg – BOD 5/m^3.Day*)
MLSS – Mixed liquor suspended solids (*kg*)
MLVSS – Mixed liquor volatile suspended solids (*kg – BOD 5/Day*)
R_t – Detention time (*h*)
H – Aeration Tank depth (*m*)
Q – Influent flow rate (*m3/h*)
V_S – sedimentation velocity (*m/sec*)
F/M coefficient is the food to Micro - Organism ratio
SMLSS – Sludge Mixed liquor suspended solids
MLSS – Mixed liquor suspended solids (*kg/m^3*)
V – Aeration tank volume (*m^3*)
SSE – Effluent suspended solids density (*kg/m^3*)
Q – Influent (*m3 /day*)
MLSS _ Mixed liquor suspended solids

RS – Return sludge
Q_{Max} – Max influent in without rain case.
SLS – Effluent sludge due to suspended solids (*kg*)
A – Constant (.35-.55) average ~ .5
B – Constant (.07 -.15) average ~ .1
y – Aeration tank efficiency
ON-Need of oxygen for Nitrogen zings
B_V – Tank volume rate (*kg-BOD5/m3.day*)
MLSS – Mixed liquor suspended solids (*kg /m^3.Day*)
Q_V – amount of oxygen per 24 hours for one cubic meter of aeration lagoon (*Kg/m3.day*).
C_X – Dissolved oxygen density (*mg/lit*)
C_S – amount of water dissolved oxygen in saturation condition (*mg/lit*)
Q_C – oxygen required per .1 (*m^3*)of aeration tank per 24 hours (*kgO2 / m^3 day*)
F –atmospheric pressure coefficient=1(at sea level) andfor every 1000 (*m*) height the "*F*" value decrease .1

1. INTRODUCTION

Miscible fluid mixtures that formed in the interpenetration between wastewater flows at parallel between plates and turbulent moving in pipe are common phenomena in our daily life. While two miscible fluids are mixed together, their appearances in terms of colors and shapes will be changed due to their mixing and interpenetration. The interpenetration between the mixture components could be regarded as a combination of the diffusing process and remixing process. This paper presents the computational performance of numerical methods for modeling of Pure Oxygen diffusion in wastewater transient flow. This Model was defined by method of the Eulerian based expressed in a method of characteristics *"MOC"based on* finite difference form. Pure Oxygen has been diffussed in the activated sludge wastewater treatment process (pipeline and aeration tank). This method needs to low detention time and low structural space for wastewater treatment plant[1]. In model studies and analysis of prototype problems, similarity law for flow in hydraulic machine is generally valid for one-phase flow. Present guidelines and standards for equivalent model and prototype analysis accept that similarity of flow in model and prototype turbomachines exist before the critical cavitation coefficients are reached. In between the zone of incipient cavitation and critical cavitation, similarity is considered to be satisfactory for the analyses of internal flow in machines. However, for flow in the downstream of the hydraulic machine and pipe system, a small quantity of Oxygen gas bubbles will change system wave speed characteristics to the extend that similarity for the model study of hydraulic transient and hydraulic vibration problems may be difficult to satisfy. When hydraulic machines operate in the zone of cavitation, air bubbles or Oxygen will flow through the machine resulting in two-phase flow in the downstream of the fluid system. Though deaeration devices may be used to minimize the air content, some air will still remain in the fluid system. Since it is almost impossible to predict the quantity of air getting through the machine and remain downstream of the system, systematic analysis has to be carried out with various assumed amount of Oxygen content in the model analysis of the transient fluid flow problem. The first study on the effects of air content on wave speed in a transient fluid system was conducted by Whiteman and Pearsall (1959). Detail survey on effects of air on wave speed in fluid systems was given by Lee (1991) [2].

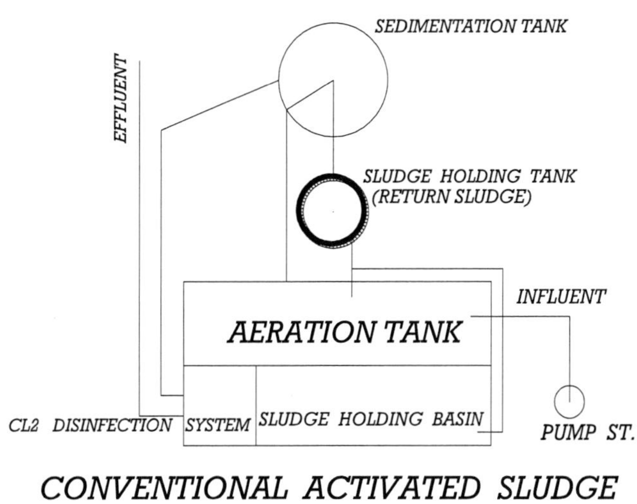

Figure 1. wastewater flow diagram in conventional activated sludge system (A.T. =Aeration Tank, S.T. = sludge Tank).

2. Formulation

Present work used the method of characteristics *"MOC"* to solve virtually any hydraulic transient problems of wastewater flow in conventional activated sludge system. The method of characteristics *"MOC"* is based on a finite difference technique, where pressures are computed along the pipe for each time step. Two cases are considered for modeling:

1- The combined elasticity of both the wastewater and the pipe walls is characterized by the pressure wave speed (Arithmetic method -combination of Joukowski formula and Allievi formula): [3]

$$H_2 - H_1 = (C/g)(v_2 - v_1) = \rho C(v_2 - v_1),\ \text{(Joukowski formula)} \tag{1}$$

$$c = 1/[\rho((1/k) + (d.C_1/E.e))]1/2,\ \text{(Allievi formula)} \tag{2}$$

With combination of Joukowski Formula and Allievi Formula[3]:

$$\begin{aligned} &\lambda\begin{bmatrix}(\partial v/\partial t) + (1/\rho)(\partial p/\partial s) + \\ g(dz/ds) + (f/2D)v|v|\end{bmatrix} + \\ &c^2(\partial v/\partial s) + (1/p)(\partial p/\partial t) = 0, \\ &\lambda =^{+} c \,\&\, \lambda =^{-} c \end{aligned} \tag{3}$$

Hence, water hammer pressure or surge pressure (*ΔH*) is a function of independent variables (*X*) such as:

$$\Delta H \approx \rho, K, d, C1, Ee, V, g, \tag{4}$$

2-The method of characteristics (*MOC*) based on a finite difference technique where pressures are computed along the pipe for each time step,

$$H_P = 1/2\begin{pmatrix} C/g(V_{Le} - V_{ri}) + (H_{Le} + H_{ri}) \\ -C/g(f\,\Delta t/2D)(V_{Le}\ |V_{Le}| - V_{ri}|V_{ri}|) \end{pmatrix}, \tag{5}$$

$$V_P = 1/2\begin{pmatrix} (V_{Le} + V_{ri}) + (g/c)(H_{Le} - H_{ri}) \\ -(f\,\Delta t/2D)(V_{Le}\ |V_{Le}| + V_{ri}|V_{ri}|) \end{pmatrix}, \tag{6}$$

Transient analysis results that are not comparable with actual system measurements are generally caused by inappropriate system data (especially boundary conditions) and inappropriate assumptions.

Behavior of the wastewater fluids flow state as a combination of the diffusing process and remixing process have been studied.In this process high speed treatment has been achieved.Also $CO2$ is released and $O2$ is utilized by micro – organisms as result of their activity Figure (1). In this work the fomulation of proccess in wastewater transmission line and aeration tank are as follows[1-2]:

$$V = LO/BV, \tag{7}$$

$$BV = MLSS \times MLVSS, \tag{8}$$

$$R_t = H / VS, \tag{9}$$

$$R_t = V / Q$$

In the case of low *F/M*, Micro - organisms feed by organic material in wastewater or feed by other Micro -Organisms Figure (1).

In order to save balancing condition, we need to high return sludge and high *MLSS*.

On the base of European standard, Tables(1-2) for Days without rain[4], the ratio of *RS* per *max* in- fluent become 100 percent and can be achieved from the flowing relations:

$$RS / Q_{max} = [MLSS / (SMLSS - MLSS)] \times 100, \tag{10}$$

3. Materials and Methods

There was a problem in the treatment process when Sludge volume index *S.V.I* was more than 200. In this condition, sedimentation was failed. By decreasing *S.V.I*, aeration time was decreased.

Sludge volume index – *S.V.I*

The amount of surplus sludge that must be removed from the settling tank was related to Biochemical oxygen demand «*BOD5*» of influent entrance to aeration tank and settling tank output.

Sludge age

$$\text{Sludge age} = (MLSS) \times (V) / (SS_e \times Q + SLS), \tag{11}$$

Oxygenation Calculation:

A Portion of absorbed pure Oxygen spent informs of energy consumption and multiplying of Bacteria. The other part of *O2* spent for oxidation of organic carbonate and organic nitrogenous material [4].

$$Q_V = A_Y B_V + B(MLSS) + 3 - 4(ON), \tag{12}$$

Process Design

$$O_C = [C_S / A(F)(C_S - C_X)] \times [.5YB_V + .1MLSS + 3.4ON], \tag{13}$$

Design criteria:

Influent production per person = 200 (*lit/day*)

Max influent factor = 1.71 (per 14 hours)

$$BV = .5(KG.BOD5 / M3DAY),$$ From table (2)

$$BV / MLVSS = F / M = .15 \text{ and } BOD5 = 60gr / p-d = .06kg / p-d,$$

Wastewater temperature=20 ◦*C*.

4. Results

In this work pure oxygen was introduced in to the wastewater from a 40 lit pure *O2* container. By submerged porous diffusers and by air nozzles pure *O2* was entered to aeration tank. Then micro – organisms by received pure *O2* was growed quickly.

Research approach:

Pollution Decreasing Figure (2-3) to 20 (*mg/lit*) for *BOD 5*

- Influent pollution calculation:
- Fixed population=700
-Office workers=500
-Total population=1200

$$Q_m^d = (700)(.2) + 500(.05) = 165(m^3 / day),$$

Max Influent per hour:

$Q^d{}_m$ = 165/14 = 12 (m^3/h) = 3.3 (*lit/sec*)

Total amount of Influent pollution :

.06(1200) =72(*kg/p-day*)

Average amount of influent pollution:

72/165 = .45 (kg/m^3) = 450(*mg/lit*)

Due to pollution-detention curves, Decrease of Influent pollution relation with detention time is

Detention time: R_t= *V/Q* = Tank Volume / influent flow rate

Rt +200/12 =17(*h*)

From :

Pollution Decreasing percent = .35

450 (1-0.35) = 293 (*mg/lit*)

Average amount of influent pollution

(Per24 hours) =.293 (165) = 48.5 (*kg BOD/day*)

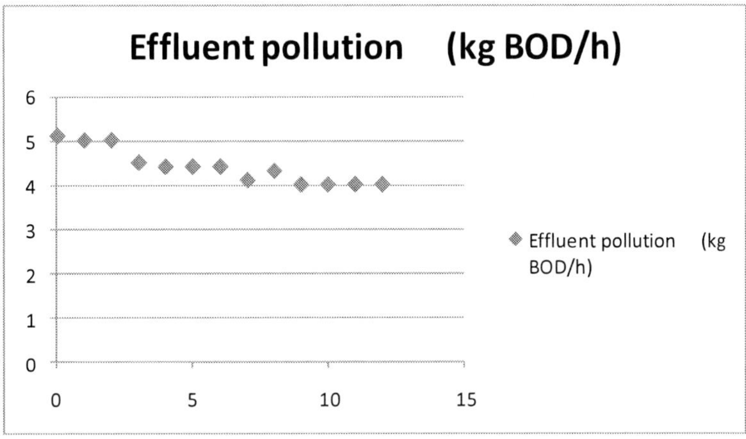

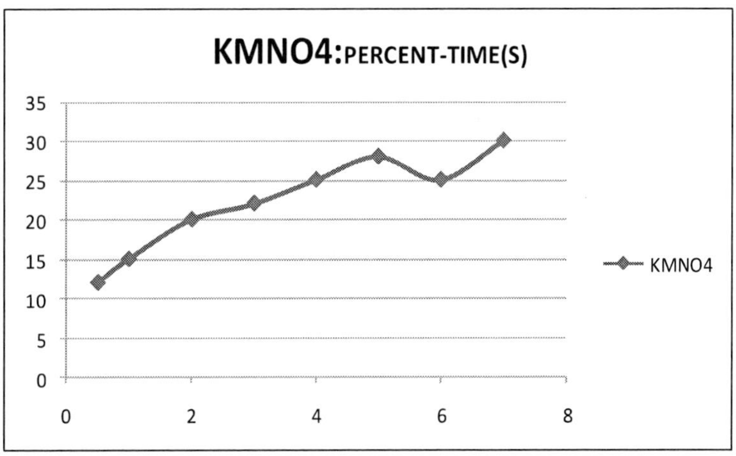

Figure 2. Effluent pollution-detention time curve.

Oxygen Required Calculation Figure (3):

From Table (1) andTable (2) :

C_S= 9.17 and A = .9

and Y= .925and F= 0.85 and $MLSS$ = 3.3 (kg/m^3) and 3.4(ON)=.23

$$O_C = [C_S / A(F)(C_S - C_X)] \times [.5(Y)B_V + .1(MLSS) + 3.4ON] \quad (14)$$

$$O_C = [9.17/.9(.85)(9.17 - 1.5)] \times [.5(.925).5 + .1(3.3) + 3.4(ON)]$$

$O_C = 1.2366$ ($kg\ O2/m^3.day$) , $\sum O_{C} = [200\ (1.2366)]\ /24 = 10.3$ ($kg\ O2/h$)

Detention time in aeration tank (h) Figure (3)

Research's Results:

Aeration Tank Volume (real value) =200($M3$)

Detention time (real value): $R_t = V/Q = 200/12 = 17(h)$

Theoretical required volume (existence data)

$V = Lo/BV = 72/.5 = 144m^3$

Theoretical detention time $R_t = V/Q = 144/12 = 12\ (h)$

Air(pure O2) Entrance Approaches

Modeling of air or pure *O2* injection in wastewater (Figure 4), influenced on hydraulic similarity. Two different types of air content models have been proposed in the literature in predicting the transient pressure behavior: the concentrated vaporous cavity model (Brown 1968; Provoost, 1976) [6] and the discrete air release model (Lee, 1991 — 95; Wylie, 1993) [6]. The concentrated vaporous cavity model produces satisfactory results in slow transients, but produces unstable solutions for rapid transients, such as pump's stoppage with reflux valve closure. The discrete air release model produces satisfactory results in pump shut down cases, but is susceptible to long term numerical damping (Ewing, 1980; Jonsson, 1985) [6]. Typically, in the discrete air release model, the wave speed distribution along a pipeline (with node points i =0, 1, . . . N) was given by Lee (1991). [6]

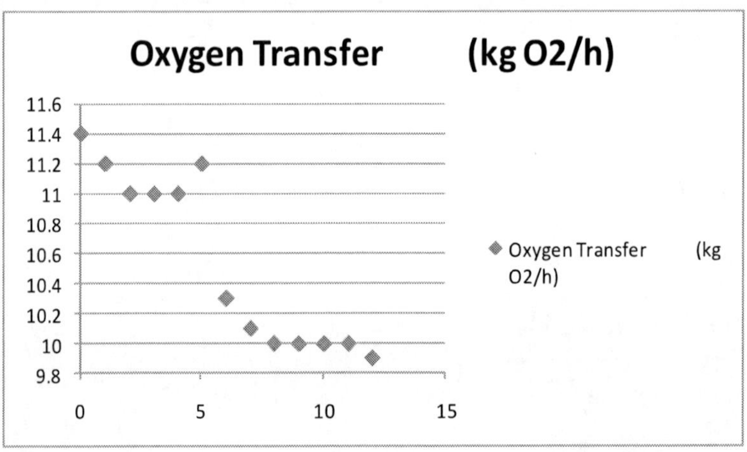

Figure Continued

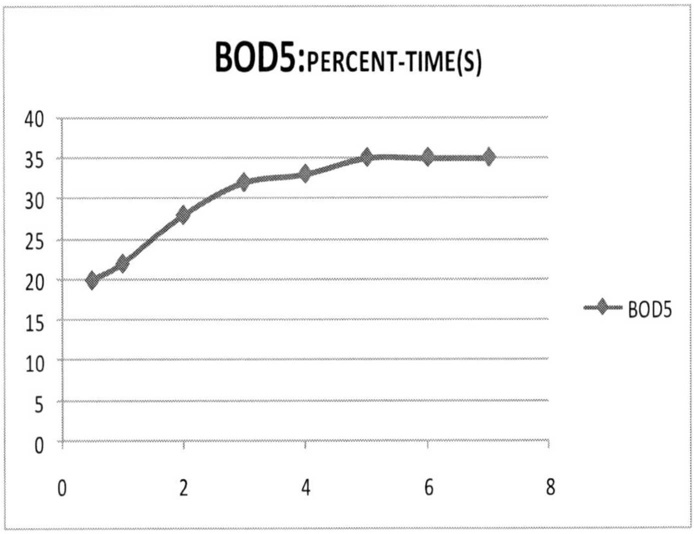

Figure 3. Oxygen Required for pollution decreasing due to detention time.

In this work ,at first step it was a simulated transient pressure in the system due to an emergency power failure without any protective equipment in service. After a careful examination of results, it was selected protective equipment and simulated the system again using modeling to assess the effectiveness of the devices which selected to control transient pressures. The formulated models can accurately describe complex flow features {such as wastewater vacuum system-pressurized flow interfaces, interface reversals, and open-pressurized pipeline surges. much of the complex dynamics in unsteady sewer flow is due to the air phase; it is suggested to incorporate the air-phase flow component to the mixed flow model.

Table 1. Saturation Dissolved oxygen (Cs) in water – At 760 mm-hg (mg/lit) [4]

Sea water			Brackish water	Water	Water temperature degree
Dissolved salts in water (*mg/lit*)					
20000	15000	10000	5000	0	
11.32	12.14	12.97	13.74	14.62	0
10.01	10.7	11.39	12.09	12.8	5
8.98	9.55	10.13	10.73	11.33	10
8.14	8.63	9.14	9.65	10.15	15
7.42	7.86	8.3	8.73	9.17	20
6.74	7.15	7.56	7.96	8.38	25
6.13	6.49	6.86	7.25	7.63	30

Figure 4. pure *O2* injection to aeration tank of wastewater treatment plant.

Table 2. Various aeration types Technical data[4]

Efficiency (%)	Air volume $m^3/kgBOD$	Returned Sludge *RS* (%)	Suspended solid weight in tank $MLSS(kgSS/m^3)$	Pollution load Sludge *F/M(kgBOD5/ kg MLVSS.d)*	Pollution load Tank Volume *BV (kgBOD5/ M^3.d)*	Type and Method of aeration
75 - 95	90 – 125	50 – 150	3 – 6	0.5 – 0.15	0.1 – 0.4	Extended aeration
85 - 95	45 - 90	15 – 50	1.5 – 3	0.2 – 0.4	0.3 – 0.6	Conventional activated sludge
85 - 95	45 - 90	15 – 50	1.5 – 3	0.2 – 0.4	0.3 – 0.6	Tapered aeration
85 - 95	45 - 90	20 – 75	2 – 3.5	0.2 – 0.4	0.6 - 1	Step aeration
80 - 90	45 - 90	25 – 100	2 – 5	0.2 – 0.4	1 - 2	Contact stabilization A.S.
75 - 90	25 - 90	100 - 500	4 - 10	0.4 - 1.5	1.6 - 6	Two stage activated sludge

CONCLUSION

As it was mentioned in the introduction, a robust and efficient numerical model able to reproduce unsteady gravity flows, unsteady pressurized flows and the simultaneous occurrence of gravity and pressurized flows in sewers was developed by the authors . The method of characteristics *"MOC"* approach transforms the water hammer partial differential equations into the ordinary differential equations along the characteristic lines defined as the continuity equation and the momentum equation are needed to determine V and P in a one-dimensional flow system. Solving these two equations produces a theoretical result that usually corresponds quite closely to actual system measurements if the data and assumptions used to build the numerical model are valid.

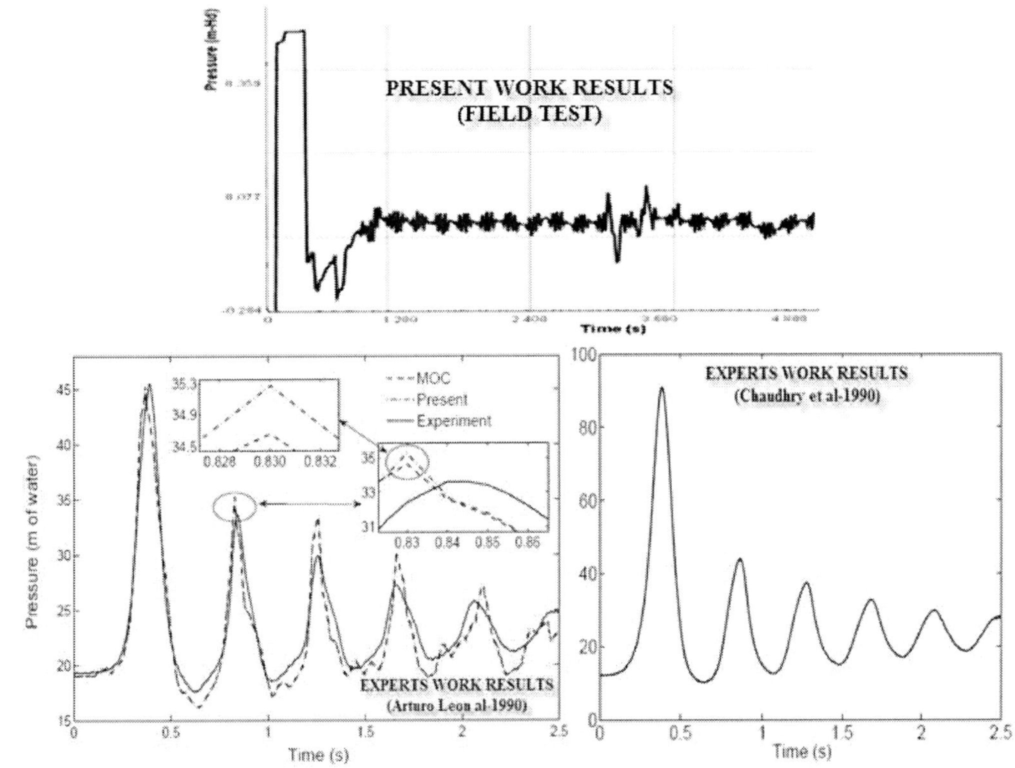

Figure 5. wastewater depth versus time , present work compartion with other experts[5-7].

The test procedure was as follows: A steady state flow of an O2- wastewater mixture was established in the wastewater pipe by controlling the exit valves and the pressure of the injected O2 at the inlet. The flow velocity of the O2- wastewater mixture was maintained at a high enough rate so that slug flow could be avoided by limiting the rate of O2 injection. Transient flow was created by a rapid control valve closure at the downstream end of the wastewater pipe. Transient-state pressures were monitored by high-frequency-response pressure detectors at two locations. By comparing the simulated results using the MOC and the proposed scheme, it can be seen that the pressure traces computed using the MOC are lower than the proposed scheme for all the simulations. This means that the MOC is more dissipative than the proposed scheme. However, the MOC agrees with the experiments slightly better than the proposed scheme, when the physical dissipation is estimated using only a steady friction formulation (as used in present work).The important aim in this research was to decrease of *BOD5* to amount of 80-90 percent. This work was a part of long term project at Rasht city in the north of Iran, which aims at developing a hydraulic model for real-time simulation of unsteady flows in sewers ranging from gravity flows, to partly gravity-partly surcharged. Numerical tests showed that to achieve a given level of accuracy, when a Courant number close to one and a large number of grids was used. Comparison between laboratory observations on fluid streamlines in pipes for three cases including: laminar flow-transient flow and turbulent flow showed the little error in the prediction of fluid flow software analysis results Figure (5).

REFERENCES

[1] Asli, K.H., Nagiyev, F.B., Hagi, A.K. Physical modeling of fluid movement in pipelines, Nanomaterials Yearbook - , USA, 2009, (has been accepted for publication in 2010, 1st quarter), https://www.novapublishers.com/catalog/product_info.php?products_id=11587.

[2] Lee, T.S. and Pejovic, S.: "Air Influence on Similarity of Hydraulic Transients and Vibrations". ASME Journal of Fluid Engineering, Vol.118, No.4, pp.706-709 (1996).

[3] Asli, K.H., Nagiyev F.B. , Haghi A.K. Three-dimensional Conjugate Heat Transfer in Porous Media, International Journal of the Balkan Tribological Association, Thomson Reuters Master Journal List, ISSN: 1310-4772, Sofia, Bulgaria, 336-346,2009.

[4] Monzavi, M.T. wastewater treatment, Tehran, Iran, 7-162, 1991.

[5] Arturo, S. Leon, Mohamed, S., Ghidaoui, M., Arthur R., Schmidt, M., Marcelo, H. Garc, a., M. Godunov-Type Solutions for Transient Flows in Sewers, 20-44, Illinois 61801, USA., (2005).

[6] Arturo Leon, S. Improved modeling of unsteady free surface, pressurized and mixed flows in storm-sewer systems, Submitted in partial fulfillment of the requirements for the degree of Doctor of Philosophy in Civil Engineering in the Graduate College of the University of Illinois at Urbana-Champaign, 57-58 (2007).

[7] Asli, K.H., Nagiyev, F.B., Hagi, A.K. "Some Aspects of Physical and Numerical Modeling of water hammer in pipelines", Nonlinear Dynamics An International Journal of Nonlinear Dynamics and Chaos in Engineering Systems, ISSN: 0924-090X (print version) ISSN: 1573-269X (electronic version) Journal no. 11071 Springer , USA,http://nody.edmgr.com/,2009, (has been accepted for publication in 2009).

In: Recent Progress in Chemistry… ISBN: 978-1-61668-501-0
Editor: A. K. Haghi, pp. 47-57

Chapter 4

THE EFFECTS OF MELT-SPINNING TEMPERATURE AND TAKE-UP SPEED ON THE PHYSICAL AND MECHANICAL PROPERTIES OF AS-SPUN POLYPROPYLENE FILAMENTS

Shahram Arbab*[*,1,2], Ehsanollah Nosratian [2], Ahmad Zeinolebadi[3], Mohamadreza Mohades Mojtahedi[2] and Ahmad Mosavi Shoshtari[2]

[1]Engineering Faculty, Department of Textile Engineering, U niversity of Guilan , P.O. Box 41635-3756 Rasht, Iran
[2]Textile Engineering Department, Amirkabir University of Technology, Tehran, Iran,
[3]Polymer Engineering Department, Amirkabir University of Technology, Tehran,Iran,

ABSTRACT

Three kinds of polypropylene resins with different melt flow index (MFI) values were melt-spun at different spinning temperatures and take-up speeds. The extrusion rate was set to result in fibers with similar linear density. Thus the effects of fiber diameter on the mechanical properties were omitted. The effects of different spinning conditions on the physical (birefringence, density and shrinkage) and mechanical properties (tenacity, modulus and elongation at break) of the fibers were investigated. It was shown that increasing spinning temperature caused an increase or a decrease in the properties of the fibers due to different effects of temperature on the melt viscosity, mobility of polymer chains, chains orientation and nucleation and growth rate of crystallites. Moreover, it was found that the mechanical properties of the fiber is mainly controlled by the characteristics of the starting polymer and variation of the spinning conditions could only adjust these properties in the limits set by the initial polymer properties.

[*] Corresponding author: E-mail: a7928949@aut.ac.ir , Tel: ++98 21 64542699, Fax: ++ 98 21 66400245

INTRODUCTION

The demand for polypropylene in the textile industry continues to increase due to the superior properties of this polymer such as low density, chemical resistance, mildew resistance and smooth touch to the skin [1]. Polypropylene fibers are usually produced via melt spinning which consists of imposing uniaxial deformation on the extruded melt while it is rapidly cooled down [2]. The final physical and mechanical properties of the fibers are mainly affected by the inherent characteristics of the polymer (e.g. molecular weight, polydispersity index and branching content) and processing conditions such as spinning speed and temperature, take-up speed, cooling rate and drawing tension and ratio [2-6].

Numerous investigations have been carried out to characterize the structure of the fiber and elucidate the relationship between structure formation and processing conditions on one hand and structure-property relationship (e.g. the relationship between crystalline structure and the final physical and mechanical properties of the fibers) on the other hand [5,6]. Capuccio et al. [7] found that the structure and mechanical properties of as-spun polypropylene yarns are mainly affected by spinning temperature and molecular weight distribution. Sheehan and Cole [8] and wang et al. [9] studied the conditions leading to the para-crystalline structure. Spruiell et al. [10] analyzed the kinetic of crystallization along the quenching zone and found a firm correlation between the orientations and thread-line tension. Nadella et al. [11] also showed that higher molecular weight and lower spinning temperatures increased the stress and molecular orientation. In addition, mechanical properties of fibers have been found to be mainly determined by the overall orientation and especially by the orientation of chains in the amorphous regions [12]. Moreover, Prost et al. [13] studied the effects of molecular weight of polypropylene on the relaxation time and elongational viscosity of its melt which affected the orientation of chains in the time frame of the cooling process. Heat setting, structure deformation during the heating process and their effects on the physical properties and dimension stability of the fibers were widely discussed by Ahmed [14].

In this study, we tried to shed light on the effects of the spinning temperature and take-up speed on the structure and properties of as spun polypropylene fibers in the melt spinning process. Commercial grade polypropylene resins with different melt flow index (MFI) values were selected to give generality to the results. As-spun polypropylene filaments were produced with nearly same linear densities under various take-up speeds and spinning temperatures. Structural, physical and mechanical properties of the filaments were characterized and their variations with spinning conditions were analyzed.

EXPERIMENTAL

Spinning: The melt spinning was carried out using an Automatic pilot-plant spinning machine with a screw 35 mm in diameter and aspect ratio (length over diameter) of 25. The spinneret consisted of 36 holes, each 0.25 mm in diameter. The take-up speed and polymer throughput were tuned such that the linear density of as-spun yarns was maintained approximately at a constant value. The starting materials were commercial polypropylene granules supplied by Bandar Imam Petrochemical Co. with melt flow index (MFI) of 12, 16 and 20 g/10min at 190 °C under 2.16 kg load.

Fibers characterization: A gradient column containing water/methanol mixture at 23 °C were used to assess fiber density. An interference microscope (Amplival Interphako, Germany) was used to measure fiber birefringence. Glycerin oil was used as the liquid phase. Tensile properties of fibers (Young's modulus, tenacity and elongation at break) were measured using an Instron tensile tester (Instron 5565). Fibers with gauge length of 2.5 cm were clamped and stretched with a rate of 25 mm/min up to rupture and their mechanical characteristics were determined. For each sample at least 15 replicates were characterized and the average values were reported. Fiber shrinkage was obtained after leaving yarn skins in an oven with a temperature of 130 °C for 15 min. All values are reported within a maximum 10% error.

Results and Discussion

Spinning conditions and Melt flow index values (MFI) in this study are given in table 1. The take-up speed and polymer throughput were adjusted such that the linear density of as-spun yarns was approximately maintained at a constant value.

Figures 1a and b present the difference in optical birefringence of all fibers at different take-up speeds and spinning temperatures. It is seen that higher take-up speed led to higher birefringence values for most of the samples at different spinning temperatures, Figure 1a. This implied that molecular orientation along the fiber axis and chain extension increased at higher take-up speeds regardless of the molecular characteristics of the polymer, namely molecular weight and polydispersity index which can be inferred from different MFI values. Indeed, at higher take-up speeds a higher deformation rate, which enhances molecular orientation in both amorphous and crystalline regions, is imposed on the fibers [2]. Thus, the birefringence value which is a measure of molecular orientation, increased by increasing take-up speed, Figure 1a. Increasing spinning temperature, however, resulted in different behaviors, Figure 1b. Generally, increasing temperature causes a reduction in polymer melt viscosity due to lower molecular friction coefficients and enhances thermomechanical degradation of polymer chains at the same time. The lower viscosity reduces the stress on the fiber and consequently weakens chains extension.

Table 1. Spinning conditions MFI values

<table>
<tr><td>Resin Code</td><td>MFI (g/10min)</td><td>Extrusion temperature (°C)</td><td>Take-up speed (m/min)</td><td>Linear density (Tex)</td></tr>
<tr><td>M12</td><td>12</td><td>260, 270, 280, 290</td><td rowspan="3">For all samples 1600-2000-2400-2800</td><td rowspan="3">For all samples 11.5-12.3</td></tr>
<tr><td>M16</td><td>16</td><td>260, 270, 280</td></tr>
<tr><td>M20</td><td>20</td><td>230, 240, 250, 260</td></tr>
</table>

On the other hand, higher mobility of polymer chains at higher temperatures facilitates orientation of chains in amorphous/crystalline regions parallel to the fiber axis. Thus, the interplay between the lower viscosity and enhanced mobility of chains at higher temperatures led to a non-monotonic pattern in the fiber birefringence which showed a decrease, an increase or even a maximum (minimum), Figure 1b.

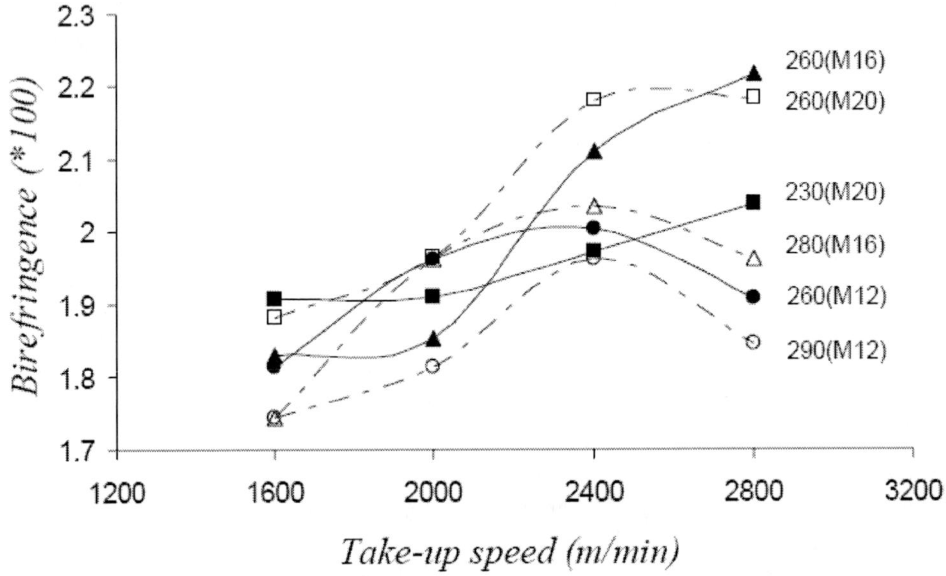

a

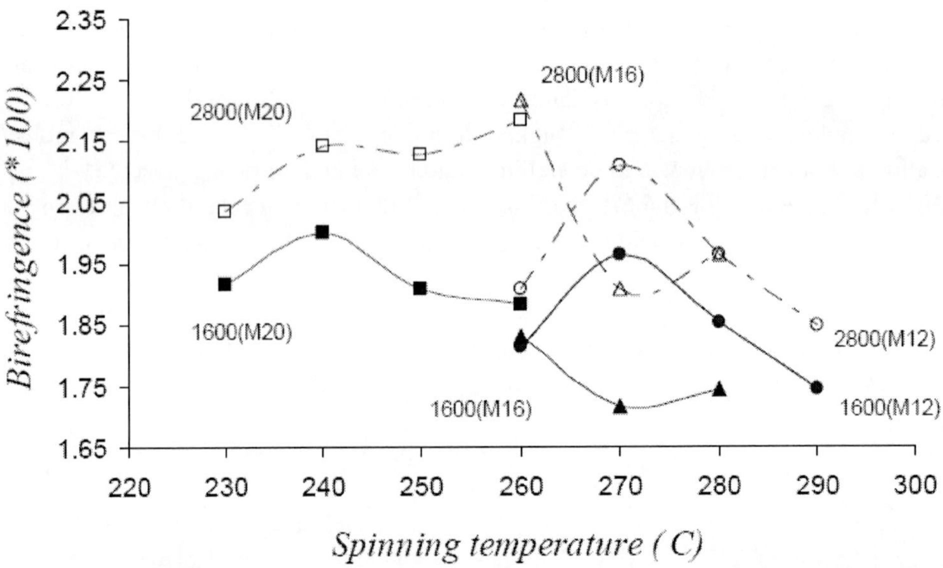

b

Figure 1. Fiber birefringence versus a) Take-up Speed, the first number on each curve indicates spinning temperature and the number in parentheses indicates sample code. b) Spinning temperature, the first number on each curve indicates take-up speed and the number in parentheses indicates sample code.

Increasing the take-up speed caused an increase in the fibers' density at all temperatures and for all samples, Figure 2a. Density of semi-crystalline polymers depends on their crystallinity percent [17], thus the higher density of the fibers at higher take-up speeds implies the higher crystallinity percent of the fibers. This is due to the fact that higher take-up speed imposed higher orientation and extension on the molecular chains which enhanced in part the crystallinity.

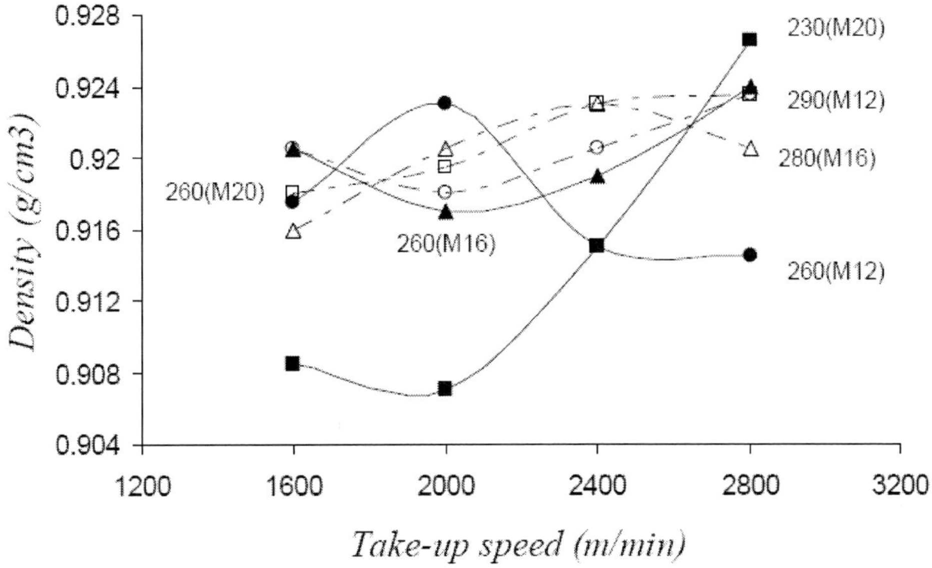

a

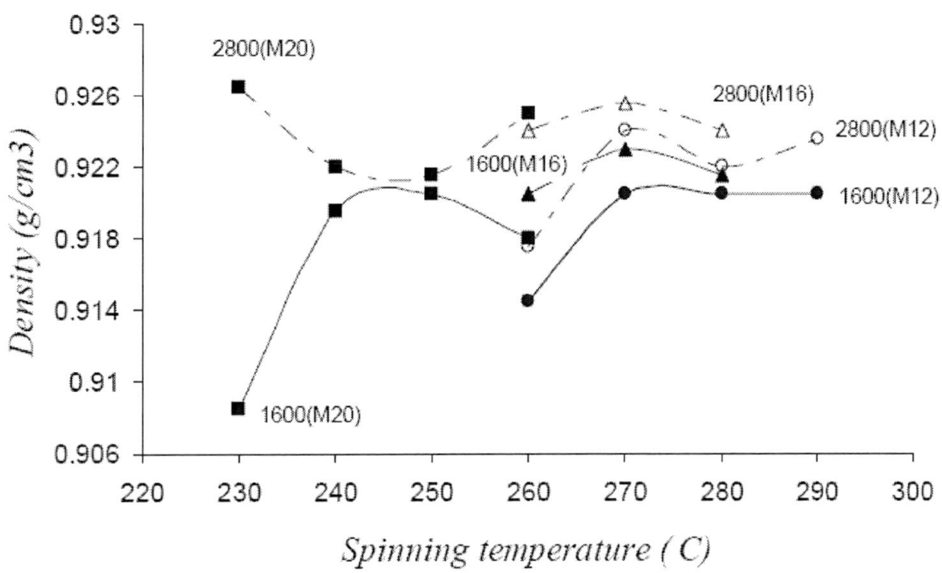

b

Figure 2. Fiber density versus a) Take-up Speed, the first number on each curve indicates spinning temperature and the number in parentheses indicates sample code. b) Spinning temperature, the first number on each curve indicates take-up speed and the number in parentheses indicates sample code.

This behavior was observed for polymers with different MFI values, hence it can be inferred that enhancement of crystallinity by increasing take-up speed was a general trend. Increasing temperature, however, led to an increase or a decrease in the fiber density, Figure 2b. At higher temperatures the nucleation rate is lower whereas the growth rate of the crystallites is higher [17]. Thus, increasing the temperature may increase or decrease the overall crystallinity of the fiber and consequently its density.

Tenacity, modulus and elongation at break of the fibers are presented in figures 3-5, respectively. Generally, fibers' tenacity increased by increasing the take-up speed. This can be attributed to the higher crystalinity degree at higher take-up speeds.

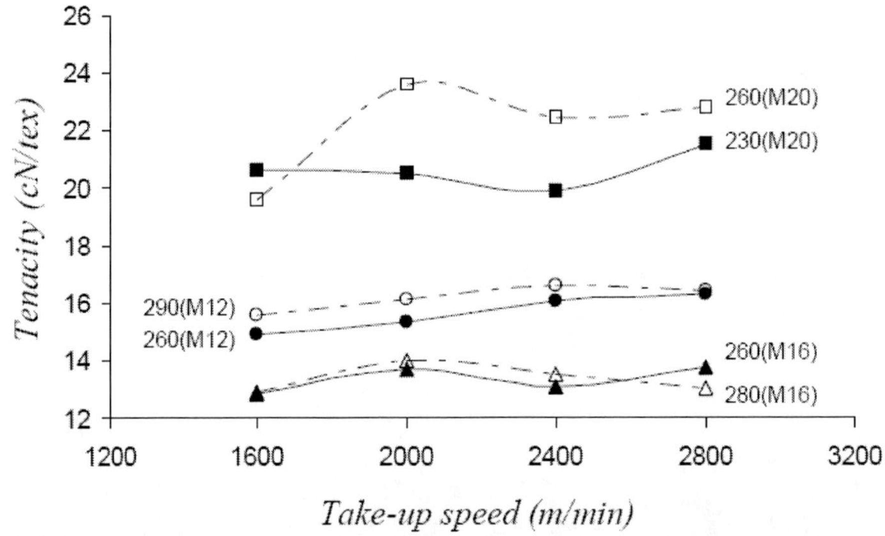

a

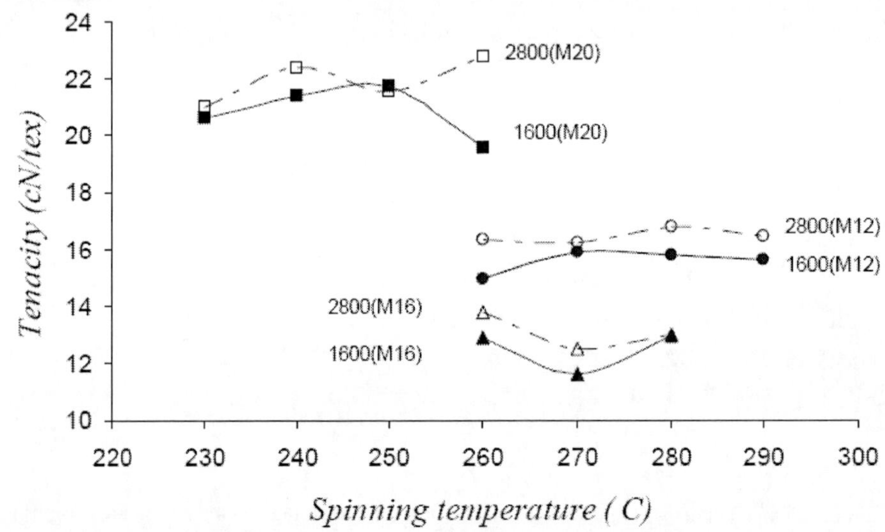

b

Figure 3. Fiber tenacity versus a) Take-up Speed, the first number on each curve indicates spinning temperature and the number in parentheses indicates sample code. b) Spinning temperature, the first number on each curve indicates take-up speed and the number in parentheses indicates sample code.

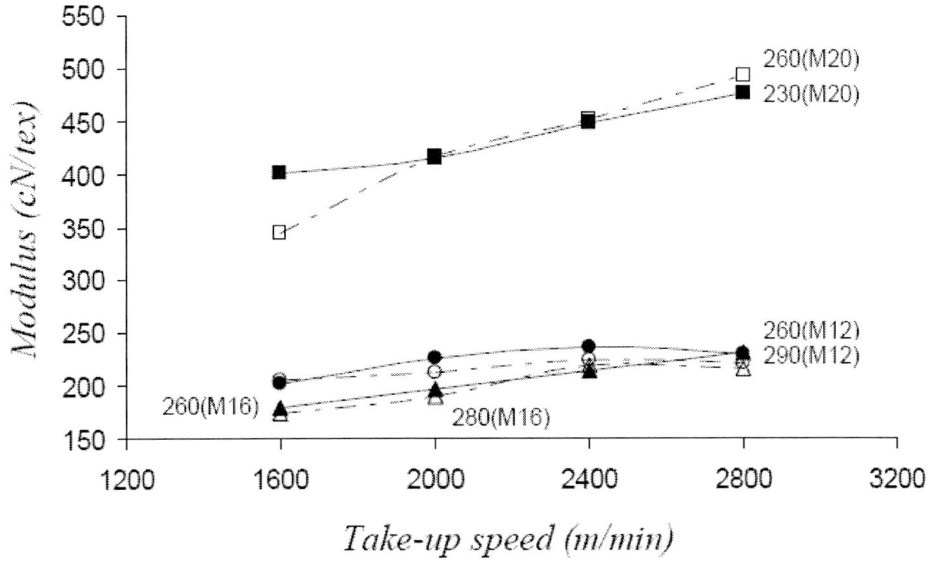

a

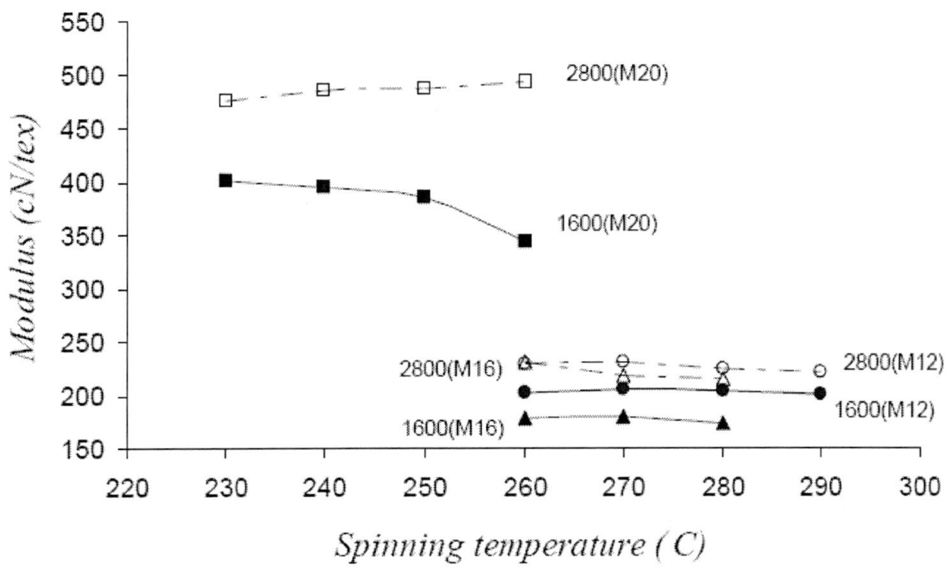

b

Figure 4. Fiber modulus versus a) Take-up Speed, the first number on each curve indicates spinning temperature and the number in parentheses indicates sample code. b) Spinning temperature, the first number on each curve indicates take-up speed and the number in parentheses indicates sample code.

On the other hand, fiber's tenacity increased or decreased by increasing spinning temperatures and in this case also no monotonic behavior was seen. Moreover, it is seen that increasing take-up speed or temperature did not affect the fibers' tenacity significantly. In other words, fiber tenacity is not mainly controlled by the spinning conditions, but the molecular characteristic of the polymer played the major role in determining the mechanical behavior of the fibers.

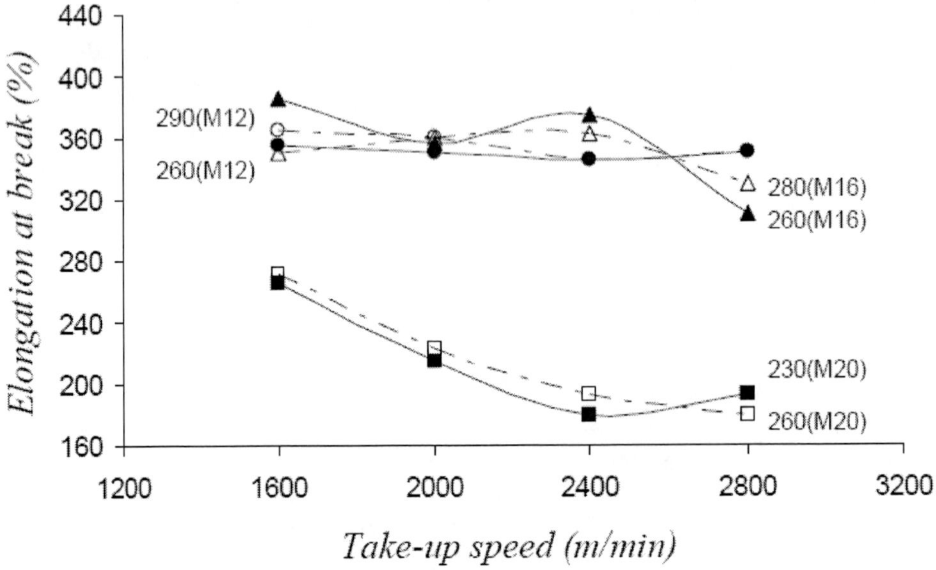

a

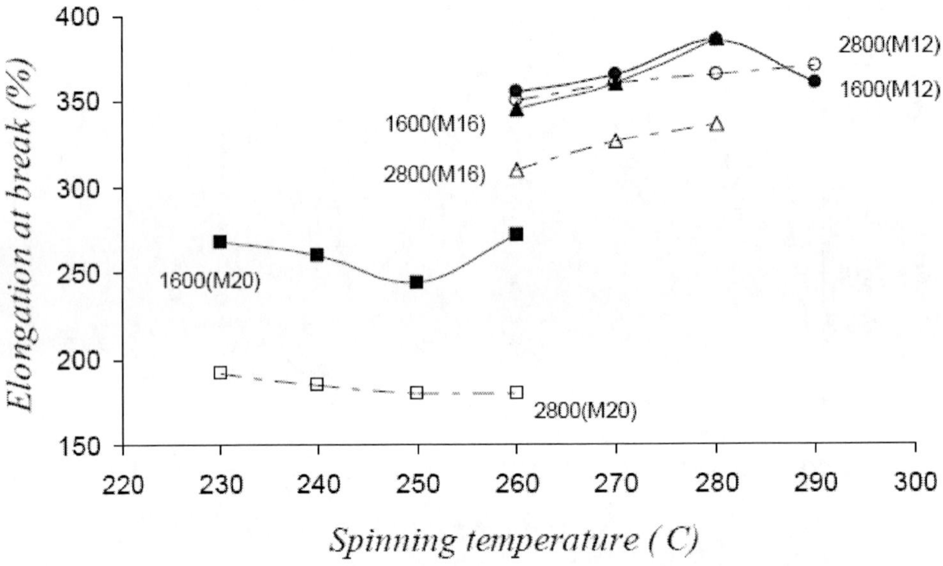

b

Figure 5. Fiber elongation at break versus a) Take-up Speed, the first number on each curve indicates spinning temperature and the number in parentheses indicates sample code. b) Spinning temperature, the first number on each curve indicates take-up speed and the number in parentheses indicates sample code.

It is seen that fibers spun from the M20 resin had higher tenacity values than the fibers spun from the other two resins. Thus, molecular weight, polydispersity index and also molecular structure (e.g. degree of branching, comonomer content, tacticity…) are the main controlling parameters of the mechanical properties and spinning conditions seemed to adjust the mechanical properties of the fibers just in the limits set by the molecular characteristics of the polymer.

The effect of increasing the take-up speed on the fibers' modulus is similar to its effect of the fibers' tenacity. As expected, fibers' modulus increased by increasing the take-up speed, but this effect was more significant for M20 resin. This also confirms the conclusion that the final mechanical properties of the fibers are mainly determined by the characteristics of the starting material.

Increasing spinning temperature did not affect the fibers' modulus. This is of practical importance, in the point of view of energy consumption, as it indicates that it is not necessary to spin fibers at higher temperatures to obtain a high modulus.

Increasing take-up speed or spinning temperature led to a decrease in elongation at break of the fibers. The variation of the fibers' elongation at break is consistent with the changes of modulus. The general behavior of polymeric fibers is that elongation at break decreases as fiber modulus increases.

Another important physical property of the fiber, which has great practical importance especially in post processing and drying process, is the shrinkage. Increasing the spinning temperature caused an increase or decrease in the fiber shrinkage due to the different effects of temperature on the orientation and extension of polymer chains, Figure 6a.

On the other hand, increasing the take-up speed led to an increase in the shrinkage at all spinning temperatures and for all samples, Figure 6b. Shrinkage is related to the degree of chain orientation in the amorphous region.

Thus, the increase in the shrinkage percent is an indication that the degree of orientation increased significantly at higher take-up speeds.

Therefore, it can be concluded that by increasing the take-up speed the chains in the amorphous region are oriented/extended to a higher degree regardless of the chain structure or molecular weight.

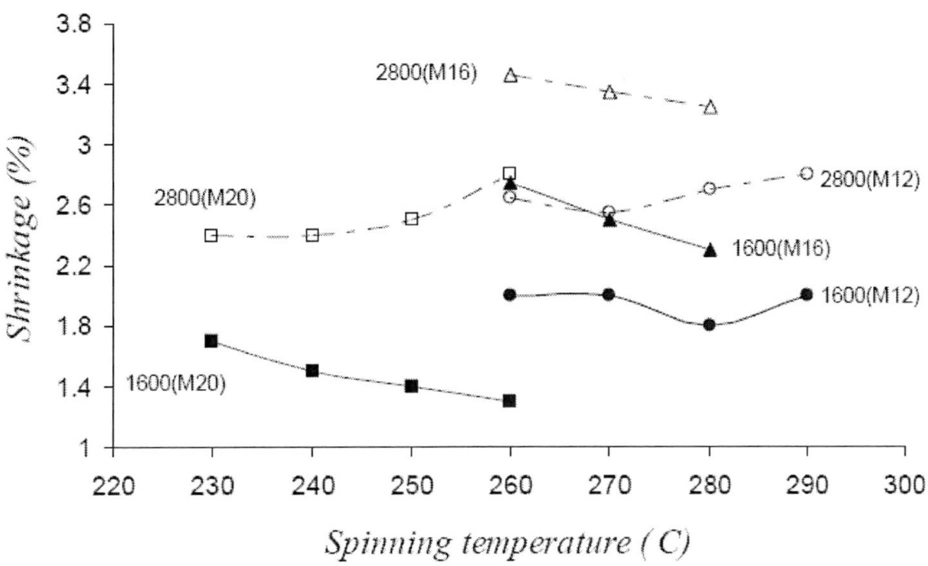

a

Figure 6. (Continued).

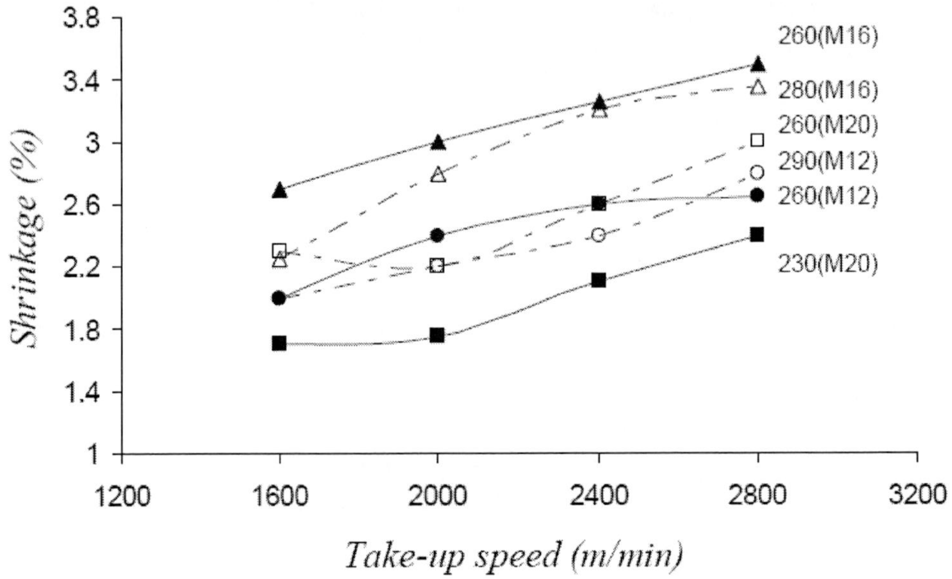

b

Figure 6. Fiber shrinkage versus a) Spinning temperature, the first number on each curve indicates take-up speed spinning temperature and the number in parentheses indicates sample code. b) Take-up Speed, the first number on each curve indicates spinning temperature and the number in parentheses indicates sample code.

Conclusions

Increasing spinning temperature caused an increase or decrease in the physical properties (birefringence, density and shrinkage) of the fibers due to the different affects of temperature on the viscosity of the molten polymer, mobility of chains and nucleation and growth rate of crystallites. The variation of the spinning temperature, however, had little influence on the mechanical properties of the fiber. On the other hand, increasing the take-up speed led more or less to a monotonic change in the physical or mechanical properties of the fibers. Birefringence, density, tenacity, modulus and shrinkage percent increased whereas elongation at beak decreased by increasing the take-up speed. It was also found that variation of the spinning conditions (temperature or take-up speed) did not influenced the mechanical properties of the fibers significantly. The inherent characteristic of the starting material such as molecular weight and its distribution or branching content determined the final properties of the fiber and spinning conditions could adjust the fibers properties just in the limits set by the characteristic of the polymer.

References

[1] Peckstadt JP (2000) *Chemical Fibers International* 50:412-413.
[2] Kolb R, Seifert S, Stribeck N, Zachmann HG (2000) *Polymer* 41:1497-1505.
[3] Van Parys M (1998) *Chemical Fibers International* 48:317-323.

[4] Broda J (2003) *J. Appl. Polym. Sci* 89:3364-3370.
[5] Paradkar RP, Patel RM, Knickerbocker E, Doufas A (2008) *J. Appl. Polym. Sci* 109:3413-3420.
[6] Patel RM, Doufas A, Paradkar RP (2008) *J. Appl. Polym. Sci* 109:3398-3412.
[7] Capuccio V, Coen A, Bertinotti F, Conti W (1962) *Chim End* 44:463-471.
[8] Sheehan WC Cole TB (1984) *J. Appl. Polym. Sci*, 8:2359-2388.
[9] Wang IC, Dobb MG, Tomka JG (1995) *J. Text. Institute*, 86:383-392.
[10] Spruiell JE, White JL, (1975) *Appl. Polym. Symp* 27:121-127.
[11] Nadella HP, Henson HM, Spruill JE, White JL (1997) *J. Appl. Poly. Sci,* 21:3003-3022.
[12] Samules, RJ (1974) Structured Polymer Properties. Wiley, New York.
[13] Prost C, Nemoz G (1989) *Macromol Chem, Macromol. Symp*, 23:16-20.
[14] Ahmad, (1982) Polypropylene Fibers, Science and Technology. Elsevier, Amsterdam.
[15] Lu FM, Spruiell JE (1987) *J. Appl. Polym. Sci* 34:1541-1556.
[16] Shimizu J, Jinan C, Kikutani T, Takaku A (1989) *J. Appl. Polym. Sci*, 37:2683-2697.
[17] Sperling LH (2006) Introduction to physical polymer science, John Wiley, New York.

In: Recent Progress in Chemistry...
Editor: A. K. Haghi, pp. 59-66
ISBN: 978-1-61668-501-0

Chapter 5

TRANSPORT PROPERTIES OF BULKY PAN NANOFIBER MATS

Sedigheh Borhani*[*1], Seyed Abdolkarim Hosseini Ravandi[1] and Seyed Gholamreza Etemad[2]

[1]Department of Textile Engineering, Isfahan University of Technology
Isfahan 84156-83111, Iran

[2]Department of Chemical Engineering, Isfahan University of Technology
Isfahan 84156-83111, Iran

ABSTRACT

In this study a non-conventional electrospinning technique was designed for production of high bulky polyacrylonitrile (PAN) nanofiber mats. Optimum nanofiber mats are achieved with 15 wt% solution of PAN in DMF. Such mats result in a bulk porosity which is as high as 99.9 and density as low as 0.84×10^{-3} g/cm^2. Effect of nanofiber mats porosity on the air permeability and moisture transfer was investigated. Based on the results, high bulky nanofiber mats possess high moisture transfer. Experimental data reveals that with a slightly decrease in the bulk porosity, air permeability decreased noticeably, while moisture transfer changed slightly.

INTRODUCTION

Fibrous materials used for filter media provide advantages of higher filtration efficiency and lower air resistance, which are closely associated with fiber fineness [1, 2]. Filtration efficiency is one of the most important concerns for filter performance [1- 5]. There are various methods to produce ultra fine fibers [1, 2]. Recently, much attention is being directed towards electrospinning as a unique technique for the fabrication of nanofibers [1- 9]. In electrospinning, a high voltage is applied to a capillary containing polymer solution. At a

* Email: sborhani@cc.iut.ac.ir

sufficient voltage to overcome surface tension forces, a charged fluid jet is ejected from the needle tip. The jet is stretched and elongated before it reaches the target, then dried and collected as randomly oriented structures in form of nonwoven mat. Electrospinning provides an ultra thin mat of extremely fine fibers with very small pore size and high porosity, which makes them unique candidates for use in filtration, and possibly protective clothing applications [1- 5].

These nanowebs have good aerosol particle obstruction and comparatively low air resistance. Recently, the filtration properties of electrospun mats have been studied [3- 8]. To provide appropriate mechanical properties, nanofiber webs have been applied onto various substrates. Substrates are often chosen to resemble conventional filter materials [8].

Gibson et al [6, 7] have reported some properties of electrospun mats. They compared performances of electrospun fiber mats with properties of textiles and membranes currently used in protective clothing systems and showed that electrospun layers are extremely efficient for trapping airborne particles. Also, they reported that the air flow resistance and aerosol filtration properties are affected by the coating weight. It was shown that an extremely thin layer of electrospun nanofibers eliminated particle penetration through the layer [6-7]. Transport properties of electrospun nylon6 mats were investigated by Ryu et al [9]. They found that concentration of polymer solution affected the fiber diameter, pore size, Brunauer-Emmett-teller (BET) surface area and gas transport properties of mats. It was shown that the filtration efficiency of Nylon6 nanofilters is superior to the commercialized high efficiency particulate air (HEPA) filter for 0.3 micrometer test particles [4]. Li et al [10] found that the pore size and pore size distribution of electrospun polylacticacid (PLA) membranes are strongly associated with fiber mass, fiber diameter and fiber length.

Researchers found that the electrospun mats provide good aerosol particle protection, without a significant change in moisture vapor transport. It was shown that materials used in protective clothing must provide a combination of high barrier performance and thermal comfort [6, 11]. It has been recognized that the heat and moisture transport behavior of textile materials is one of the most important factors influencing the dynamic comfort and performance of clothing in normal use [12]. Significant theoretical and experimental investigation has been done in this field by various researchers [12, 13]

Despite the many excellent studies which have been done on the properties of nanofiber mats, due to insufficient information on heat and moisture transfer of nanofiber mats, more studies in these fields are required.

In this study, bulky nanofiber mats were produced by using a modified electrospinning method. Bulk porosity of nanofiber mats was measured and the effect of bulk porosity on the air permeability and moisture transfer of nanofiber mats was evaluated.

EXPERIMENTAL

Polymer solution with concentration 15wt. % was prepared by dissolving PAN in DMF. PAN solution in DMF was loaded into a 1ml syringe. A modified electrospinning system consists of two syringe pumps was used to produce bulky nanofiber mats. The experimental setup of electrospinning has been shown in Figure 1.

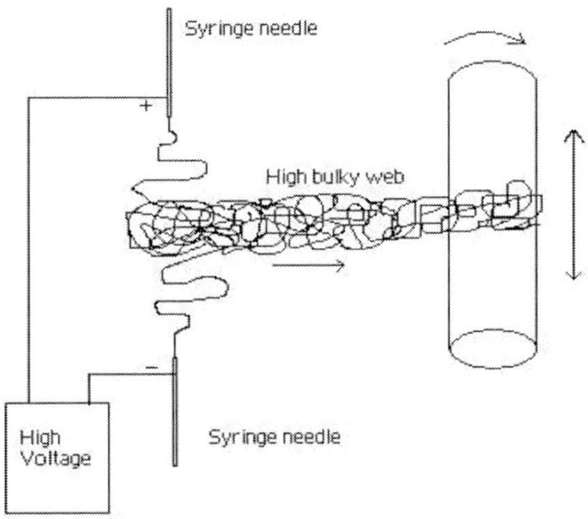

Figure 1. Schematic illustration of the modified electospinning system setup.

Polymer solution fed at speed of 2.8 $\mu l/\min$ through needles with an outer diameter of 0.7 mm. A 9 kV voltage applied between the two needles while nozzle distance was 10cm. The tip to collector distance was 15 cm.

The rotation speed of the drum was changed from 31.5 to 103.5 RPM. The bulk porosity of nanofiber mats was measured by the following equations:

$$\varepsilon = 1 - \frac{\rho_{mat}}{\rho_f} \quad (1)$$

$$\rho_{mat} = \frac{W_{mat}}{L \times A} \quad (2)$$

where ε, ρ_{mat}, ρ_f, W_{mat} ,L *and* A are bulk porosity, nanofiber mat density (g/cm^3), nanofiber density (1.16g/cm^3), weight of nanofiber mat (g), nanofiber mat thickness(0.2cm), and nanofiber mat area(cm^2), respectively.

The morphology of electrospun mats was observed by a scanning electron microscope (SEM, Philips XL-30).

Thickness of nanofiber mats was measured by measuring of collector diameter before and after electrospinning using digital images as shown in Figure 2. Microstructure measurement software was used to calculate the diameter of PAN nanofibers from SEM images at magnification of 5000. A minimum of 100 fibers were used to calculate the mean values of the fibers diameter.

Air permeability of PAN nanofiber mats was measured by using a Shirley permeation analyzer at 65 % relative humidity (RH), 25 $^{\circ}C$, and 100Pa.

In order to investigate the dynamic moisture transfer of nanofiber mats, an experimental apparatus was designed and constructed which was able to simulate the sweating of human body (Figure 3).

It consists of a controlled environmental chamber, sweating guarded hot plate and data acquisition system. The guarded hot plate, maintained at 37°C and used as a heat source, was housed in a chamber with ambient conditions of 25°C, and 65%RH. The diffusion cell consists of a water container, a piece of animal skin for simulating human skin, humidity sensors. One side of nanofiber mats faces the sweating skin but does not contact it, whereas another side is exposed to the controlled environment. The driving forces for the movement of moisture vapor are the temperature and vapor gradients maintained between the points where the moisture vapor emerges from the simulated skin (37°C, 90%RH) and the ambient environment controlled at 25°C and 65% RH.

For comparison between different samples, each sample is tested for 10 min. The one–way analysis of variance at the 0.05 level of significance was used to comparisons among the treatments means.

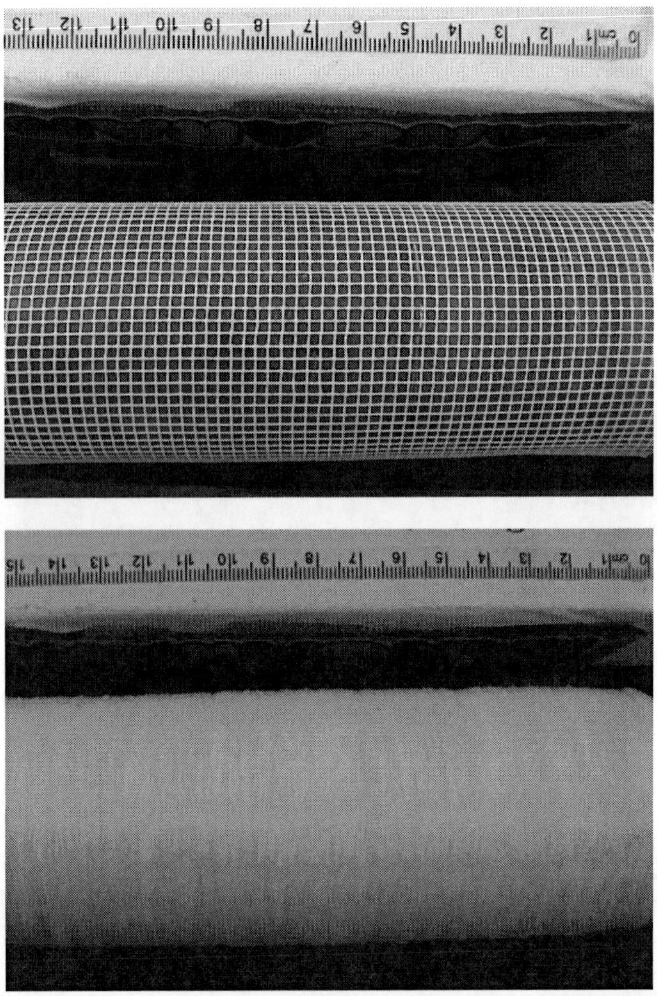

Figure 2. Digital photos of Collector, a- before electrospinning b- after electrospinning.

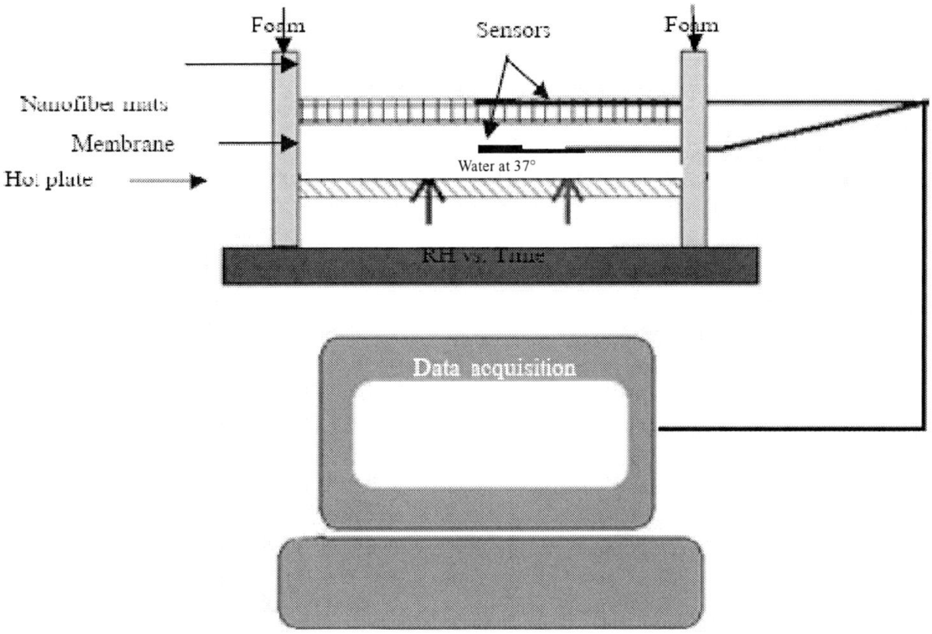

Figure 3. Schematic design of the instrument for measurement of moisture transfer.

RESULTS AND DISCUSSION

Present experimental investigation was carried out for a wide range of rotation speed of collector. Table 1 represents the characteristics and transport properties of bulky nanofiber mats.

Results indicate that bulk nanofiber mats have high porosity. Increase in speed rotation of collector increases the density of nanofiber mats and decreases bulk porosity of nanofiber mats. Statistical analysis stresses on the significant reduction of the bulk porosity of nanofiber mats.

Based on the results in Table 1and Figure 4, the air permeability of nanofiber mats increases by increasing in porosity and decreasing density of nanofiber mats. Reduction in mat density results in increase of the bulk porosity and pore size of the mat. Also results show that the moisture transfer of nanofiber mats is high which reduces by increasing the density and decreasing the bulk porosity of nanofiber mats as shown in Figure 5.

Statistical analysis shows the transport properties significantly reduce as bulk porosity of nanofiber mats decreases.

Comparing the nanofiber mats with different porosities, shows that the nanofiber mat with higher porosity and air permeability has the higher moisture transfer. This result indicates that the high resistance to the air permeability of nanofiber mats dose not impeded the moisture transfer through the bulk nanofiber mats. This difference is related to the size of water vapor and air molecules.

Table 1. Characteristics and transport properties of bulk nanofiber mats

rotation speed of collector *(RPM))*	Density of nanofiber mats *(g/cm^3)*	Bulk porosity (%)	Air permeability *(l/m^{2}sec)*	Moisture transfer (%)
substrate	0	~100	4000 >	100
31.5	0.84×10^{-3}	99.92	477.5	98.4
49.5	1.65×10^{-3}	99.86	245.5	82.5
67.5	3.1×10^{-3}	99.73	126.7	74.7
85.5	4.62×10^{-3}	99.6	58.7	64.8
103.5	6.64×10^{-3}	99.43	46.9	60.6

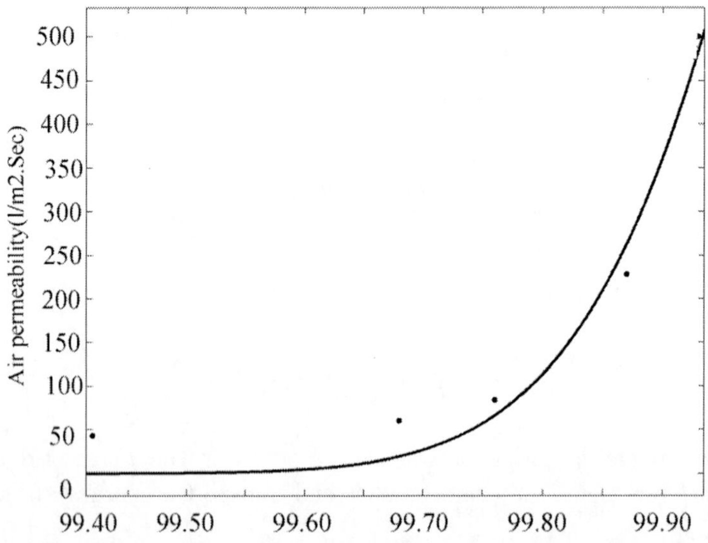

Figure 4. Air permeability versus bulk porosity of nanofiber mats.

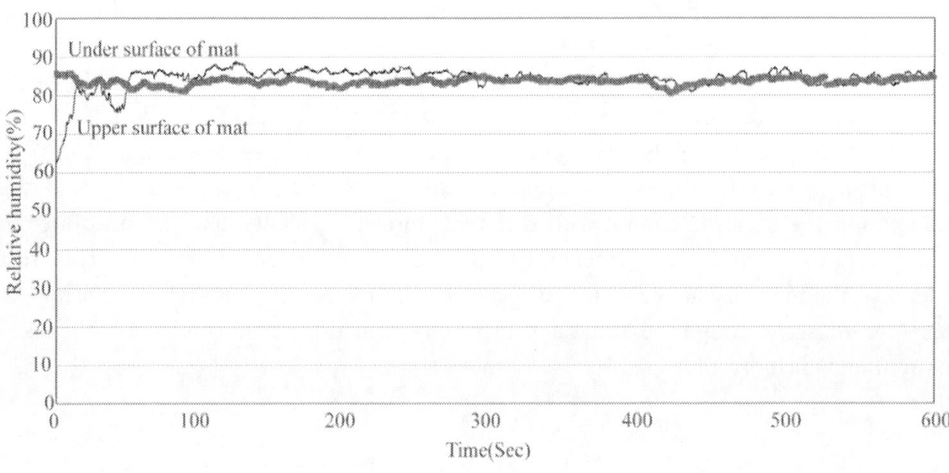

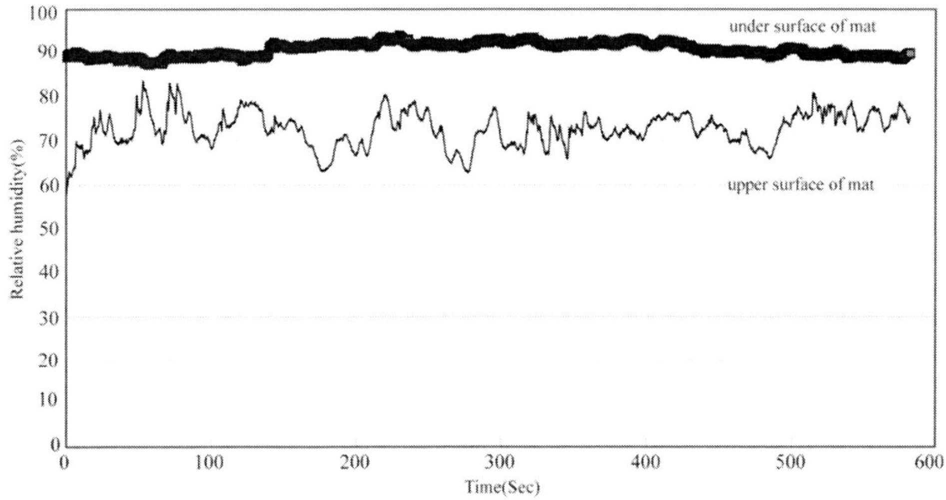

Figure 5. Experimental results of the relative humidity of the nanofiber mats with different porosity, a-99.92%, b-99.42%.

CONCLUSIONS

Electrospun nanofiber mats by modified electospinning system have high porosity. Results show that the porosity of nanofiber mats increases by decreasing speed rotation of drum. Air permeability and moisture transfer of nanofiber mats were reduced with reduction the porosity of nanofiber mats. Based on experimental results, moisture transfer behavior is influenced by nanofiber mats porosity. Bulk nanofiber mats with high rates of water vapor diffusion and low air permeability are promising candidates for protective clothing applications.

REFERENCES

[1] Huang Z. M., Zhang Y.Z., Kotaki M., Ramakrishna S., A Review on Polymer Nanofibers by Electrospinning and their Applications in Nanocomposites, *Composites Science and Technology*, 63, 2223-2253, 2003.

[2] Ramakrishna, S.; Fujihara, K.; Teo, W.E.; Lim, T.C.; and Ma, Z., Introduction to Electrospinning and Nanofibers, *World Scientific Publishing Co. Pte.Ltd. ISBN* 981-256-415-2, 2005.

[3] Gibson P., Gibson H.S., Use of Electrospun Nanofibers for Aerosol Filtration in Textile Structures, *Proceeding of 24th Army Science Conference (ASC2004) Orland, FL,* 20 November-December 2004.

[4] Ahn Y.C., Park S.K., Kim G.T., Hwang Y.J., Lee C.G. , Shin H.S., Lee J.K., Development of High Efficiency Nanofilters Made of Nanofibers, *Current Applied Physics,* 6, 1030-1035, 2006.

[5] Graham K., Ouyang M., Raether T., Grafe T., McDonald B., Knauf P., Polymeric Nanofibers in Air Filtration Application, *Fifteenth Annual Technical conference and*

Expo of the American Filtration and Separations Society, Galveston, Texas, April 9-12, 2002.

[6] Gibson P., Gibson H.S., Rivin D., Transport Properties of Porous Membranes Based on Electrospun Nanofibers, *Colloids and Surfaces A,* Vol.187-188,469-481, 2001.

[7] Gibson P., Gibson H.S., Rivin D., Electrospun Fiber Mats: Transport Properties, *AIChE Journal*, Vol.45, No.1, 190-195, 1999.

[8] Grafe T., Gogins M., Barris M., Schaefer J., Canepa R., *Nanofibers in Filtration Applications in Transportation, International Conference and Exposition of the INDA, Chicago, Illinois, December 3-5,* 1-15, 2001.

[9] Ryu Y.J., Kim H.Y., Lee H.K., Park H.C., Lee D.R., *Transport Properties of Electrospun Nylon6 Nonwoven Mats*, European Polymer Journal, Vol.39, pp.1883-1889, 2003.

[10] Li D., Frey M.W., Joo Y.L., Characterization of Nanofibrous Membranes with Capillary Flow Porometry, *J. of Membrane Science,* 2006, 286, 104.

[11] Graham K., Gogins M., Gibson H.S., Incorporation of Electrospun Nanofibers into Functional Structures, *INJ, 21-27, summer,* 2004.

[12] Li Y., Wong A.S.W., Clothing Biosensory Engineering, *The Textile Institute, England,* 2006.

[13] Wang Z., Li.Y., Zhu Q.Y., Lue Z.X., Radiation and Conduction Heat Transfer Coupled with Liquid Water Transfer, Moisture Sorption, and Condensation in Porous Polymer Materials, *Journal of Applied Polymer Science*, Vol.89, pp.2780-2790, 2003.

In: Recent Progress in Chemistry...
Editor: A. K. Haghi, pp. 67-76
ISBN: 978-1-61668-501-0

Chapter 6

INFLUENCE OF COMPATIBILIZER AND PROCESSING CONDITIONS ON THE DISPERSION OF NANOCLAY IN A FIBER-FORMING POLYPROPYLENE MATRIX

Samira Aslanzadeh, Mohammad Haghighat Kish and Ali Asghar Katbab
Amirkabir University of Technology, Tehran, Iran

ABSTRACT

Polypropylene (PP)/Montmorillonite (MMT) nanoclay based composite was prepared by melt compounding with maleic anhydride grafted polypropylene (MA-g-PP) in a twin screw extruder and an internal mixer. Test specimens were prepared by compression molding. Mechanical properties such as tensile modulus, tensile strength and maximum percent strain were measured. The clay dispersion was investigated using wide angle X-ray diffraction and scanning electron microscopy. The XRD patterns and SEM images show that introducing the proper compatibilizing agent ratio and suitable processing condition like a twin screw extruder decrease the immiscible clay component in the matrix. However, the increase of tensile strength is not achieved.

1. INTRODUCTION

Conventionally, micro and macro-composites have been filled with synthetic or natural inorganic particles in order to improve the material properties or simply to reduce the costs. The addition of these particles sometimes imparts disadvantages to the resultant composites such as weight increase, brittleness and opacity. Besides, nanocomposites are a new type of composites that are particle-filled polymers for which at least one dimension of the dispersed particles is in the nanometer range. In recent years polymer/layered silicate nanocomposites (PLSN) have obtained particular attraction, both in industry and in academia, as they often reveal notable enhancement in the materials properties at very low loading levels comparing to virgin polymer or conventional composites [1-3].

In order to obtain good interfacial adhesion and mechanical properties the hydrophilic clay needs to be modified prior to its introduction in most polymer matrices which are organophilic. When nanometric dispersion of primary clay platelets is obtained, the aspect ratio of the filler particle is increased and the reinforcement effect is improved [4].

Polypropylene (PP) is one of the most interesting thermoplastic materials because of its low price and balanced properties. PP is the lightest polymer and has wide range applications such as packaging, fiber, automotive industry, nondurable goods and construction materials. However, due to the low polarity of PP, obtaining the exfoliated and homogenous dispersion of the silicate layer at the nanometer level in the polymer matrix is hard. This is principally due to the fact that the silicate clays layers have polar hydroxyl groups and are compatible only with polymers containing polar functional groups [4].

Due to the nonpolar nature of PP, third component known as compatibilizer such as maleic anhydride-grafted polypropylene copolymer are introduced to compensate the difference of polarity between polymer matrix and clay particles [5-8].

The exfoliation and dispersion of nanoclays in polyolefins matrices prepared by melt compounding depend on the organic modifier of the nanoclay, the initial interlayer spacing, the concentration of functional groups in the compatibilizer and its overall concentration in the composite, the viscosity of the plastic resin, and the operational conditions, such as screw configurations of extruders, rpm, temperature, residence time, etc [9].

Since nylon-6/clay nanocomposites (NC's) with excellent thermal and mechanical properties were reported by the Toyota Research Center, polymer/clay NC's have attracted much attention. The improvements in thermal, mechanical and flammability properties of polymer/clay NC's are significantly higher than those achieved in traditional filled polymers. To date, polymer/clay NC's have experienced some success in several kinds of polar polymers. However, for polymers with low polarity, such as polyolefins, the current results are not satisfactory due to the low compatibility between the clay and the polyolefins [10]. Polypropylene is a highly hydrophobic and nonpolar polymer and cannot be used as a matrix for the dispersion of inorganic hydrophilic clays in order to prepare NC's without a compatibilizing modification [11]. As a result, a complete exfoliation in polypropylene NC's has not obtained yet, and moreover, the decrease of tensile strength of polypropylene NC's still cannot be satisfactorily avoided [12].

The aim of the present paper is to evaluate the influence of compatibilizer and processing conditions on the dispersion behavior of nanoclays in a fiber-forming polypropylene matrix as well as the tensile strength of polypropylene NC's.

2. Experimental

2.1. Raw Materials

Polypropylene homopolymer, Moplen HP552R, having melt flow rate of 25 g/10 min, was obtained from Arak Petrochemical Co. (Arak, Iran), and was used as-received. A polypropylene-graft-maleic anhydride (PP-g-MA) with 0.1% of MA, PRIEX 20 070 from Priex Co. with a melt flow rate of 64g/10 min) was used as a compatibilizing agent. The clay or pristine OMMT used in the study is Dellite 67G from Laviosa Chimica Mineraria S.p.A.

(Livorno, Italy), which is an organo-modified montmorillonite (dimethyl dihydrogenated quaternary ammonium montmorillonite).

2.2. Preparation of PP/Clay Composite Samples

Preparation of the nanocomposite samples was done using a twin screw extruder (Werner and Pfleiderer, ZSK25) and an internal mixer or Rheomix made by Haake. The temperature of the twin screw extruder was maintained at 160, 170, 180, 180, 190 and 190 °C from hopper to die, and the screw speed was about 200 rpm. The reaction of an internal mixer was carried out at 180 °C and 100 rpm for 15min. Table 1 shows the specification of prepared samples.

Table 1. Specification of prepared samples

Sample	Clay (wt %)	PP-g-MA (wt %)	Method of Preparation	Number of Extrusion
PP	0	0	Direct Melt Mixing in an Internal Mixer	---
PPC	1	0	Direct Melt Mixing in an Internal Mixer	---
PPCG	1	3	Direct Melt Mixing in an Internal Mixer	---
PP1	0	0	Direct Melt Mixing in a Twin Screw Extruder	1
PP2	0	0	Direct Melt Mixing in a Twin Screw Extruder	2
PPS1-3	1	3	Direct Melt Mixing in a Twin Screw Extruder	1
PPS2-3	1	3	Direct Melt Mixing in a Twin Screw Extruder	2
PPS1-9	1	9	Direct Melt Mixing in a Twin Screw Extruder	1
PPS2-9	1	9	Direct Melt Mixing in a Twin Screw Extruder	2

2.3. Characterization

The structure of layered silicates and the morphology of extruded composites specimen were analyzed by wide-angle X-ray scattering (WAXD) and scanning electron microscopy (SEM). X-ray diffraction (XRD) data were performed with a Siemens Diffractometer D5000. The X-ray beam was nickel-filter CuKa radiation of wavelength 1.540560 °A operated at 40kV and 30 mA. Diffraction spectra were obtained over a 2θ range of 2–10° at a scanning rate 0.02°/s and the interlayer spacing (d001) was calculated using Bragg's equation $\lambda=2d\sin\theta$, where λ is the wavelength. The sample for WAXD measurement was prepared in film by compression molding at 190 °C and 1.5 MPa. The mechanical properties were evaluated by the tensile modulus, tensile strength and maximum percent strain.

RESULTS AND DISCUSSION

XRD is a rather simple and widely employed technique for the characterization of layered-clay dispersion in a polymer matrix. WAXS analysis was employed to quantify the gallery height between adjacent silicate platelets, and subsequently to prove the widening of this distance as matrix polymer intercalates between the galleries. Changes in the value of 2θ reflect changes in the gallery distance of the clay and allow distinguishing the intercalated silicate fractions [13]. Therefore, Figure 1 shows the diffraction pattern for pristine OMMT, PPC and PPCG composites prepared in an internal mixer and PPS1-3, PPS2-3, PPS1-9 and PPS2-9 composites prepared in a twin screw extruder with constant clay content, i.e., 1 wt.%. As expected, the layered-structure of nanoclay (Dellite 67G) is evident. The resulting interlayer distances are shown in Table1. The pristine OMMT exhibits a sharp peak (d_{001}) at around 2.48°, corresponding to an interlayer d-spacing of 3.56 nm. The very broad feature in the 4-8° 2θ region possibly indicates ion-exchange, possibly due to high molecular weight of the surfactant [14].

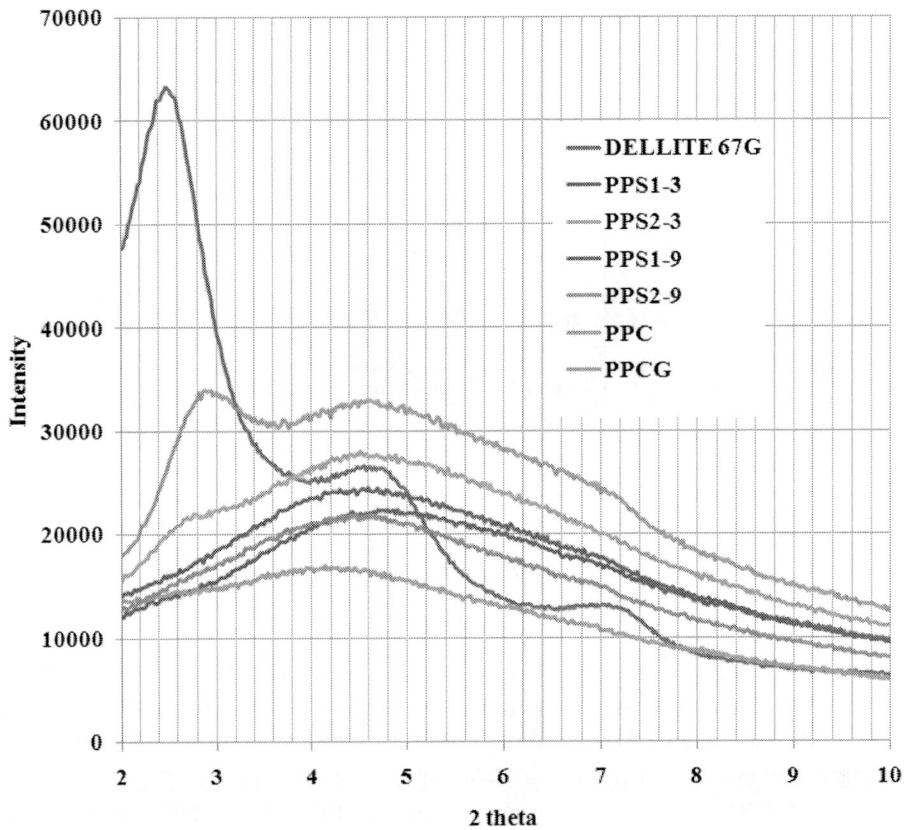

Figure 1. XRD patterns of Dellite 67G and composite samples.

However, after melt mixing and extrusion molding, the characteristic peak is substantially weakened, denoting that delaminating or disordering of the clay may have occurred. Thus, the XRD traces suggest that the system contains both an immiscible component and a nano-dispered component. Introducing the compatibilizing agent to the composites decreases the (d_{001}) diffractions of the clay. However, the characteristic peak is not weakened by increasing the ratio of compatiblizing agent to clay.

It is clearly observed that the (001) base peak position shifts into higher angles passing from 2.48° for the pristine OMMT to 2.93° and 2.78° for the PPC and PPCG composites, which indicates that PP does not intercalate into OMMT, even for the composite sample containing maleic grafted PP and the broad feature in the 4-8° 2θ regions remains. It means that the clay internal spacing decreased. In these two samples, no dispersion of clay particles occurs. It suggests that nanocomposite structure is not achieved. This is also confirmed by SEM images showing the agglomeration of clay particles in the surface of PPC and PPCG composites, as shown in Figure 2. It is also seen that there is no significant effect of compatibilizer on the dispersion of clay particles in the polymer matrix. However, the size of agglomerated particles decreases with the introducing the compatibilizing agent.

Table 2. XRD results of Dellite 67G and composite samples

Sample	2 θ (°)	d value (nm)
PP	2.48	3.56
PPC	2.93	3.01
PPCG	2.78	3.17
PPS1-3	---	---
PPS2-3	---	---
PPS1-9	---	---
PPS2-9	---	---

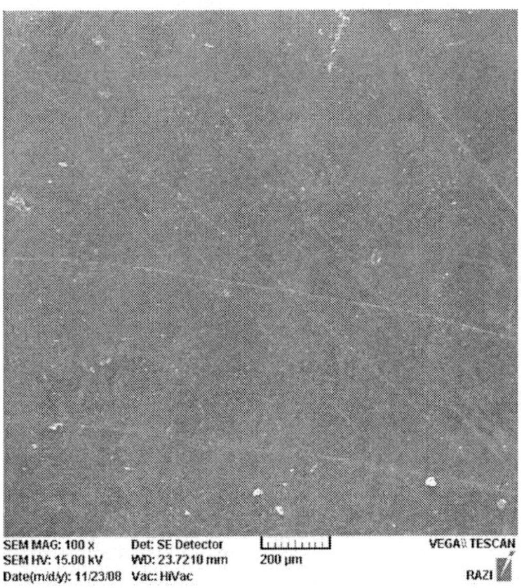

PP

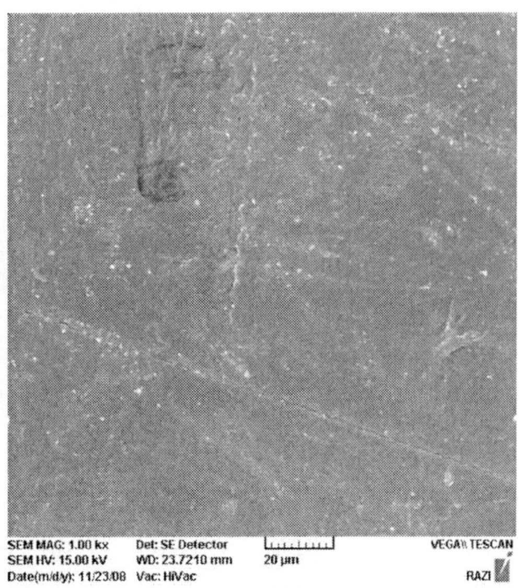

PP

Figure 2. (Continued).

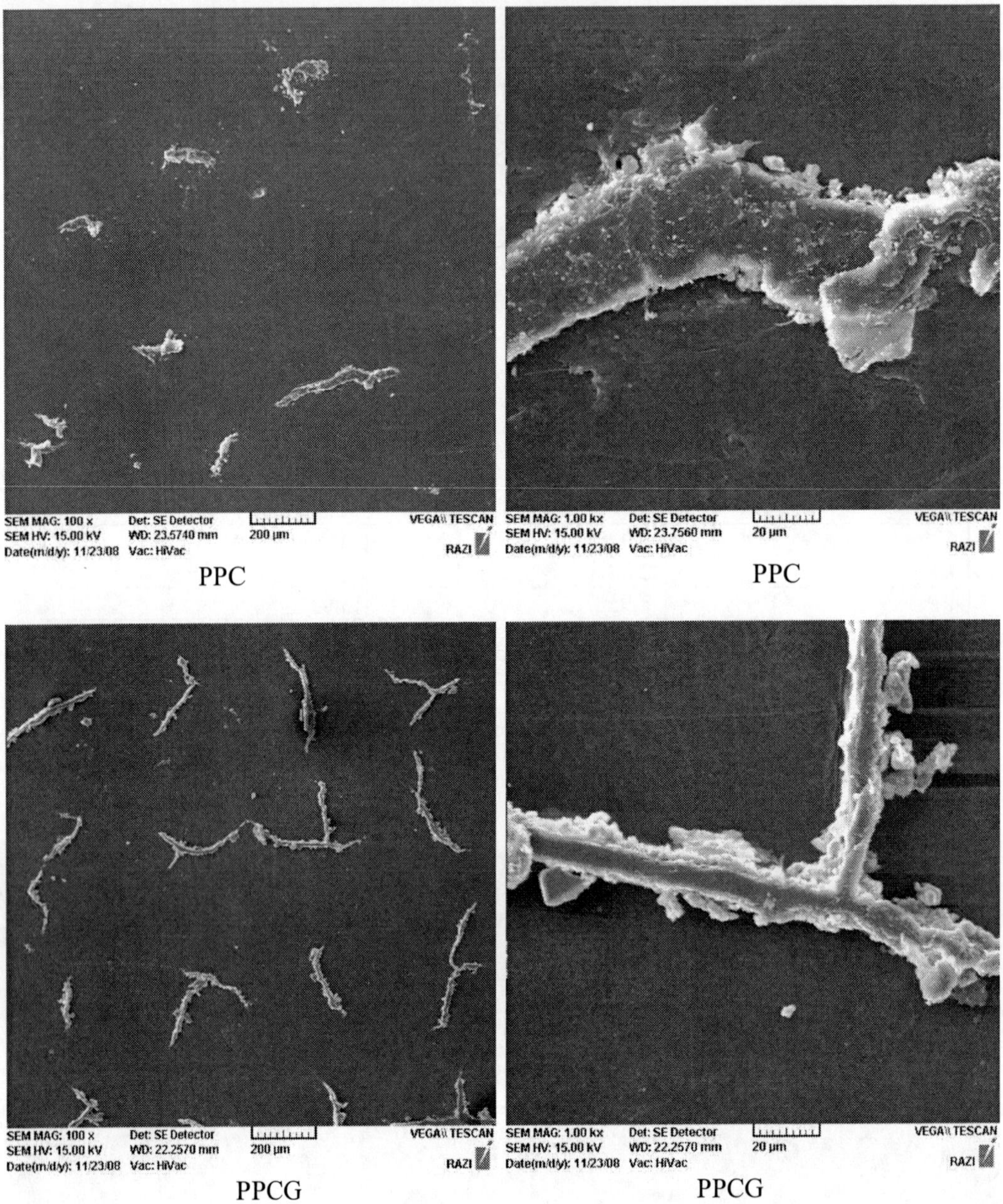

Figure 2. Surface SEM images of pristine polypropylene and composite samples prepared in an internal mixer.

For the XRD patterns of PPS1-3 and PPS2-3 composites prepared in a twin screw extruder, the broad feature in the 4-8° 2θ regions broadens with low intensity and shifts to lower angles.

The d_{001} peaks weakens, denoting that exfoliation or disordering of the clay may have occurred. The diffuse nature of d_{001} peaks makes it difficult to estimate accurately their d-spacing which suggest either a formation of intercalated nanocomposites or a mixed intercalated-exfoliated system (or simply a disordered system).

Thus the XRD traces suggest that the PPS1-3 and PPS2-3 composites contain an immiscible component (from the broad feature at 4-8°) and a nano-dispersed component. Figure 3 shows the SEM images of agglomerated clay particles in the surface of PPS1-3 composite which is much less than microcomposites prepared in an internal mixer. Figure 4 demonstrates the SEM images of agglomerated clay particles in the surface of the PPS2-3 composite which is less than PPS1-3. It suggests that 2 time extrusion process is able to disperse the immiscible component part which the XRD patterns indicates this matter since occurrence of diminution of peak intensity.

The ability to exfoliate and disperse the layered silicate into a polymer matrix is related to a number of factors such as the exchange capacity of the silicate, the polarity of polymer and the chemical nature of the organic modifier.

Processing conditions are the one of the factors affecting the dispersion of the silicate layers. The disappearance of the peak in the XRD patterns and the decrease of agglomerated particles in the SEM images indicate that the dispersibility of clay particles is more pronounced in the composite samples prepared in a twin screw extruder in comparison to an internal mixer.

XRD investigations and SEM imaging was carried out to study the dispersion of clay in the matrix. Melt compounding in the twin screw extruder and using the compatibilizing agent decrease the large-scale particle agglomeration.

However, there is no significant change in the tensile strength of composites as shown in table 3.

It can be concluded that tensile strength remains the same because of the existence of immiscible clay component, as also mentioned in [12]. Conversely, it is worthy to note that the increase of the strength observed in the literature [15].

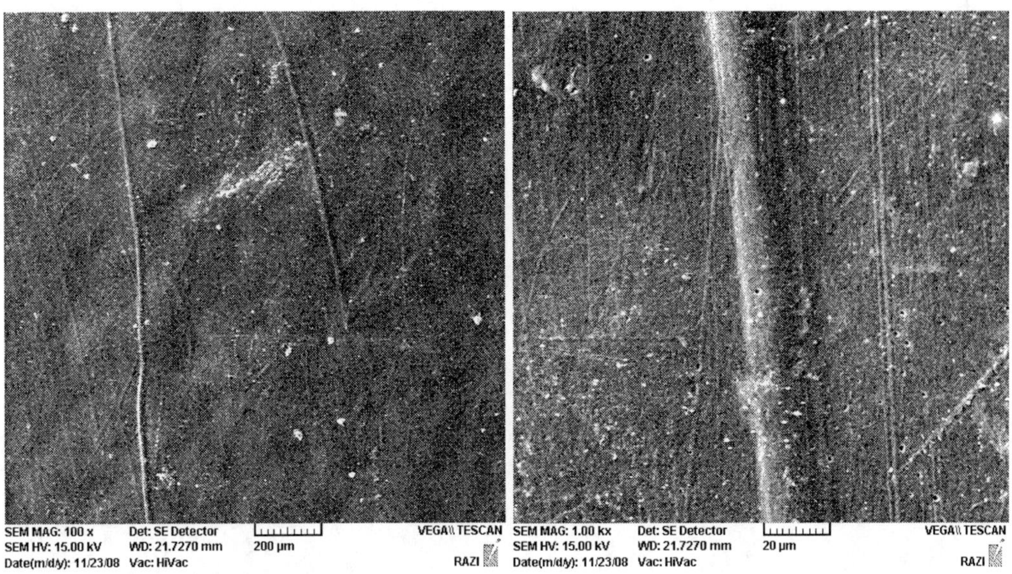

Figure 3. (Continued).

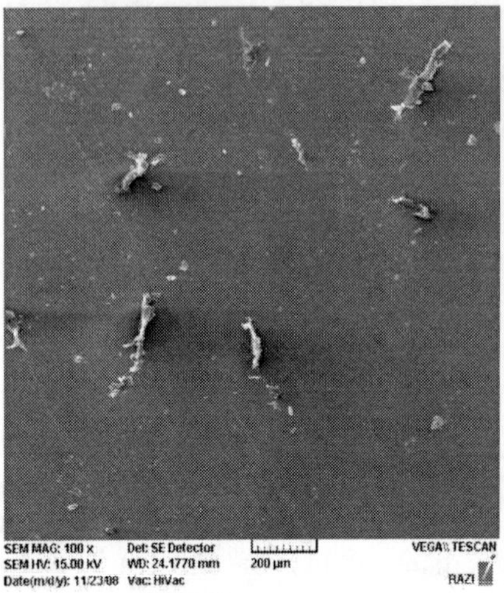

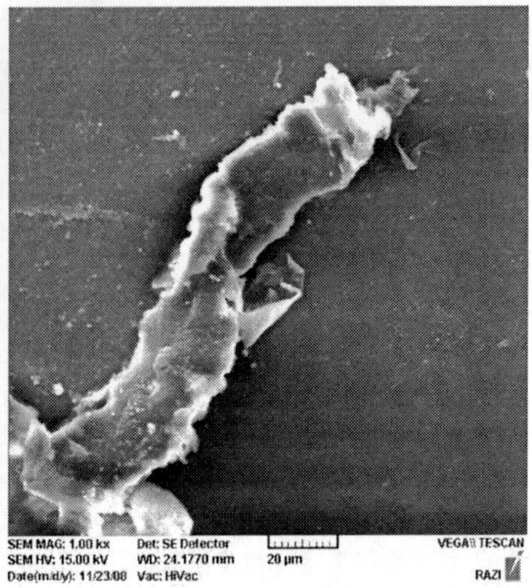

PPS1-3 PPS1-3

Figure 3. Surface SEM images of pristine polypropylene and composite samples prepared in a twin screw extruder.

Table 3. Mechanical properties of the samples prepared in an internal mixer

sample	Module (MPa)	Tensile (MPa)	Strain (%)
PP	1850 ± 54	39.9 ± 1.8	4.3 ± 0.5
PPC	1828 ± 98	38.8 ± 0.5	4.4 ± 0.3
PPCG	1780 ± 144	37.5 ± 2.8	3.8 ± 0.1
PP1	1772 ± 159	40.6 ± 1.9	4.8 ± 0.1
PPS1-3	1690 ± 228	40.2 ± 3.5	4.8 ± 0.6
PP2	1736 ± 176	38.6 ± 2.0	4.5 ± 0.1
PPS2-3	1872 ± 52	40.3 ± 1.2	4.4 ± 0.2

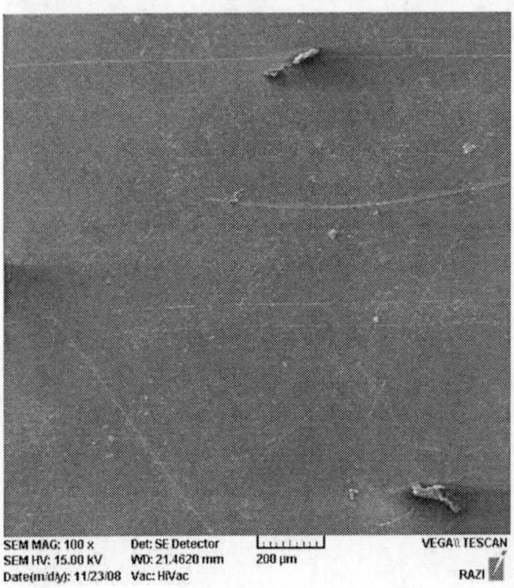

PP2 PP2

Figure 4 Continued

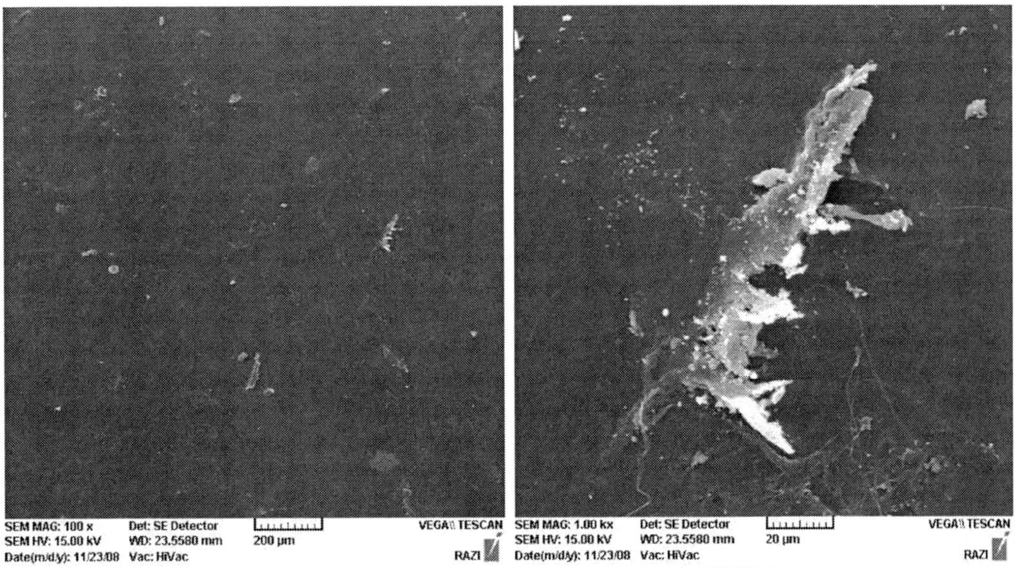

PPS2-3 PPS2-3

Figure 4. Surface SEM images of pristine polypropylene and composite samples prepared in a twin screw extruder.

Conclusions

In this research the XRD patterns and SEM images show that introducing the proper compatibilizing agent ratio and suitable processing condition like a twin screw extruder decrease the immiscible clay component in the matrix. However, the increase of tensile strength is not achieved.

References

[1] M. Alexandre, P. Dubois, *Mat. Sci. Eng. R*, 28, 1-63, 2000.

[2] S. S. Ray, M. Okamoto, *Prog. Polym. Sci*, 28, 1539-1641, 2003.

[3] S. Pavlidou, C. D. Papaspyrides, *Prog. Polym. Sci*, 33, 1119-1198, 2008.

[4] D. Garcia-Lopez, O. Picazo, J. C. Merino, J. M. Pastor, *Eur. Polym. J.*, 39, 945-950, 2003.

[5] M. Kawasumi, N. Hasegawa, M. Kato, A. Usuki, A. Okada, *Macromolecules*, 30, 6333-6338,1997.

[6] E. Manias, A. Touny, L. Wu, K. Strawhecker, B. Lu, T. C. Chung, *Chem. Mater.*, 13, 3516-3523, 2001.

[7] B. Morgan, J. D. Harris, *Polymer*, 44, 2313-2320, 2003.

[8] A. P. Reichert, H. Nitz, S. Klinke, R. Brandsch, R. Thomann, R. Mulhaupt, *Macromol. Mater. Eng*, 275, 8-17, 2000.

[9] S. Sanchez, J. G. Martinez Colunga, M. L. Lopez-Quintanilla, I. Yanez Flores, M. L. Garcia-Salazar, C. Gonzalez Cantu, *Polym. Bull*, 60, 829-836, 2008.

[10] C. Ding, D. Jia, H. He, B Guo, H. Hong, *Polym Test*, 24, 94-100, 2005.

[11] Y. Tang, B. M. Lewin, *Polym. Degrad Stab.*, 93, 1986-1995, 2008.

[12] D. Pospiech, B. Kretzschmar, M. Willeke, A. Leuteritz, D. Jehnichen, A. Janke, M. Omastova, *Polym. Eng. Sci.*, 47, 1262-1271, 2007.

[13] C. Remili, M. Kaci, S. Kachbi, S. Bruzaud, Y. Grohens, *J. Appl. Polym. Sci.*, 112, 2868-2875, 2009.

[14] J. Zhang, C. A. Wilkie, Polymer, 47, 5736-5743, 2006.

[15] W. Wenyi, Z. Xiaofei, W. Guoquan, C. Jianfeng, *J. Appl. Polym. Sci.*, 100, 2875-2880, 2006.

In: Recent Progress in Chemistry...
Editor: A. K. Haghi, pp. 77-86
ISBN: 978-1-61668-501-0

Chapter 7

PREPARATION AND CHARACTERIZATION OF POLYPROPYLENE NANOFIBRE CONTAINING AG NANOPARTICLES VIA MELT ELECTRO SPINNING

Mohammad Zarrebini Esfahani[1], Behnaz Baghaei*[1], Masoumeh Arabian[2], Mahvash Keivani[2] and Massoud Salehi[3]

[1]Department of Textile Engineering, Isfahan University of Technology, Isfahan, 84156-83111, Iran
[2]Isfahan University of Technology, Isfahan Science and Technology Town, Cara Palayeh Co., Isfahan, 84155-666, Iran
[3]Department of mechanical Engineering, Isfahan University of Technology, Isfahan, 84156-83111, Iran

ABSTRACT

Polypropylene nanofibres containing silver nanoparticles were successfully prepared via melt-electrospinning process. The electrospun nanofibres were characterized using SEM and XRD. AATCC 100 test, a quantitative method, was employed to evaluate the antibacterial properties of samples. The results of the intensity pattern of X-ray diffraction indicated that the crystallinity of the samples containing silver nanoparticles was slightly less than those of pure polypropylene nanofibres. In order to determine fibres diameter and diameter of the silver nanoparticles, two sets of SEM micrographs along and perpendicular to fiber axis were taken. It was found that while fibre diameter lies in the range of 110 to 120 nm. The diameter of the imbedded silver nanoparticles in the nanofibres was 40 nm. It was also noticed that some aggregation of the silver nanoparticles had occurred. The results of antibacterial evaluation showed that the nanofibres, which contained silver nanoparticles exhibited excellent antibacterial properties.

---

* Email: b.baghaei@tx.iut.ac.ir

INTRODUCTION

Polypropylene (PP) fibre is one of the most widely used synthetic fibres in textile industry. PP has some advantages: it is cheaper and stronger than other synthetic fibre. Besides, it is well applicable for various field, for example carpets, automotive interior trim, films, package, cover stock, cable, napkins, and so on. In particular, it is used for sanitary applications such as surgical masks, diapers, filters, hygienic bands, etc, which need to display antibacterial effect [1, 2].

In fiber science related literature, fibers with diameters below 100 nm are generally classified as nanofibres [3]. Researchers are interested in nanofibres due to their micro and nano structural characteristics which provide the development of smart materials. More considerably; large surface area, small pore size and the production possibility of three dimensional structures increase the interest in nanofibres.

Recently production of nanofibres using nanocomposites has attracted attention [4, 5]. This is due to the fact that, this type of nanofibre combines the unique properties of nanocomposites [6] with the outstanding characteristics of nanofibres I.e. high specific surface area plus high interpenetrating capacity in other materials [7]. Therefore, metal/polymer nanocomposites have not only the potential to meet the requirements of applications such as photonic, electric sensors, filters, and artificial tissue, but also can act as catalysts. Ag nanoparticles are the most embedded metal nanoparticles used in conjunction with polymers. This is because Ag nanoparticles exhibit remarkable catalytic activity, surface-enhanced Raman scattering activity, high electrical conductivity and antimicrobial activity etc [8, 9].

The use of electrostatic forces to produce fibres is known as electrospinning. This is in fact an extension of advance polymer research. The generated forces, due to the electric field, to which the polymer, either as a melt or solution, is subjected to, are in some instances capable of drawing the fiber diameter into the nano region. The potential of making fibers finer than those previously made through conventional fibre making processes, allows for a more detailed understanding of polymers spinning in a wide of fiber diameters. It has been shown by Jäger [10] that as the fiber diameter is decreased, the mechanical properties of the fiber increase exponentially as a result of the decreased probability of surface flaws. Utilization of this trend through electrospinning, allows for the potential for polymers to be made of a lighter weight and smaller diameter that are capable of mechanically outperforming those fibers that currently exist. Electrospinning has now been used for the better part of a century, however commercialization and a full understanding of the process is a goal yet unattained. To date, the fundamental goal of electrospinning has been to produce fibers with diameters less than 100 nanometers. As a result of the polymeric parameters that are needed to obtain such fibers, nearly all research has been focused on electrospinning of a polymer solution. Electrospinning from the melt, despite environmental, economical, and safety benefits, has been, for the most part, ignored [11].

In melt-electrospinning, Ag nanoparticles are doped in molten polymer to prepare Ag/polymer composite. It is extremely difficult to disperse Ag nanoparticles uniformly into the polymer matrix. This problem arises mainly due to the ease by which nanoparticles agglomerates and also high viscosity of molten polymer. In the past few years, compounding of polypropylene with nano Ag using compounding extruder has gained popularity [12].

Melt-electro spinning is a process by which fibers with diameters in the range of 10 to 500 nm are formed when a suspended droplet of molten polymer is charged to high voltage. When voltage is sufficiently high to overcome surface tension forces, fine jets of the melt are shoot out toward a grounded collector. The jet is stretched, elongated before it reaches the collector where it is cooled. The cooled filaments are collected as nanofibres [13]. This novel nanofibre spinning technique has been mainly explored to achieve pure polymer nanofibres.

In this paper, production of PP nanofibre containing silver nanoparticles using the above technique together with their characterizations using XRD and SEM analysis are presented. Additionally antibacterial properties of nanofibres are evaluated using quantitative AATCC 100 test. Inclusion nanosilvers into polymers to form a nanocomposite have been demonstrated to have a profound effect on the crystallization of the polymer, which in turn affects the properties of nanofibres, and antibacterial properties of the nanofibres.

EXPERIMENTAL

Materials

Nanoparticulate spherical silver powder with purity greater than 99.9%, and density of 10.49g/cm³ supplied by Nano CID Co, Iran were used in this research. Polypropylene chips (isotactic PP as shown in Figure 1) with the following specifications were provided by Bandar Imam 2 Integrated Petrochemicals Complex, Iran. M_n: 5.0 $\times 10^4$ g/mol, Melt Flow Index: 16.0 g/10 min, density: 0.90 g/ml, and polydispersity: 4.6.

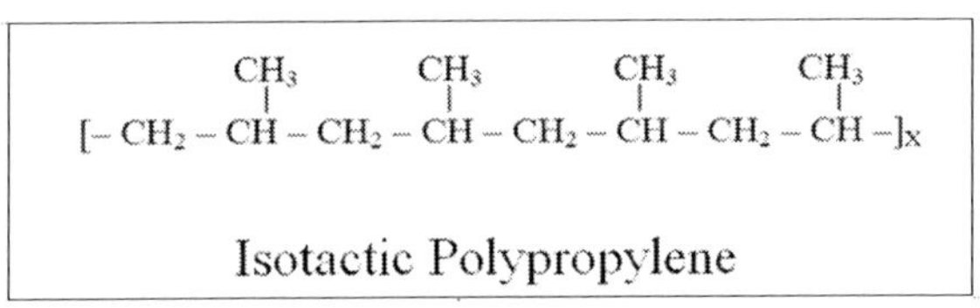

Figure 1. Structure of polymer utlized through the melt-electrospinning experimentation.

Spinning Process

In order to facilitate the spinning process, Polypropylene/Ag master-batch were prepared at two Ag nanoparticles concentration, using a compounding extruder as shown in Figure 2 [11]. The master-batches were prepared at temperature of 185-195°C, using a screw rotation of 120 rpm.

Thus, a mixing time of 12 min was achieved. Total moisture-free state was obtained by drying both Polypropylene chips and Polypropylene/Ag master-batches for at least 2h at 100°c in vacuoa. Polypropylene /Ag master-batch and Polypropylene chips at various ratios, as shown in Table 1 and 2, were fed to a conventional twin-screw extruder. Melting temperature of about 200°C, screw speed of 120 rpm giving a mixing time of 15 min was employed.

In melt-electrospinning system is shown in Figure 3 [11]. A metallic capillary having a diameter of 1.7 mm and a volume of 3cc was loaded with the Polypropylene /Ag nanocomposite.

The capillary acted as anode was positioned at a distance of 2 cm from cathode which itself was stuck on a drum. The rotational speed of the drum was positively controlled. The drum surface area was covered by an Aluminum foil. Polypropylene /Ag nanofibre formed using a voltage of 20.0 kV was collected on the covered drum surface. 1 mA constant current was supplied by a high voltage transformer.

Spinning temperature and flow rate through extruder is considered to be the two main parameters that control the amount of polymer that exits from the spinneret. These were set at 180-200°C and 1-2 rev/min respectively.

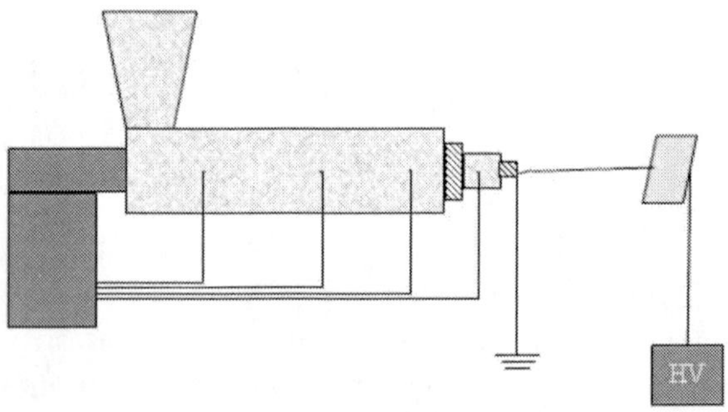

Figure 2. Schematic drawing of an extruder based melt-electrospinning system.

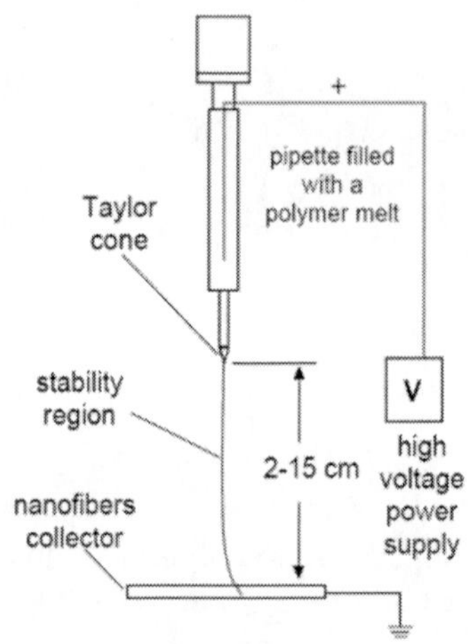

Figure 3. Schematic drawing of melt-electrospinning system.

Table 1. Materials ratio in masterbatch samples

Sample Code	Polypropylene (Wt %)	Silver nanoparticles (Wt %)
Masterbatch 1/9	90	10
Masterbatch 1/32	97	3

Table 2. Materials ratio in PP nanocomposite samples

Sample Code	Polypropylene (Wt %)	Master-batch	
		Polypropylene (Wt %)	Silver nanoparticles (Wt %)
PP nano 1	97	2.7	0.3
PP nano 2	85	13.5	1.5
PP nano 3	90.1	9.6	0.3
PP nano 4	50.5	48	1.5
PP nano 5	100	-	-

SEM and XRD Observation

For cross-sectional morphology and determination the average diameter of the produced nanofibres, micrographs along and perpendicular to the nanofibres axis were taken using Philips XL30 Oxford 6650.

The nanofibres were analyzed using an X-ray diffractometer (Philips, mode E "Pert"-MPD). The angles (2Θ) of diffraction of all samples were measured from 5° to 50° in the equatorial direction.

Test of Antibacterial Properties

Antibacterial test were carried out to measure antibacterial properties of polypropylene/Ag nanofibres by the quantitative antibacterial test method (AATCC 100 Test Method 100-1993) [14]. The tests were performed with one bacillus. Staphylococcus aureus (American Type culture Collection, ATCC No 6538) is used as the Gram-positive organism. The degree of antibacterial effect was presented as the reduction ratio for the bacilli. The equation for quantitative antibacterial evaluation is given by (equation (1)):

$$R(\%) = \frac{B-A}{B} \times 100 \qquad (1)$$

where:

R is the percentage reduction ratio,

A is the number of bacteria recovered from the inoculated treated test specimen swatches in a jar incubated over the desired contact period, and B the number of bacteria recovered from the inoculated treated test specimen swatches in the jar immediately after inoculation.

Results and Discussion

Fibres and Particles Diameter

Fibre Diameter

Figure 4 show visual representation with information about the collected Polypropylene/Ag nanofibres diameters. The micrographs revealed that the average diameter of PP nanofibres ranges from 110nm to 120nm. It was found that, not only fiber diameter decreases as the amount of nano Ag in nanocomposite increases, but also fibers appear to achieve a smoother surface. The latter can be attributed to the enhancement of electrical conductivity when higher amount of nano Ag are used [15, 16, and 17].

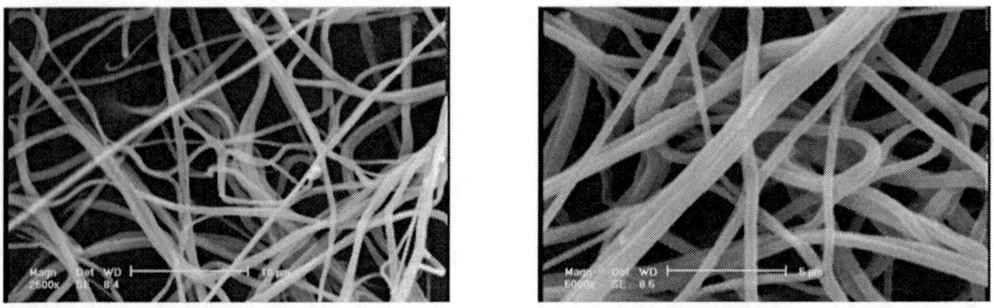

a) b)

Figuew 4. SEM micrograph of polypropylene/silver fibres melt-electrospun containing a) 0.3 %wt nanosilver (AVG: 120 nm)) 1.5 %wt nanosilver (AVG: 110 nm).

Particles Diameter

Silver nanoparticles embedded in the cross-section of the polypropylene nanofibres are shown in SEM micrographs of Figure 5. As can be seen in Figure 5 the embedded silver nanoparticles are uniform spheres with a diameter of 40 nm. It must be emphasized that an acceptable degree of uniform dispersion of the particles in the polypropylene nanofibres is achieved. This is despite of a mild aggregation of the particles that can be observed.

Structural Analysis

Figures 6, 7 show XRD patterns of pure PP nanofibres and PP nanofibres containing Ag nano particles. The results show that, an increase in the amount of nano Ag in nanocomposite leads to reduction of the crystallinity of the nano fibers. Although no definite explanation can be offered in respect of this phenomenon, it is assumed that, Ag nano particles in the matrix can be held responsible for the reduction in crystallinity [18].

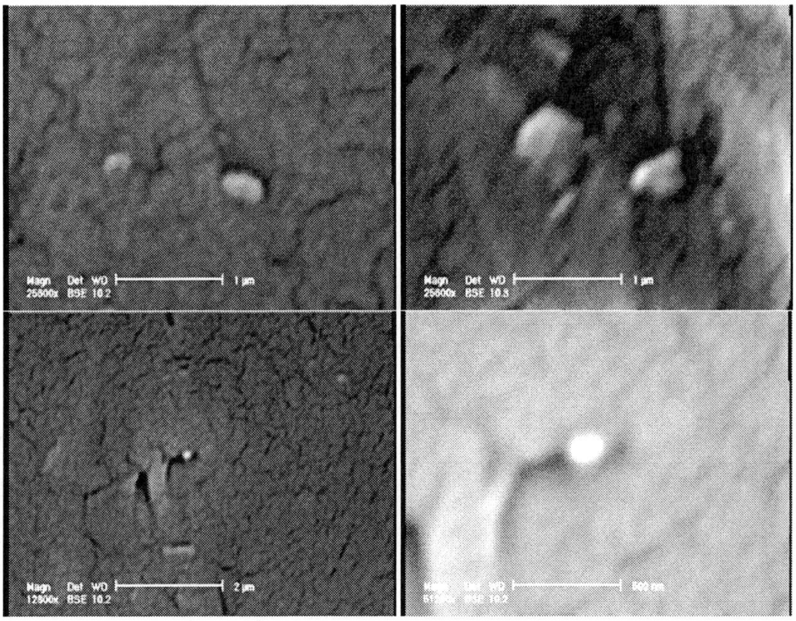

Figure 5. SEM micrograph of cross section of polypropylene/silver fibre melt-electrospun.

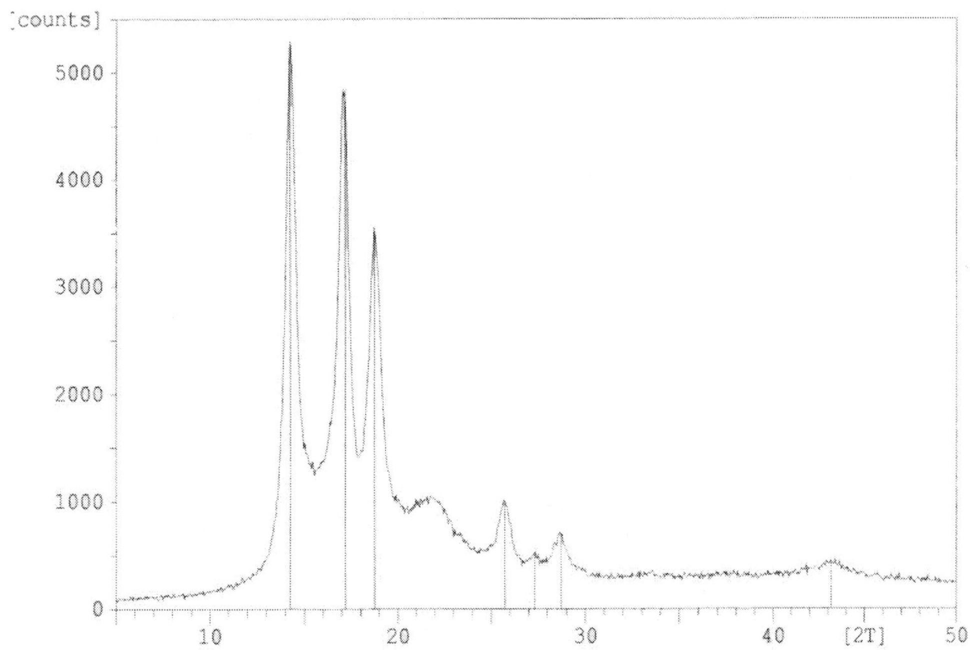

Figure 6. XRD of pure polypropylene nanofibres melt-electrospun.

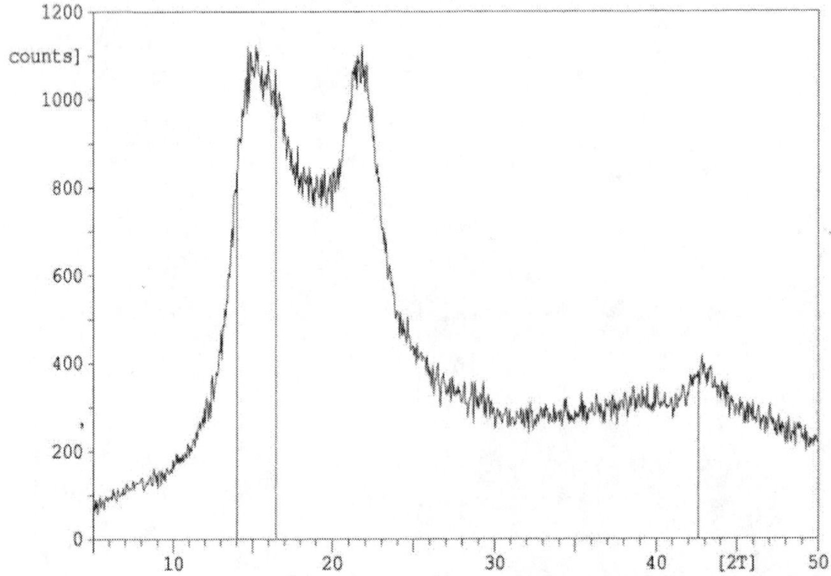

Figure 7. XRD of polypropylene/silver nanofibres melt-electrospun containing 1.5 %wt nanosilver.

Evaluation of Antibacterial Effects

Table 3 shows the antibacterial performance of the samples. Considering the result, it can be said that even at low silver nanoparticles content, the produced nanofibres exhibited excellent antibacterial properties. Additionally, it is stated that the antibacterial properties is enhanced as the finesse of the fibre is increased. This is due to the fact that as fibre diameter decreases, the contact area between nanoparticles and the bacteria is increased, thus bacteria propagation is severely restricted.

Table 3. Antibacterial evaluation of the samples by AATCC 100 test method

ample Code	Silver nanoparticles Content (Wt %)	Antibacterial evaluation (%) Staphylo coccus aureus
PP nano 1	0.3	99.9
PP nano 2	1.5	99.9
PP nano 5	0	26.3

CONCLUSION

In this study, it has been shown that polypropylene nanofibres containing silver nanoparticles can be successfully produced via melt-electrospinning process. SEM micrographs show that melt-electrospinning of polypropylene/nanosilver resulted in nanofibres with diameters in the range of 110 to 120 nm. It is also concluded that, good dispersibility of the silver nanoparticles in the matrix is obtained.

XRD diffraction patterns showed that the crystallinity of polypropylene nanofibres containing silver nanoparticles is slightly less than of those of pure polypropylene nanofibres.

It was also observed that the polypropylene nanofibres containing silver nanoparticles exhibited excellent antibacterial properties even at low silver content.

REFERENCES

[1] Moore EP Jr (1996). *Polypropylene Handbook*. New York: Carl Hanser Verlag.

[2] Mukhtar A (1982). *Polypropylene Fibers*. New York: Elsevier Science.

[3] T. Subbiah, G. S. Bhat, R.W.Tock, S. Parameswaran, S. S. Ramkumar. (2005). Electrospinning of Nanofibers, *Journal of Apply Polymer Science*, 96, 557-569.

[4] E. Ochoa-Fernández , D.Chen, Z. Yu, B. Tøtdal, M. Rønning, A. Holmen. (2004). Effect of carbon nanofiber-induced microstrain on the catalytic activity of Ni crystals. *Surface Science Letters*, 554, 1107-1112.

[5] L.V.W. Randall, M.T. Thomas and E.C. Valerie. (2001). Substrate-support interactions in metal-catalyzed carbon nanofiber growth, *Carbon*, 39, 2277-2289.

[6] Y.H. Niu and R.M. Crooks. (2003). Dendrimer-encapsulated metal nanoparticles and their applications to catalysis, *Comptes Rendus Chimie,* 6, 1049-1059.

[7] Z.M., Huang, Y.Z. Zhang, M. Kotaki, S. Ramakrishna. (2003), A review on polymer nanofibers by electrospinning and their applications in nanocomposites, *Composites Science and Technology*, 63 (15), 2223-2253.

[8] M.M. Miranda, N. neto. (2004). Surface-enhanced Raman scattering of π-conjugated "push–pull" molecules Part I. *p*-Nitroaniline adsorbed on silver nanoparticles. *Colloids and Surfaces. A, Physicochemical and Engineering Aspects,* 249, 79-84.

[9] R. A. Ganeev, M. Baba, A. I. Ryasnyansky, M. Suzuki, H. Kuroda. (2004). Characterization of optical and nonlinear optical properties of silver nanoparticles prepared by laser ablation in various liquids, *Optics Communications*, 240 (46), 437-448.

[10] Jäger, H. (1986). *Die Natur der Haftung zwischen Carbonfasern und Polymermatrizen und Schlagzähigkeitsverhalten carbon-faserverstärkter Polymerverbundkörper bei 298 K und 77 K.*, University of Karlsruhe.

[11] S.H.Jeong, S.Y Yeo, S.C Yi, (2005). The effect of filler particle size on the antibacterial properties of compounded polymer/silver fibers .*Journal of Materials Science*, 40 (20), 5407-5411.

[12] L. Larrondo, R. St. John Manley, (2003). Electrostatic fiber spinning from polymer melts. I. Experimental observation on fiber formation and properties. *Journal of Polymer Science: Polymer Physics Edition,* 19, 909-920.

[13] J. M. Lyons (2004). Melt-electrospinning of Thermoplastic Polymers: An Experimental and Theoretical Analysis. *A Thesis Submitted to the Faculty of Drexel University.*

[14] *American Association of Textile Chemists and Colorists*, AATCC Technical Manual (2000). 75.

[15] Y. wang, Q. Yang, G. Shan, C.Wang, J. Du, sh wang, y. Li, X. Chen, X. jina, Y. Wei. (2005). Preparation of silver nanoparticles dispersed in polyacrylonitrile nanofiber film spun by electrospinning. *Materials letters*, 59, 3046-3049.

[16] H. Lee, E. jeong, ch. beak, j. youk. (2005). One step preparation of ultra fine polyacrylonitrile fibers containing silver nanoparticles. *Materials letters*, 59, 2977 - 2980.

[17] Q. Yong, D. li, Y. Hong , Z. Li, C. Wang, S. Qiu, Y. Wei. (2003). Preparation and characterization of a pan nanofibre containing Ag Nanoparticles via electrospinning. *Synthetic metals*, 137, 973-974.

[18] S. Y. Yeo and S. H. Jeong (2003). Preparation and characterization of polypropylene/silver nanocomposite fibers. *Polymer International*, 52, 1053-1057.

In: Recent Progress in Chemistry…
Editor: A. K. Haghi, pp. 87-97
ISBN: 978-1-61668-501-0

Chapter 8

MODIFICATION OF PCL NANOFIBROUS SCAFFOLDS FOR NERVE TISSUE ENGINEERING

Laleh Ghasemi-Mobarakeh[1], Molamma P Prabhakaran[2], Mohammad Morshed[3], Mohammad Hossein Nasr-Esfahani[4], Hossein Baharvand[5] and Seeram Ramakrishna[2]

[1]Islamic Azad University, Najafabad branch, Isfahan, Iran
[2]NUS Nanoscience and Nanotechnology Initiative, Division of Bioengineering, National University of Singapore, 9 Engineering Drive 1, Singapore 117576
[3]Department of Textile Engineering, Isfahan University of Technology, Isfahan, Iran
[4]Department of Stem Cells, Cell Science Research Center, Royan Institute, Esfahan Campus, ACECR, Esfahan, Iran
[5]Department of Stem Cells, Cell Science Research Center, Royan Institute, ACECR, Tehran, Iran

ABSTRACT

One limitation of electrospun scaffolds fabricated with synthetic polymers such as PCL is low cell affinity to them due to low hydrophilicity and lack of surface cell recognition sites. To overcome this drawback of the synthetic materials, natural polymers have been widely used to modify the synthetic material to improve the cell adhesion properties.

In this study polycaprolactone (PCL) was blended with gelatin for improvement the cell adhesion and biocompatibility of PCL nanofibrous scaffolds. Fiber diameter and hydrophilicity of PCL and PCL/gelatin nanofibrous scaffolds were evaluated. *In vitro* cell culture study was carried out using nerve stem cells (NSCs). Proliferation and morphology of NSCs on PCL and PCL/gelatin nanofibrous scaffolds were investigated.

Fiber diameter was found to decrease and hydrophilicity was found to improve with incorporation of gelatin in electrospinning solution. The results of MTS assay and SEM micrographs showed that PCL/gelatin nanofibrous scaffolds serve as better substrate for NSCs compared to PCL nanofibrous scaffolds.

INTRODUCTION

Tissue engineering is a new approach to reconstructing and regenerating lost or damaged tissue [1]. It is a multidisciplinary field that applies the principles of engineering and the life science by combining synthetic and living components in appropriate configurations and environmental conditions [2, 3]. The ultimate goal of tissue engineering as a treatment concept is to replace or restore the anatomic structure and function of damaged, injured, or missing tissue or organs following any injury or pathological process by combining biomaterials, cells or tissue, biologically active molecules, and/or stimulating mechanical forces of the tissue microenvironment[4]. Among the numerous attempts to integrate tissue-engineering concepts into strategies to repair nearly all parts of the body, neuronal repair stands out. This is partially due to the complexity of the nervous system anatomy, functioning and the inefficiency of conventional repair approaches [4]. The design and preparation of scaffolds suited to tissue engineering is critical for the success of tissue regeneration [5]. For successful tissue engineering, it is essential to have as many biomimetic scaffolds as possible [6]. Collagen is a major natural extracellular matrix component and has a fibrous structure with fiber bundles varying in diameter from 50-500nm. In order to mimic the structure of native extracellular matrix, nanofibrous scaffolds have been an interesting candidate in tissue engineering [7]. Nanofibrous scaffolds mimic natural extracellular matrix (ECM) and its nanoscale fibrous structure [6]. Polymeric nanofibers can be processed by a number of techniques such as drawing, template synthesis, phase separation, self-assembly and electrospinning [3, 7]. Electrospinning is one of the most important techniques to manufacture nanofibers. An attractive feature of electrospinning is the simplicity and inexpensive nature of the setup [8]. It yields continuous fibers, and is versatile enough to be used for the production of nanofibers from a variety of materials [6]. A high voltage power supply is required to create an electrically charged jet of polymer solution or melt. [8]. Many parameters can influence the transformation of polymer solutions into nanofibers through electrospinning. These parameters include (a) the solution properties such as viscosity and conductivity (b) governing variables such as voltage and distance between the tip and the collecting screen, and (c) ambient parameters such as solution temperature and humidity [9-11].

Synthetic biodegradable polymers are now widely used in tissue engineering due to their generally good strength and adjustable degradation rate. However, cell affinity towards synthetic polymers is poor as a consequence of their low hydrophilicity and lack of surface cell recognition sites [12]. Natural polymers are generally favored over synthetic polymers due to their biocompatibility and biodegradation in biomedical applications. However, their mechanical properties and their processability are often poor and to fabricate ultrafine fibers from natural polymers is usually more difficult than from synthetic polymers. Therefore, to overcome these problems, solution blends of synthetic and natural polymers have been prepared and spun into nanofibers [13]. Blending synthetic and natural polymers can improve cell adhesion of blended scaffold. PCL is a biodegradable and biocompatible polymer that has been successfully electrospun. Furthermore, it has been shown to be capable of supporting a wide variety of cell types [14-18], although its poor hydrophilicity, slow degradation kinetics, and lack of natural cell recognition sites greatly limit its application in the biomedical field [19]. Gelatin is a natural biopolymer derived from collagen by controlled hydrolysis. Because of its many merits, such as its biological origin, biodegradability, biocompatibility and commercial availability at relatively low cost, gelatin has been widely used in the pharmaceutical and medical fields [20-21].

This study is aimed at fabricating of PCL/gelatin nanofibrous scaffolds for nerve tissue engineering application. We fabricated nanofibrous scaffolds containing PCL and gelatin and studied the effect of incorporation of gelatin in PCL nanofibrous scaffold, evaluating the morphology and hydrophilicity of the resultant nanofibers. Further, Morphology and proliferation of NSCs on PCL and PCL/gelatin nanofibrous scaffolds were studied for understanding the effect of gelatin incorporation towards nerve tissue engineering.

EXPERIMENTAL

Materials

PCL (Mw= 80,000), gelatin type A from porcine skin and hexafluoro-2-propanol (HFP) were all obtained from Sigma-Aldrich (St.Louis, MO). Hexamethyldisilazane (HMDS) was purchased from Fluka. Dulbecco Modified Eagle's Medium (DMEM), fetal bovine serum (FBS), phosphate buffered saline (PBS), trypsin-EDTA and hourse serum were purchased from Gibco, Singapore. CellTiter 96® AQueous One solution reagent (MTS), used in the cell-adhesion assay, was purchased from Promega.

Methods

Fabrication of PCL and PCl/Gelatin Nanofibrous Scaffolds

PCL and PCL/gelatin nanofibrous scaffolds were produced by electrospinning method. In brief, a polymer solution with a concentration of 6 wt % was prepared by the dissolution of PCL: gelatin mixture with a weight ratio of 70:30 in HFP. The viscosity of the PCL and PCL/gelatin solutions were measured at 25C and rate of 5rpm using Brookfield digital viscometer (model DV-II Pro). Electrospinning was carried out from a 5-mL syringe with a needle diameter of 0.4 mm and by the application of a high voltage of 12 kV with a mass flow rate of 1 mL/h toward a target that was placed 12 cm from the syringe tip. Figure 1 shows schematic illustration of electrospinning set-up used in this research.

Characterization of Nanofibrous Scaffolds

Morphology of electrospun PCL and PCL/gelatin nanofibrous scaffolds were observed by scanning electron microscopy (SEM) after gold coating. The diameter of the fiber was measured from the SEM micrographs using image analysis software (Image J, National Institutes of Health, USA).

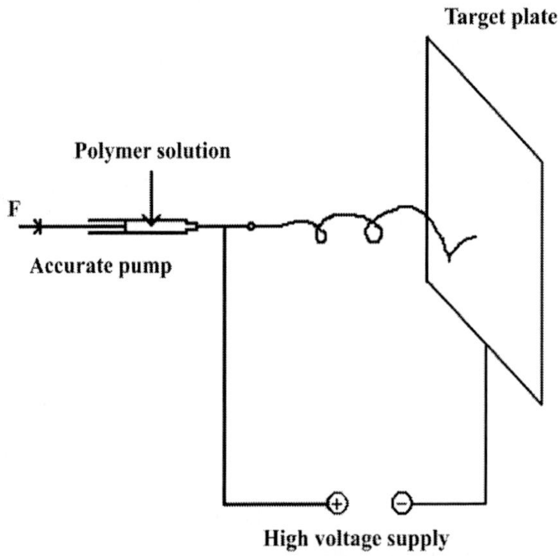

Figure 1. Schematic of electrospinning setup.

In this study, to determine the wettability (or hydrophilicity) of PCL and PCL/gelatin nanofibrous scaffold, the contact angle was measured by a video contact angle system (VCA Optima, AST Products). The droplet size was set at 0.5 μl. Five samples were used for each test and the average value was reported with standard deviation (±SD).

Cell Culture Study

The nanofibrous scaffolds were exposed to UV radiation and washed with PBS before cell seeding. NSCs were cultured in DMEM supplemented with FBS, horse serum and penicillin/streptomycin.

After reaching 70% confluency, the cells were detached by trypsin–EDTA and viable cells were counted by trypan blue assay. Cells were further seeded onto nanofibrous scaffolds, placed in a 24-well plate and tissue culture polystyrene (TCP as control) and cultured with DMEM/F12 1:1 mixture containing N2 supplement at 37 °C, 5% CO2 and 95% humidity.

6 days after cell seeding, the samples were fixed with 3% glutaraldehyde. Specimens were then rinsed with water and dehydrated with a graded concentration (50%, 70%, 90%, 100% v/v) of ethanol. Subsequently the samples were treated with HMDS and kept in a fume hood for air drying and sputter coated with gold before being examined on a scanning electron microscope.

MTS assay was carried out to compare the proliferation of nerve cells on PCL and PCL/gelatin nanofibrous scaffolds and TCP as the control. After 2, 4 and 6 days of cell seeding in 24-well plate, cells were washed with PBS and incubated with 20% of MTS reagent containing serum free medium. After 3 hr of incubation at 37°C in 5% CO_2, aliquots were pipetted into a 96-well plate. The absorbance of the content of each well was measured at 492 nm using a spectrophotometric plate reader (Fluostar Optima, BMG Lab Technologies, Germany).

Immunocytochemistry was used for investigation of differentiation of NSCs to nerve cells on nanofibrous scaffolds. Immunocytochemistry is a common lab practice which uses antibodies that target specific peptides or protein antigens in the cell via specific epitopes. In this study immunofluorescent labeling for neurofilament 200 kD (NF200) was performed to investigate differentiation of NSCs seeded on the nanofibrous scaffolds to nerve cells using laser scanning confocal microscopy (LSCM, Leica TCS SP2, Germany). Briefly, after 4 day of culture, the samples were rinsed with PBS solution twice and fixed in 4% paraformaldehyde for 30 min at RT followed by permeation with 0.5% Triton X-100 for 5 min. Non-specific labeling was blocked by incubating with goat serum (diluted in PBS at 1:20 ratio).

Then the samples were incubated in rabbit anti-NF200 (diluted as 1:200, Sigma, USA) overnight and kept in fluorescein isothiocyanate (FITC) conjugated goat anti-rabbit secondary antibody for 1 h. Finally, the samples were mounted on to glass slides and viewed under LSCM [22].

Statistical Results

All data presented are expressed as mean ± standard deviation (SD). Statistical analysis was carried out using single-factor analysis of variance (ANOVA). A value of $p \leq 0.05$ was considered statistically significant.

RESULTS AND DISCUSSION

Morphology and Characterization of Electrospun PCL and PCL/Gelatin Nanofibrous Scaffolds

Previous researches showed that solution properties are the main factors influencing the transformation of polymer to nanofibers. The effects of solution properties and processing parameters on the structure and morphology of the electrospun nanofibers have been investigated by previous researches. One of the most significant parameters influencing the fiber diameter is the solution viscosity. A higher viscosity results in a larger fiber diameter [8-11].

Table 1 shows the viscosity of PCL and PCL/gelatin solution. As can be seen in table 1, the solution viscosity decreases with addition of gelatin in electrospinning solution.

Table1. solution viscosity of PCL and PCL/gelatin solution

Solution	Viscosity(CP)
PCL	101±4
PCL/gelatin	75±6

Figure 2 shows the SEM micrographs of electrospun PCL and PCL/gelatin nanofibers fabricated in this study.

Figure 2. Morphology of (A) PCL and (B) PCL/gelatin nanofibrous scaffold.

Table 2 shows the results of fiber diameter and contact angle measurement of PCL and PCL/gelatin nanofibrous scaffolds.

Table 2. fiber diameter and contact angle of PCL and PCL/gelatin nanofibrous scaffolds

Substrate	Fiber diameter(nm)	Contact angle(°)
PCL	431±118	118±6
PCL/gelatin	189±56	32±8

As depicted in figure 2 and table 2, fiber diameter decreased with incorporation of gelatin in polymeric system which could be due to decrease of solution viscosity by incorporation of gelatin in electrospinning solution. As can be seen in table 2, the contact angle of PCL/gelatin nanofibrous scaffolds was lower than the contact angle of PCL nanofibrous scaffolds, showing an improved hydrophilicity after blending of PCL and gelatin. Fig 3 shows structural unit of gelatin. Improvement of hydrophilicity of PCL/gelatin nanofibrous scaffolds might be attributed to the amine and carboxylic functional groups in gelatin structure whereas such functional groups are not present in PCL structure.

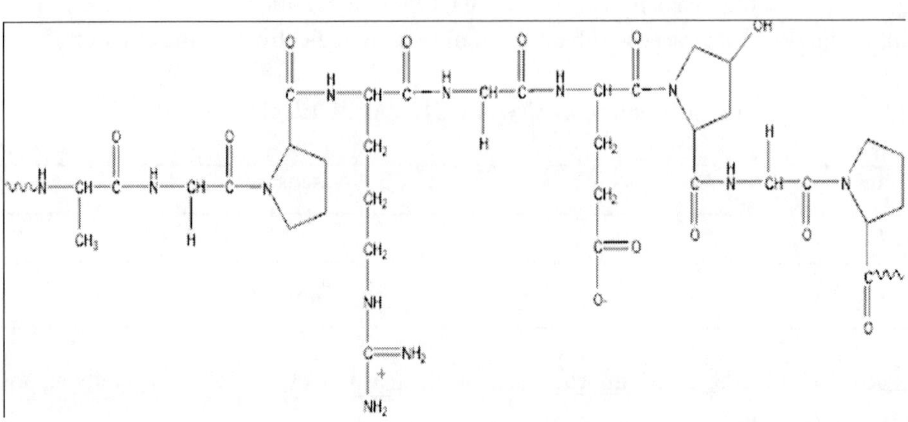

Figure 3. structural unit of gelatin.

Gelatin is a water soluble polymer and it is susceptible to *in vivo* degradation by enzymes such as gelatinase, which hinders its application as a scaffold for tissue regeneration/ drug delivery [23-28].

For long-term biomedical applications, electrospun gelatin membranes must be crosslinked to improve the water-resistant property of the resultant nanofibrous membrane [26]. However crosslinking of gelatin might be cytotoxic to cellular growth during *in vitro* or *in vivo* experiments due to the residues left behind from the unreacted crosslinking agent [29-31].

Blending gelatin with synthetic polymers such as PCL is a feasible approach that may not only reduce the potential problem of cytotoxicity of cross-linking reagents, but also provide a new biomaterial with good biocompatibility and improved mechanical, physical and chemical properties [32].

In Vitro Cell Culture Study

Proliferation of NSCs

MTS assay was carried out to evaluate the proliferation of NSCs on PCL and PCL/gelatin nanofibrous scaffolds. MTS assay is a colorimetric and an indirect measurement of cell proliferation that serves as an indicator of cell viability after seeding cells on different scaffolds.

The assay is based on the reduction of yellow tetrazolium salt in MTS to form purple formazan crystals by dehydrogenase enzymes secreted from the mitochondria of metabolically active cells [19].

The assay measures the formation of a soluble formazan product which is directly proportional to the number of live cells in culture. Figure 4 shows MTS results of NSCs on PCL, PCL/gelatin nanofibrous scaffolds and TCP as control. As shown in figure 4, the NSCs proliferation was significantly ($p<0.05$) higher in PCL/gelatin nanofibrous scaffolds than PCL nanofibrous scaffolds and TCP as control.

The hydrophilic-hydrophobic characteristic of scaffold is important in tissue culture and this can influence the initial cell adhesion and migration to a higher extent. According to previous literatures, hydrophobic surfaces lead to lower cell adhesion in the initial step of cell culture.

It can be concluded that presence of functional groups such as amino group on gelatin imparts more hydrophilic sites on PCL/gelatin nanofibrous scaffolds, resulting in greater attachment and proliferation of NSCs while poor hydrophilicity together with the lack of functional group results in low cell adhesion and proliferation of NSCs on PCL nanofibrous scaffolds.

TCP is normally treated by oxygenated gas plasma to create a more hydrophilic oxidized polymer surface and provides a surface chemistry that absorbs sufficient amounts of trace ECM proteins from serum-supplemented media to promote cell attachment [33].

Therefore, it is not surprising that proliferation of NSCs on PCL nanofibrous scaffolds is less than that on the TCP as TCP has higher hydrophilicity compared to PCL nanofibrous scaffolds.

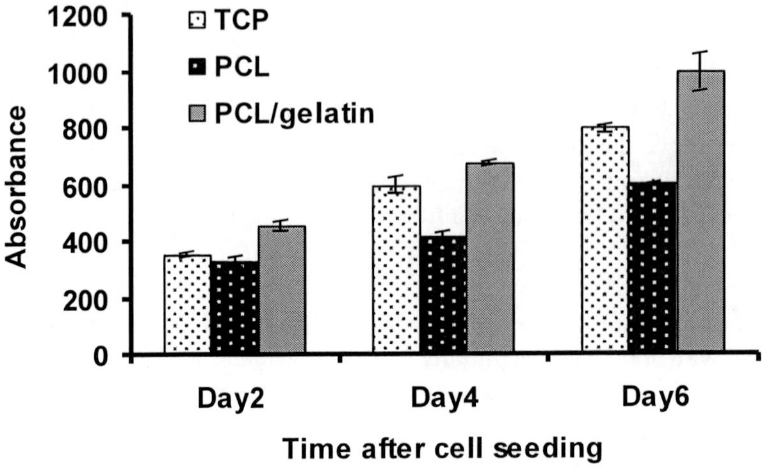

Figure 4. MTS results of C17.2 on PCL and PCL/gelatin nanofibrous scaffolds and TCP after 2, 4 and 6 days of cell seeding.

Cell –Scaffold Interaction

Figure 5 shows the morphology of NSCs on PCL and PCL/gelatin nanofibrous scaffold. A normal phenotypic morphology of nerve cells was observed on different nanofibrous scaffolds in this study. As can be seen in fig5, longer filapodial extensions and processes with more spreading of NSCs were observed for cells seeded on PCL/gelatin nanofibrous scaffolds compared to PCL nanofibrous scaffolds. These results are consistent with MTS results indicating inclusion of gelatin into PCL nanofibrous scaffold provides better substrate for nerve cells.

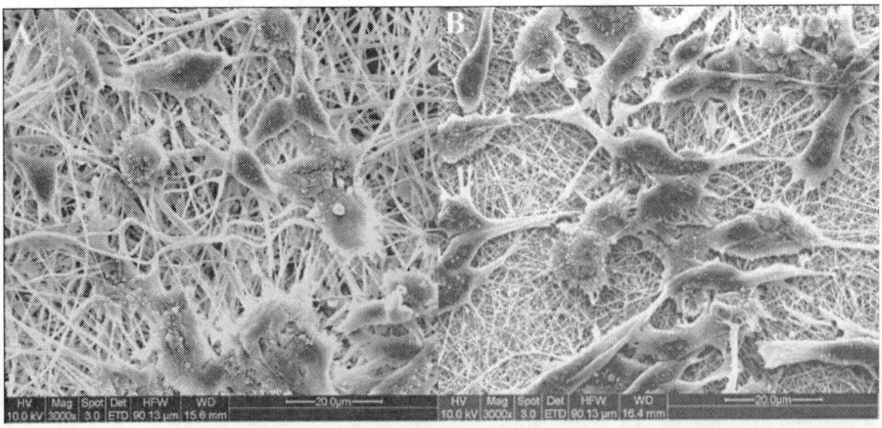

Figure 5. Morphology of nerve cells on (A) PCL nanofibrous scaffold (B) PCL/gelatin nanofibrous scaffold.

Immunocytochemistry

To observe the expression of the cytoskeleton proteins of nerve cells, neurofilament immunocytochemistry staining and LSCM observation was performed on the cell cultured scaffolds at day 4. Figure 6 shows the LSCM micrographs of differentiated NSCs on the PCL/gelatin nanofibrous scaffolds. Figure 6 shows the neurofilament expression indicating differentiation of NSCs to nerve cells on PCL/gelatin nanofibrous scaffolds. A similar result was also observed for differentiation of NSCs to nerve cells on PCL nanofibrous scaffolds (data not shown).

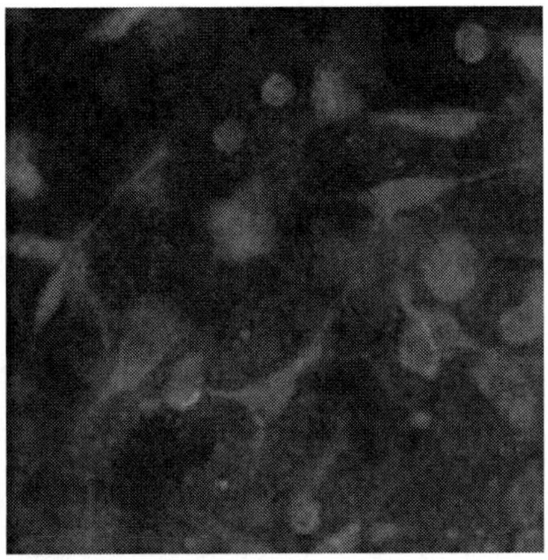

Figure 6. LSCM of NSCs after 4 day seeding on PCL/gelatin nanofibrous scaffolds.

Conclusion

In this study PCL and PCL/gelatin nanofibrous scaffolds were fabricated. The results of this study showed that the properties of nanofibrous scaffolds influenced by incorporation of gelatin in the electrospinning solution. Fiber diameter was found to decrease and hydrophilicity of PCL/gelatin nanofibrous scaffolds was found to improve compared to PCL nanofibrous scaffolds. The PCL/gelatin nanofibrous scaffold fabricated in this study has both the advantages of utilizing a synthetic polymer where the PCL domain provides mechanical strength and natural polymer where gelatin enhancing cell adhesion and proliferation. *In vitro* cell culture study showed that neurite extension and cell proliferation enhanced on the PCL/gelatin nanofibrous scaffolds compared to PCL nanofibrous scaffolds.

References

[1] Yang, S., Leong, K., Du , Z., Chua ,C.(2001). The design of scaffolds for use in tissue engineering. Part I: Traditional factors, Tissue Eng, Vol.7, 679-89.

[2] Bhattarai, SR., Bhaattarai, N., Yi, HK., Hwang, PH.(2004).Novel biodegradable electrospun membrane:scaffold for tissue engineering, *Biomaterials*, Vol. 25, 2595-2602.

[3] Ma PX. (2004). Scaffolds for tissue fabrication, *Materials today*, Vol. 7, 30-40.

[4] Zhang ,N., Yan ,H., Wen,X.(2005). Tissue-engineering approaches for axonal guidance, *Brain Research Reviews*, Vol.49, 48-64.

[5] Lee, CH., Shin, HJ., Cho, IH., Kang, YM., Kim, IA., Park, KD., Shin, JW., Nanofiber alignment and direction of mechanical strain affect the ECM production of human ACL fibroblast.(2005). *Biomaterials*, Vol.26, 1261-70.

[6] Ashammakhi, A., Ndreu, Y., Yang ,H., Nikkola, L.(2008). Nanofiber-based scaffolds for tissue engineering, *European Journal of Plastic Surgery*, Published online.

[7] Smith, LA., Ma, PX. (2004). Nano-fibrous scaffolds for tissue engineering, *Colloids Surf B Biointerfaces*, Vol.39, 125-131.

[8] Quynh,P.P.,Upma,S.(2006). Electrospinning of polymeric nanofibers for tissue engineering applications:A re.iew, *Tissue engineering*, Vol.12, 1197-1211.

[9] Ramakrishna, S., Fujihara, K. (2005). An Introduction to Electrospinning and nanofibers, Singapore, Singapore, World Scientific publishing.

[10] Tan, S-H., Inai, R., Kotaki, M., Ramakrishna,S.(2005). Systematic parameter study for ultra-fine fiber fabrication via electrospinning process , *Polymer*, Vol. 46, 6128-6134.

[11] Deitzel, J. M., Kleinmeyer, J., Harris,D ., Beck Tan ,N. C.(2001). The effect of processing variables on the morphology of electrospun nanofibers and textiles, *Polymer*, Vol. 42, 261-272.

[12] Ma,Z., Gao,C., Gong,Y., Shen, J. (2005). Cartilage tissue engineering PLLA scaffold with surface immobilized collagen and basic fibroblast growth factor, *Biomaterials,* Vol.26, 1253-1259.

[13] Safi ,S., Morshed ,M., Hosseini Ravandi,SA., Ghiaci ,M. (2007).Study of electrospinning of sodium alginate, blended solutions of sodium alginate/poly(vinyl alcohol) and sodium alginate/poly(ethylene oxide), *Journal of Applied Polymer Science*, Vol. 104, 3245-3255.

[14] Kweon ,H.Y., Yoo,M.K .,Park,I.K., Kim,T.H., Lee ,H.C., Lee,H.S.(2003). A novel degradable polycaprolactone networks for tissue engineering, *Biomaterials*, Vol.24, 801-808.

[15] Schnell,E., Klinkhammer,K., Balzer,S., Brook,G., Klee,D., Dalton,P., Mey,J.(2007). Guidance of glial cell migration and axonal growth on electrospun nanofibers of poly-ε-caprolactone and a collagen/poly-ε-caprolactone blend, *Biomaterials*, Vol.28, 3012-3025.

[16] Li,W.J., Richard ,T., Xiaoxue,H., Laquerriere,P ., Tuan,R.(2005). Multilineage differentiation of human mesenchymal stem cells in a three-dimensional nanofibrous scaffold, *Biomaterials*, Vol. 26, 5158-5166.

[17] Venugopal,J., Ma,L.L., Yong,T ., Ramakrishna,S.(2005). In vitro study of smooth muscle cells on polycaprolactone and collagen nanofibrous matrices, *Cell Biology International*, Vol. 29, 861-867.

[18] Kang,X., Xie,Y., Heather, M. P., Lee,L.J., Martha,A. B.(2007). Adipogenesis of murine embryonic stem cells in a three-dimensional culture system using electrospun polymer scaffolds, *Biomaterials*, Vol. 28, 450-458.

[19] Prabhakaran, MP., Venugopal, JR., Chyan, TT., Hai, LB., Chan, CK., Lim, AY., Ramakrishna, S.(2008). Electrospun Biocomposite Nanofibrous Scaffolds for Neural Tissue Engineering, *Tissue engineering*: Part A, Vol.14, 1787-1797.

[20] Chong,EJ., Phan,TT., Lim,IJ., Zhang,YZ., Bay,BH., Ramakrishna, S., Lim, CT.(2007). Evaluation of electrospun PCL/gelatin nanofibrous scaffold for wound healing and layered dermal reconstitution, *Acta Biomaterialia*,Vol. 3, 321–30.

[21] Koh, HS., Yong, T., Chan, CK., Ramakrishna, S.(2008). Enhancement of neurite outgrowth using nano-structured scaffolds coupled with laminin, *Biomaterials*, Vol.29, 3574-82.

[22] Liumin He ., Susan Liao ., Daping Quan ., Michelle Ngiam ., Casey K. Chan ., S. Ramakrishna ., Jiang Lu.(2009). The influence of laminin-derived peptides conjugated to Lys-capped PLLA on neonatal mouse cerebellum C17.2 stem cells, *Biomaterials*, Vol. 30, 1578–1586.

[23] Kim, YJ., Kwon, OH.(2007). Crosslinked Gelatin Nanofibers and Their Potential for *Tissue Engineering. Key Engineering Materials*, Vol.169, 342-343.

[24] Kozlov, PV., Burdygina, GI.(1983). The structure and properties of solid gelatin and the principles of their modification, *Polymer*, Vol. 24, 651-666.

[25] Zhang, YZ., Venugopal, J., Huang, ZM., Lim, CT., Ramakrishna,S.(2006). Crosslinking of the electrospun gelatin nanofibers, *Polymer*, Vol. 47, 2911-2917.

[26] McGuigan, AP., Sefton, MV. (2007). Modular tissue engineering: fabrication of a gelatin-based construct. *J. Tissue Eng Regen Med.*, Vol. 1, 136-145.

[27] Sakai, S., Hashimoto, I., Kawakami, K.(2007). Synthesis of an agarose-gelatin conjugate for use as a tissue engineering scaffold. *Journal of Bioscience and Bioengineering*, Vol. 103, 22-26.

[28] Broderick, EP., O'Halloran, DM, Rochev, YA , Griffin, M, Collighan, RJ, Pandit, AS. (2005).Enzymatic Stabilization of Gelatin-Based Scaffolds. *Journal of Biomedical Materials Research* Vol.72B, 37-42.

[29] Yung, CW., Wu, LQ., Tullman, JA., Payne, GF., Bentley, WE., Barbari, TA. (2007). Transglutaminase crosslinked gelatin as a tissue engineering scaffold, *Journal of Biomedical Materials Research Part* A, Vol.83A, 1039-1046.

[30] Li, M., Mondrinos, MJ., Chen, X., Gandhi., MR., Ko, FK., Lelkes, PI.(2006). Co-electrospun poly (lactide-co-glycolide), gelatin, and elastin blends for tissue engineering scaffolds, *Journal of Biomedical Materials Research Part A*, Vol.79A, 963-973.

[31] Coombes, AGA., Verderio, E., Shaw, B., Li, X., Griffin, M., Downes, S. (2002). Biocomposites of non-crosslinked natural and synthetic polymers, *Biomaterials,* Vol.23, 2113-2118.

[32] Zhang, Y., Ouyang, H., Lim, CT., Ramakrishna, S., Huang,ZM.(2005). Electrospinning of Gelatin Fibers and Gelatin/PCL Composite Fibrous Scaffolds, *Journal of Biomedical Materials Research*, Vol.72B, 156-165.

[33] Yang, F., Xu, CY., Kotaki, M., Wang, S., Ramakrishna, S.(2004). Characterization of neural stem cells on electrospun poly(L-lactic acid) nanofibrous scaffold, Journal of Biomaterials Science, *Polymer Edition* , Vol.15,1483–97.

In: Recent Progress in Chemistry…
Editor: A. K. Haghi, pp. 99-106
ISBN: 978-1-61668-501-0

Chapter 9

Vapor Phase Polymerization of Polypyrrol on the Electrospun PVA Nanofibres

Mahdi Nouri**, *Javad Mokhtari and Ozra Ghamgosar
University of Guilan, Rasht, Iran

Abstract

Polymers that exhibit high electrical conductivity have successfully been synthesized in the last few decades. Polypyrrole (PPy) has been one of the most extensively studied conductive polymers. Polypyrrol was deposited on the Poly vinyl alcohol (PVA) nanofibers during electrospinning process. Polymerization of the PPy was performed in the vapor phase at various conditions such as oxidant type and oxidant content in the electrospinning solution. For this purpose, a chamber was designed and electrospinning was done in this chamber. Possibility of the chemical deposition of PPy on the surface of the electrospun PVA nanofibers using ammonium persulfate as oxidant was shown. Fiber formation and morphology of the coated nanofibers were studied using a scanning electron microscope (SEM) and conductivity of the nanowebs was determined. The electrical conductivity of the coated nanofiber mats depends on the coating conditions and increases at PVA/oxidant weight ratio of 3/1. Measurement of the electrical conductivity of the coated mats shows electrical conductivity in the range of 10^{-3} to 1 S/cm depending on the chemical coating conditions.

Introduction

Electrically conducting polymers still are of interest of many researchers due to their various applications such as rechargeable batteries, antistatic coating, shielding of electromagnetic interferences, sensors, etc [1]. Among them, polypyrrole (PPy) has attracted great attention since it exhibits a high electrical conductivity, electrochemical activity and good environmental stability. PPy is polymerized by either conventional electrochemical or chemical polymerization methods in insoluble or infusible due to the strong inter- or intra-

* Corresponding author: mnouri69@guilan.ac.ir

molecular interactions. The poor processability, brittleness, insolubility and unstable electrical properties of PPy have limited its practical applications. However, polymerization of PPy onto the textile substrates in solution or vapor phase could be used as an alternative method for fabrication of PPy based materials. Due to excellent flexibility of textile materials, chemical oxidative deposition of conducting polymers onto textile surface provides a new composite material with potentially various applications [2]. The resulting composite possess both the mechanical properties of the textiles and the electrical properties of the conducting polymers which can be used as flexible smart materials in various fields. Due to the good adhesion with different substrates, PPy is capable of producing composites with a number of nonconductive fibers to prepare electrical conductive material. Rossi et al. [3] reported a sensitized glove based on the polypyrrole coated lycra/cotton fabric. Kim et al [4] proposed the ppy coated PET/Spandex as a flexible strain sensor for a large deformation up to 50%. Li et al. [5] reported a flexible strain sensor from PPy coated fabrics prepared by a chemical vapor deposition process.

PPy offers tremendous technological potential, especially as electrodes for rechargeable batteries, sensors or supercapasitors. Since electrochemical activity of the PPy electrodes is proportional to the surface area, PPy electrode must be processed to a shape in which ratio of surface area to volume or weight is high. This feature could be obtained via converting PPy to a fiber with small diameter e.g. Nano fibres. In nano-scale, the electrical conductivity of conductive polymers is increased [6]. It should be noted that systems in nano-scale may possess entirely new physical and chemical characteristics. As an example, higher electrical conductivity arises when the size of a wire is reduced below certain critical thickness (nano-scale). This provides the wide potential applications of nanofibers. Nevertheless, it is impossible at the level of the visible world to produce fibers in nano-level with unusual properties [6]. Using nanofiber structures of conducting polymers will enhance their performance due to high surface to volume ratio of nanofibers. It has been reported that optical sensors based on electrospun nanofibers showed sensitivity up to three times higher than that obtained from thin film sensors for the detection of nitro compound, ferric and mercury ions [7]. The higher sensitivities reported for the electrospun nanofibers can be attributed to their high ratios of surface area to volume.

To date, the most successful method of producing nanofibers and nanoporous materials is through the process of electrospinning. The electrospinning process has gained much attention because it is an effective method to manufacture ultra fine fibers or fibrous structures of many polymers with diameter in the range from several micrometers down to a few nanometers.

This method seems to be the most straightforward way to produce nanofibers by forcing a polymer melt or solution through a spinneret using of a high voltage electrostatic field. The electrostatic field is subjected to a droplet of polymer solution, held at the end of a capillary tube. With increasing the electrical field, electrostatic force overcomes the solution surface tension and a charged jet of the fluid is ejected from the hemi-spherical surface of the fluid at the tip of the capillary.

The ejected polymer solution jet typically develops a bending instability and then solidifies to form fibers. Often, the fibers produced in this way have a diameter in the range of nanometers. Several authors have explained the process; for the sake of brevity a few are mentioned in the list of references [8-11].

In the current study, preparation of electrically conductive PVA nanofiber using vapor phase chemical polymerization of pyrrole onto the surface of the nanofibers are investigated. PVA was selected for this study because of its excellent chemical resistance and physical properties. PVA is highly biocompatible and nontoxic.

EXPERIMENTAL

Materials and Methods

PVA (Mw=78000) was purchased from Merck. Reagent-grade pyrrole was obtained from Aldrich and distilled prior to use. All the other reagents were laboratory grades and used as received.

Electrospinning

In order to prepare electrospinning solution, a 10% (w/w) PVA aqueous solution was first prepared.. The required amount of oxidant (ammonium persulfate or $FeCl_3$) was then added to the prepared PVA solution and stirred using mechanical stirrer to form a homogenous solution. The mixtures containing various PVA/oxidant weight ratios were prepared and used for electrospinning.

In the electrospinning process, a high electric potential (Gamma High voltage) was applied to a droplet of PVA/oxidant solution at the tip of a syringe needle. A syringe pump (New Eva Pump System Inc, USA) was used to form a constant amount of solution on the tip. The output of the injection pump was 20μl/min. A charged jet is formed and ejected in the direction of the applied field.

The electrospun nanofibers were collected on a target aluminum foil, which was placed at a distance of 10 cm from the syringe tip. A high voltage in the range from 10 kV to 20 kV was applied to the droplet of solution at the tip of the syringe and the best condition for the stable electrospinning was selected which was 15kV.

Coating of Nanofibers

Polypyrrole was deposited on the PVA nanofibers during electrospinning of the nanofibers. Polymerization of the PPy was performed in the vapor phase at various conditions such as oxidant type and oxidant content in the electrospinning solution. For this purpose, a chamber was designed and electrospinning was done in this chamber (Figure 1).

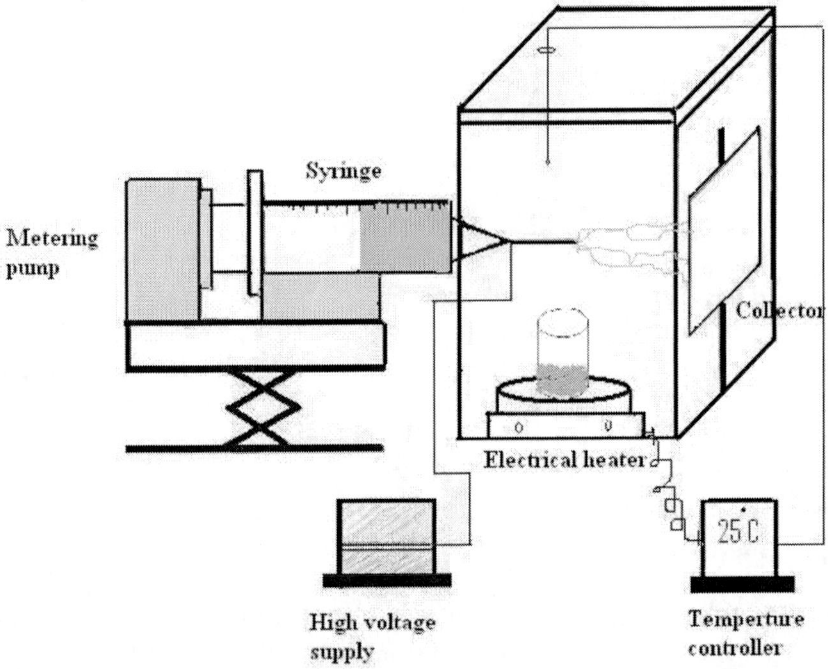

Figure 1. Schematic diagram of the electrospinning apparatus.

Characterization

Fiber formation and morphology of the coated nanofibers were determined using a scanning electron microscope (SEM), Philips XL-30. A small section of the produced web was placed on SEM sample holder and coated with gold (BAL-TEC SCD 005 sputter coater). Electrical conductivity of the coated mats was determined by employing the standard four-probe technique.

RESULTS AND DISCUSSION

Preliminary studies were performed to find optimum condition for electrospinning of PVA solution. SEM photomicrographs of electrospun PVA nanofibers from 10% solution showed that there was a uniform nanofibers with diameter ranging from 150 to 400 nm, average diameter of 350nm. As the polymer concentration increased, a mixture of beads and fibres were formed. Therefore, PVA concentration of 10% was used for preparation of electrospun nanofibers. A series of mats were prepared from the 10% PVA solution in mixture with various amount of $FeCl_3$ or ammonium persulfate ($(NH_4)_2S_2O_8$) as oxidant at the 15KV constant electric field.

Figure 2 shows SEM photomicrograph of PPy coated electrospun nanofiber using ammonium persulfate as oxidant at PVA/oxidant weight ratio of 3/1. It is observed that uniform layer of PPy with accumulations of PPy are deposited on the nanofibers surfaces. Average fibre diameter of uncoated nanofibers was 350 nm.

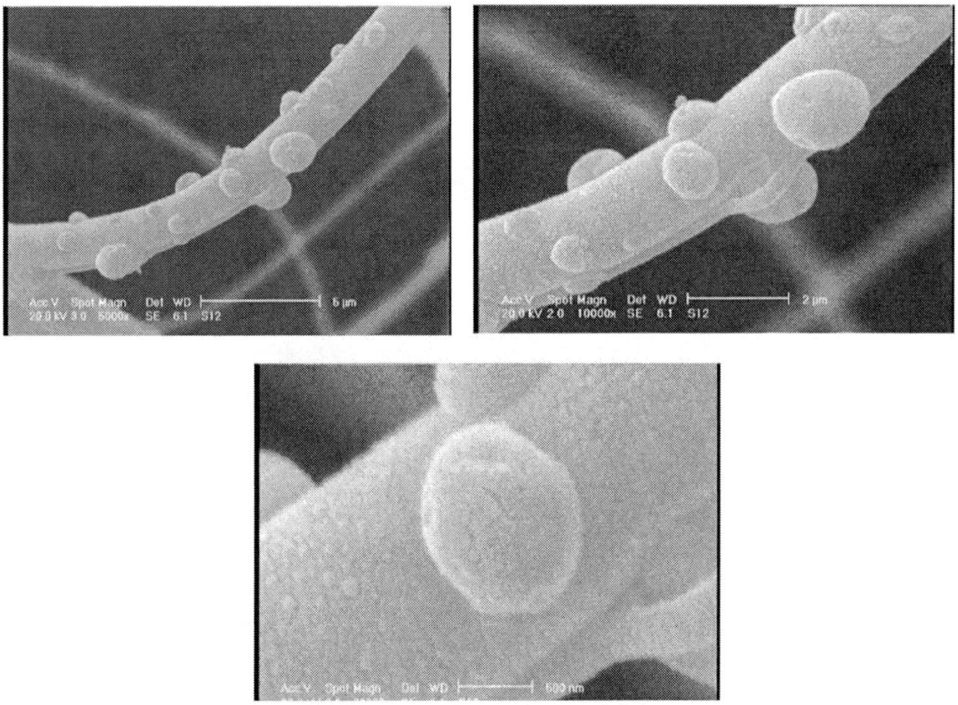

Figure 2. SEM photomicrograph of PPy coated electrospun PVA nanofibre using ammonium persulfate as oxidant (PVA/oxidant =3/1).

After coating process, average fiber diameter raised to 1300 nm. It means that a layer of PPy with thickness of approximately 1000 nm was coated onto the nanofibers. By changing the PVA/oxidant weight ratio to 2/1 and 1/1 (increase in oxidant content), deposition of the PPy onto the fibre was reduced and irregular fibers were obtained as shown in Figure 3. Electrical conductivity of the coated nanofiber mats depends upon the PVA/oxidant weight ratio. It seems that by suitable selection of the PVA/oxidant ratio and suitable coating of nanofibers increase in electrical conductivity can be achieved. Measurement of the electrical conductivity of the coated mats shows electrical conductivity in the range of 10^{-4} to 10^{-1} S/cm depending on the chemical coating conditions such as amount of the oxidant in the PVA solution. As discussed on SEM photomicrograph PVA/ oxidant ration of 3/1 is the optimum condition for preparation uniform coated nanofibers with higher electrical conductivity.

Figure 4 shows SEM photomicrograph of PPy coated electrospun nanofiber using $FeCl_3$ as oxidant. In compare to ammonium persulfate coated nanofibres, there is not a distinct sign of PPy coating onto the fiber surface. The resulted webs are not conductive which confirms that the coating of nanofibers using $FeCl_3$ was not successful.

For comparison, Figure 5 shows SEM photomicrograph of PPy coated electrospun Polyacrylonitrile nanofiber using chemical in-situ polymerization [12]. In this case, a uniform layer of PPy is deposited on the nanofibers surfaces. Average fiber diameter for uncoated nanofibers was 230nm. After coating process, average fiber diameter increased to 350nm. It means that a layer of PPy with the thickness of approximately 120 nm was coated onto the PAN nanofibers. Measurement of the electrical conductivity of the coated mats showed that electrical conductivity in the range of 10^{-3} to 1 S/cm depending on the chemical coating conditions was achieved. This is at least one order of magnitude higher than those achieved through vapor phase polymerization reported in this work.

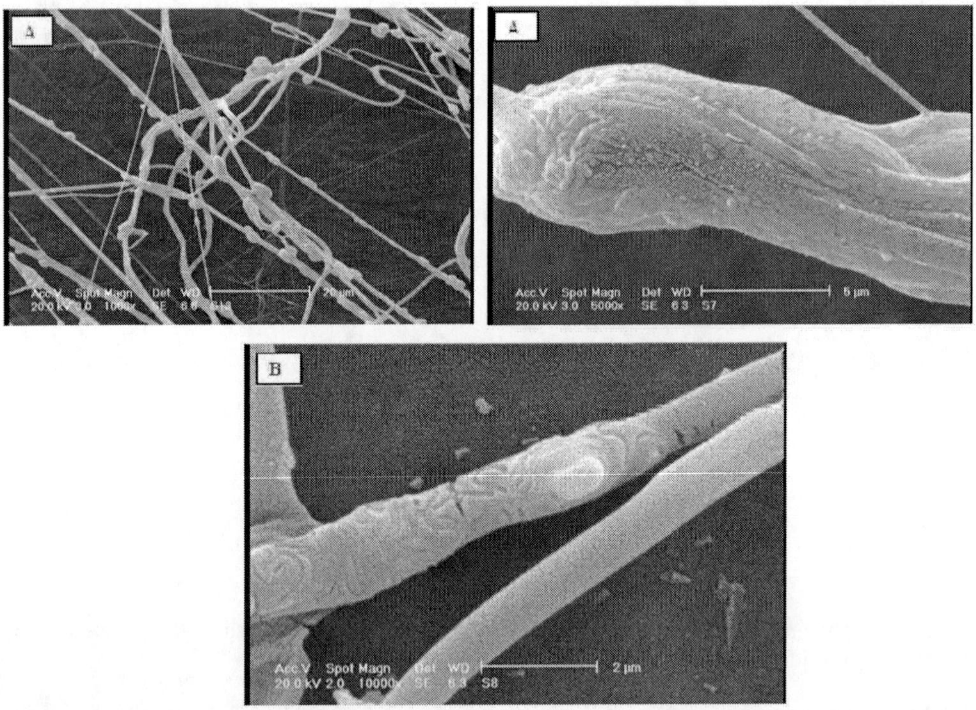

Figure 3. SEM photomicrograph of PPy coated electrospun PVA nanofibre using ammonium persulfate as oxidant ((A) PVA/oxidant =1/1, (B) PVA/oxidant =2/1).

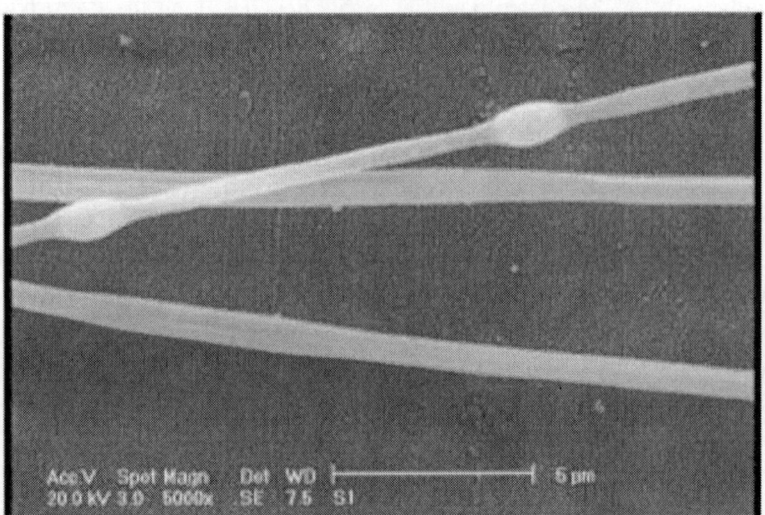

Figure 4. SEM photomicrograph of PPy coated electrospun PVA nanofibre using FeCl3 as oxidant (PVA/oxidant =1/1).

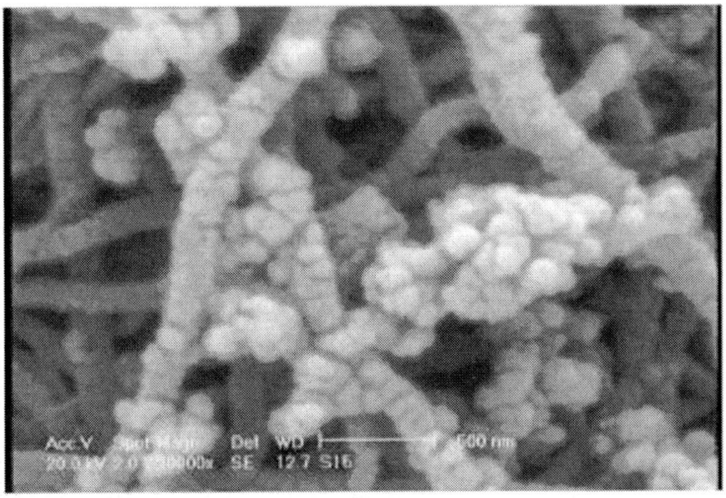

Figure 5. SEM photomicrograph of PPy coated electrospun PAN nanofiber using solution polymerization [12].

CONCLUSIONS

The vapor-phase chemical deposition of PPy on the surface of the electrospun PVA nanofibers was successfully performed. Morphology of the nanofibers and successful deposition of the PPy onto the nanofibers is governed by the oxidant type and its concentration in the electrospinning solution. At the best studied condition, a layer of PPy with the thickness of 1000 nm is formed onto the PVA nanofibers surface using ammonium persulfate as oxidant. In contrast, surface coating was not observed in the case of FeCl3 as oxidant. The electrical conductivity of the coated nanofiber mats depends on the coating conditions and reaches to 10^{-1} S/cm. Large surface area of the sensor material with the electrochemical activity is one of the most important properties for sensor application. Since it is obvious that the surface area of PPy coated nonwoven web is much greater than that of the film with same volume or weight, which makes the PPy coated nonwoven web potentially being as sensors.

REFERENCES

[1] Wallace, G.G, Spinks, G. M., Kane-maguire, L.A.P., Teasdale, P. R., (2003), " Conductive electroactive Polymers", 3th Edition, USA, C.R.C.

[2] Malinauskas , A., (2001), Chemical deposition of conducting polymers, *Polymer, 42,*3957-3972.

[3] Rossi , D.D., Sanata , A.D., Mazzoldi , A., (1999), Dressware: wearable hardware, *Mater. Sci. Eng. C7*, 31-37.

[4] Kim, H.K., Kim, M.S., Chun et al., (2003) Characterization of Electrically Counductive Polymer-Coated Textiles , *Mol. Cryst. Liq. cryst. 405,* 161-169.

[5] Li , Y., Cheng , X.Y., Leung , M.Y., Sang, J. T., Tao , K.M., Yuen, M.C.W., (2005), A flexible strain sensor from polypyrrole-coated fabrics, *Synth. Met., 155,* 89-94.

[6] He, J.H., Wan, Y.Q., Xu, L., (2007) Nano-effects, quantum-like properties in electrospun nanofibers, *Chaos, Solitons and Fractals. , 33*, 26–37.

[7] Wang, X.Y., Drew, C., Lee, S.H., Senecal, K.J., Kumar, J., Samuelson, L.A., (2002), Electrospun nanofibrous membranes for highly sensitive optical sensors, *Nano lett. , 2, 11,* 1273-1275.

[8] Huang, Z.M., Zhang, Y.Z., Kotaki, M., Ramakrishna, S., (2003), A review on polymer nanofibers by electrospinning and their applications in nanocomposites, *Composite Science and Technology, 63,* 2223-2253.

[9] Deitzel, J. M., Kleinmeyer, J., Harris, D., Beck Tan, N. C., (2001), The effect of processing variables on the morphology of electrospun nanofibers and textiles, *Polymer, 42*, 261-272.

[10] Theron, S. A., Zussman, E., Yarin, A. L., (2004), Experimental investigation of the governing parameters in the electrospinning of polymer solutions, *Polymer, 45,* 2017-2030.

[11] Tan, S. H., Inai, R., Kotaki, M., Ramakrishna, S., (2005), Systematic parameter study for ultra-fine fiber fabrication via electrospinning process, *Polymer, 46*, 6128-6134.

[12] Mirbaha, H., Nouri. M, Ghamgosar A., (2008), Polypyrrol Coated Polyacrylonitril Electrospun Nanofibers, AUTEX2008, biella, Italy.

In: Recent Progress in Chemistry…
Editor: A. K. Haghi, pp. 107-172
ISBN: 978-1-61668-501-0

Chapter 10

AN EPITOME TO THE CATALYST DEVELOPMENT

Pragnesh N. Dave*, Puneet R. Malpani and Saumya K. Pande
Institute of Technology, Chemical Engineering Department, Nirma University, Gandhinagar-Sarkhej Highway, Ahmedabad-382 481 (Gujarat), India

ABSTRACT

This chapter portrays the phenomenal development that has taken place in the field of catalyst science. It attempts to outline various catalyst preparation methods, characterization techniques, catalyst deactivation and regeneration. This chapter further delineates about the strategies for the kinetic testing of industrial catalysts (e.g. SSITKA (steady-state-isotopic-transient-kinetic-analysis and TAP technique). It explains the interaction of inorganic compounds with oxides and impact of drying of dying on supported catalyst. The salient feature of this chapter is the selection of best transition metal for the reaction using modern density functional theory (DFT) utilizing a d-band model to explain it, Electric field effect, BEP (Brønsted–Evans–Polanyi) relation and volcano curve in heterogeneous catalysis. The details regarding SMSI (strong metal support interaction) and NEMCA for the modification of catalytic activity and selectivity embodied with the future of catalysts. Within this framework, catalysis will still play a leading role in technical and economical development in the 21st century.

INTRODUCTION

The first question that comes to our mind is what is a catalyst? The simplest way to answer it is: A catalyst is a substance that facilitates a chemical reaction. A catalyst is something that makes a reaction go faster, without being consumed in the process. Because the catalyst is not consumed in the process, each catalyst molecule can participate in many consecutive cycles, so we need only a small amount of catalyst relative to the substrate. The substrate/catalyst ratio reflects the catalyst's efficiency, which is measured in terms of *turnover number* or *turnover frequency*. There are many different types of catalysts ranging

* Corresponding author: pragneshdave@gmail.com

from proton, H^+, through Lewis acids, organometallic complexes, organic and inorganic polymers, all the way to enzymes. To simplify we divide catalysis into three main categories: *homogeneous catalysis, heterogeneous catalysis, and biocatalysis.*

The catalyst turnover number (TON) and the turnover frequency (TOF) are two important quantities used for comparing catalyst efficiency. These definitions, however, vary slightly among the three catalysis field. In homogeneous catalysis, the TON is the number of cycles that a catalyst can run through before it deactivates, i.e., the number of A molecules that one molecule of catalyst can convert (or turn over) into B molecules. The TOF is simply TON/time, i.e., the number of A molecules that one molecule of catalyst can convert into B molecules in one second, minute, or hour. In heterogeneous catalysis, TON and TOF are often defined per active site, or per gram catalyst. This is because one does not know exactly how many "catalyst molecules" there are on the surface. In biocatalysis, the TON and TOF are defined by the rate measured when all the enzyme molecules are coupled with a reactant, divided by the total enzyme concentration.

In homogeneous catalysis, the catalyst is in the same phase as the reactants and products. Homogeneous catalysis mainly concentrates in the liquid phase. In the classic case, the reactant (also called the substrate) molecules and the catalyst are reacted in a solvent. For example, the trans-esterification of fatty acid triglycerides with methanol is catalyzed by hydroxide ions. This is an important process for making fatty acid methyl esters which are then used as biodiesel.

Heterogeneous catalysis covers all the cases where the catalyst and the substrate are in different phases. One important advantage of heterogeneous catalysis is the ease of catalyst separation. In gas/solid systems the catalyst is easily separated and cleaned, and in liquid/solid systems it can be simply filtered. Here we will discuss about heterogeneous catalysis in detail.

The biocatalyst is an enzyme – a complex protein that catalyzes the reactions in living cells. Enzymes are extremely efficient catalysts. An enzyme typically completes 1000 catalytic cycles in one second. Compared to this, conventional homogeneous and heterogeneous catalysts are slow and inefficient (100–10000 cycles per hour). Speed, however, is not the only advantage: enzymes specialize in converting one specific reactant into another specific product. This specialness even includes chiral centres, so an enzyme will catalyze the reaction of one substrate enantiomer, but not of the other. Such enantiospecific reactions are extremely important in the pharmaceutical and food industries, because often only one enantiomer has the desired activity. It is widely used to form any drugs, food additives, flavorings, and fragrances.

PREPARATION OF HETEROGENEOUS CATALYSTS

In any case, at the start of catalyst design consideration of desirable and undesirable reactions in the overall process as well as the fundamental properties and their mutual relationships in the different steps leads to the choice of the active species (figure 1).

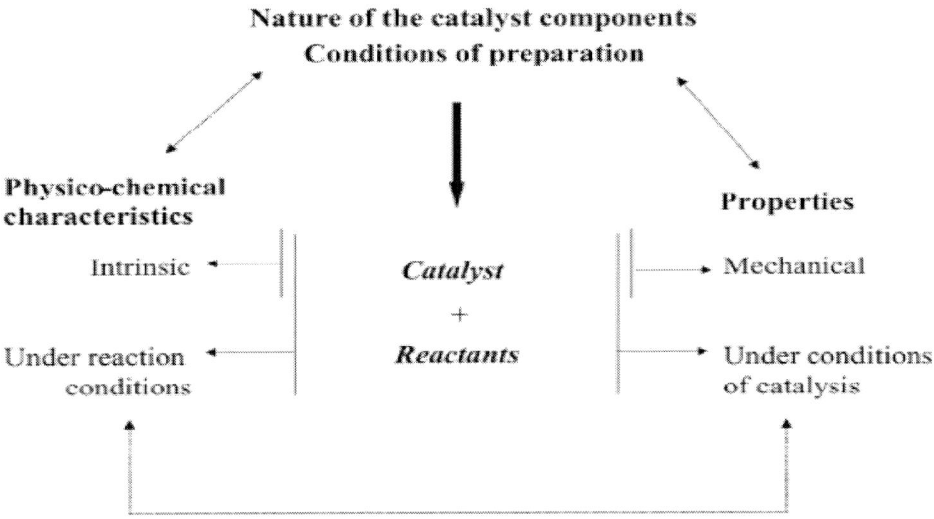

Figure 1. How the principal factors in catalyst are inter-related. [1,2].

A heterogeneous catalyst is a composite material, characterized by: (a) the relative amounts of different Components (active species, physical and / or chemical promoters, and supports); (b) shape; (c) size; (d) Pore volume and distribution; (e) surface area.

CATALYST PREPARATION

Generally, the catalysts may be classified according to the preparation procedure as: (i) bulk catalysts or supports and (ii) impregnated catalysts. On this basis the relative preparation methods are: (i) the catalytic active phase is generated as a new solid phase and (ii) the active phase is introduced or fixed on a preexisting solid by a process which intrinsically depends on the support surface.

Bulk Catalyst

The first unit operation is to obtain a solid which is achieved by *precipitation*. The aim of this step is to precipitate a solid from a liquid solution, as either a precipitate or gel. The precipitate or gel is generally a precursor, the nature of which determines the structure and properties of the final solid catalyst which formed as a result of successive operations. The precursor may be a crystalline, a gel or a flocculate precipitate.

The precipitation of a crystalline solid can be divided into three steps: *supersaturation, nucleation and growth*. In the supersaturation region the system is unstable and precipitation occurs as a result of any small perturbation figure 2. Supersaturation is reached by means of physical transformations (change in temperature or solvent evaporation) or chemical processes (addition of bases or acids, use of complex forming agents). Formation of the solid phase occurs in two elementary steps: (i) nucleation, , and (ii) growth or agglomeration of the particles.

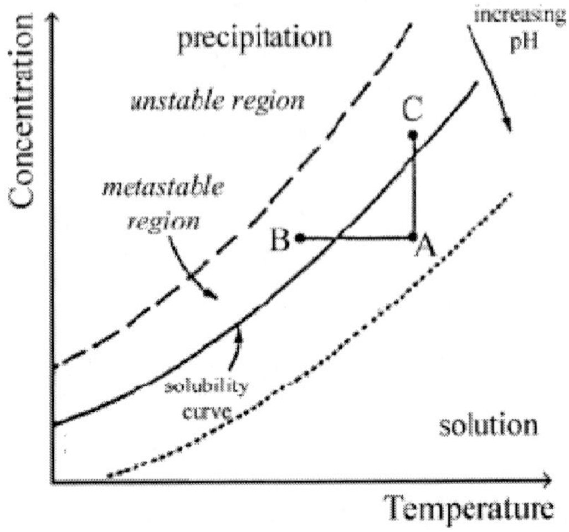

Figure 2. Concentration as a function of temperature and pH during the precipitation step (pathways to increase the concentration: A → B decrease in temperature; A → C solvent evaporation).

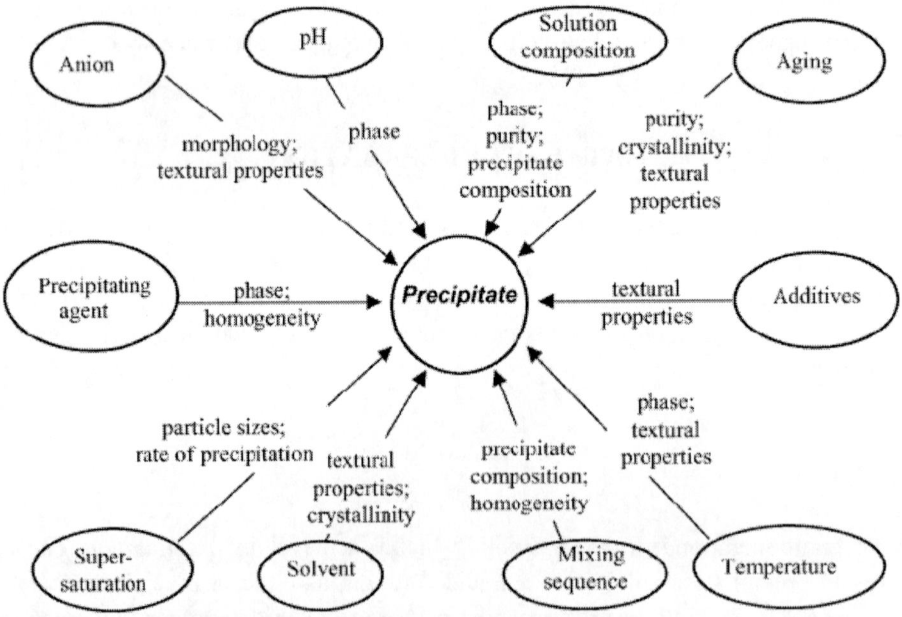

Figure 3. factors affecting the properties of precipitated catalysts [6].

Under conditions of high supersaturation the rate of nucleation is much higher than the rate of crystal growth and leads to the formation of numerous but small particles. In this condition amorphous precipitates can be obtained[3]. Usually precipitates with specific properties are desired. These properties include the nature of the phase formed, chemical composition, purity, particle size, surface area, pore size, etc., as well as the requirements of downstream processes (drying, palletizing or calcination). Basically, all process parameters influence the quality of the final precipitates, and fine tuning of the parameters is necessary in order to produce the required material (Figure 3).

The chemical and physical properties of the precipitates in contact with their mother liquor may further change, due to hydrothermal phenomena. Intimate mixing of the catalyst components can be achieved by either the formation of very small crystallites or the formation of mixed crystallites containing the constituents. During coprecipitation the pH has to be adjusted and kept constant [4]. Hydroxides and carbonates are the preferred precipitates because of their low solubility, easy decomposition and minimal toxicity and environmental problems [5].

The gelation route (sol–gel method) is a homogeneous process which results in a continuous transformation of a solution into a hydrated solid precursor (hydrogel). Sol–gel methods have several promising advantages over precipitation. In general, sol–gel synthesis have been recognised for their versatility which allows better control of the texture, composition, homogeneity and structural properties of the final solids. The nanoscale chemistry involved in sol–gel methods is a more direct way to prepare highly divided materials [3].

Four main steps may be identified in taking a precursor to a particular product via sol–gel preparation: formation of a hydrogel, its ageing, removal of solvent and heat treatment (figure 4). The versatility of this preparation method lies in the number of parameters that can be manipulated in each of these steps.

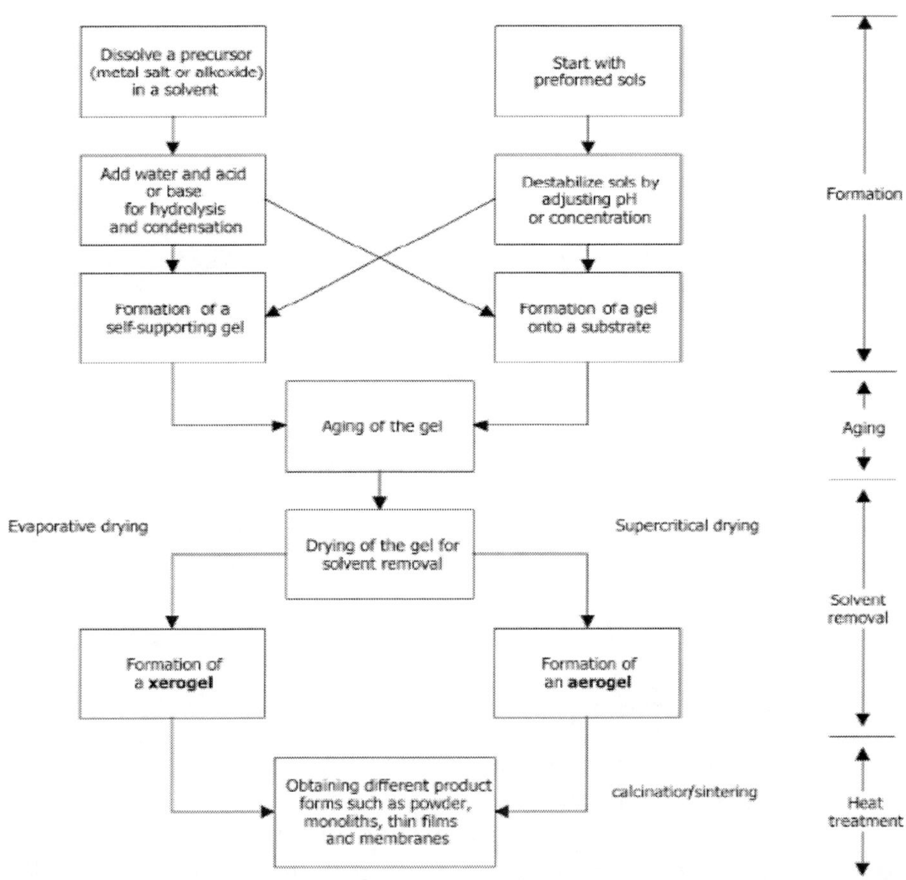

Figure 4. Schematic diagram of the different steps in the sol–gel preparation [4].

Hydrophilic colloidal solutions are formed of micelles that remain separated because of electrical charges on their surface and in the surrounding solution. These charges create repelling forces which prohibit coagulation of the micelles. The point at which the hydrogel formation, called gelation, occurs depends on the micelle concentration, temperature, ionic strength of the solution and especially the pH. The chemistry of the processes can be represented by a sequence of acid-orbase-catalysed nucleophilic additions or substitutions [hydrolysis(A) and condensation (B)].

$$\text{(A) -M-OR} + H_2O \rightarrow \text{-M-OH} + \text{R-OH} \quad (1)$$

$$\text{(B) -M-OH} + \text{XO-M-} \rightarrow \text{-M-O-M-} + \text{X-OH} \quad (2)$$

where X can be either H or an alkyl group R. However, this description simplifies the overall process because it does not represent the nature of the intermediates and end products, nor does it depict the simultaneous occurrence of the two reactions. Key factors are the relative rates of hydrolysis and condensation,due to their impact on the properties of the product [4].

The overall process produces a reticulate, metastable polymer with an open structure in which the primary units are held together by eitherchemical bonds,hydrogen bonds, dipole forces or van derWaals interactions. This framework includes almost all the solvent (Figure 5,pathA) [3,7].

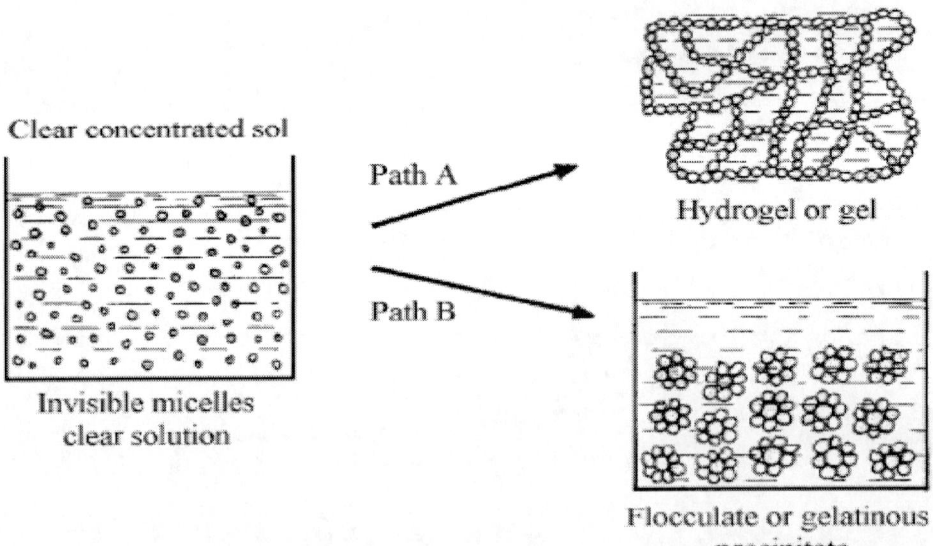

Figure 5. Mechanism of the formation of gels (A) and flocculates (B) [3, 7].

The gelation time ranges from a few minutes to a 100h and the density of the hydrogel increases with increasing concentration of the original salts in the solution and with the gelation rate. The homogeneity of the hydrogel depends on the solubility of the reagents in the solvent used, the sequence of addition of reactants, the temperature and the pH. Flocculation of a sol can be obtained through the neutralisation of the micelle charges and the flocculate precipitates more or less well. Hydrophilic sols give jelly-like flocculates that contain large amounts of water (Figure 5, pathb) [3,7]. The flocculates are denser than hydrogels and the elementary solid particles have the dimensions of the original micelles. The surface area of the final solids depends on the original micelle size and also on the ripening and drying conditions. Due to the possibility of obtaining high-surface-area solids with homogeneous compositions and controlled porosity, sol–gel methods have been widely employed for the synthesis of micro and mesoporous materials [3], mixed oxides, layer double hydroxides, hydroxyapatite, composite powders [4], etc.

Supported Catalysts

Supported catalysts consist of an active phase dispersed on a carrier [3,8,9]. The catalytic reaction takes place at the internal surface (i.e. in the pores) of the catalyst. Good supports combine relatively high dispersion with a high degree of thermal stability of the catalytic component [3,5,8,9]. Furthermore, the support should enable the production of a large shaped particle composed of very small easily sintered crystals of active phase which are prevented from coalescing by being separated by the support component. Wetting of the support with a solution or a slurry of the active phase precursors is the operation that characterizes such preparation. The most common preparation methods are *impregnation, ion-exchange, adsorption and deposition–precipitation* [3,5,7,10,11,12].

Impregnation [3,5,7,11] is the procedure whereby a certain volume of solution containing the precursor of the active phase is contacted with the solid support, which, in a subsequent step, is dried to remove the imbibed solvent. Two methods of contacting may be distinguished, depending on the volume of solution: wet impregnation and incipient wetness impregnation.

In wet impregnation (Figure 6) an excess of solution is used. After a certain time the solid is separated and the excess solvent is removed by drying. The composition of the batch solution will change and the release of debris can form a mud which makes it difficult to completely use the solution. The heat of adsorption is released in a short time.

In incipient wetness impregnation (Figure 6) the volume of the solution of appropriate concentration is equal or slightly less than the pore volume of the support. Control of the operation must be rather precise and repeated applications of the solution may be necessary. The maximum loading is limited by the solubility of the precursor in the solution. For both methods the operating variable is the temperature, which influences both the precursor solubility and the solution viscosity and as a consequence the wetting time. The concentration profile of the impregnated compound depends on the mass transfer conditions within the pores during impregnation and drying. Impact on drying on the distribution of active catalytic species on the support has been discussed in detail in the later section.

Ion exchange [3,5,7,11] consists of replacing an ion in an electrostatic interaction with the surface of a support by another ion species [3,11]. The support containing ions A is plunged into an excess volume (compared to the pore volume) of a solution containing ions B. Ions B gradually penetrate into the pore space of the support, while ions A pass into the solution, until equilibrium is established corresponding to a given distribution of the two ions between the solid and the solution [11]. For example, using a proper salt solution at ca.100 °C (to increase the exchange rate), it is possible to prepare the acid form of zeolite by exchanging NH_4^+ for Na^+ and successive calcination [13].

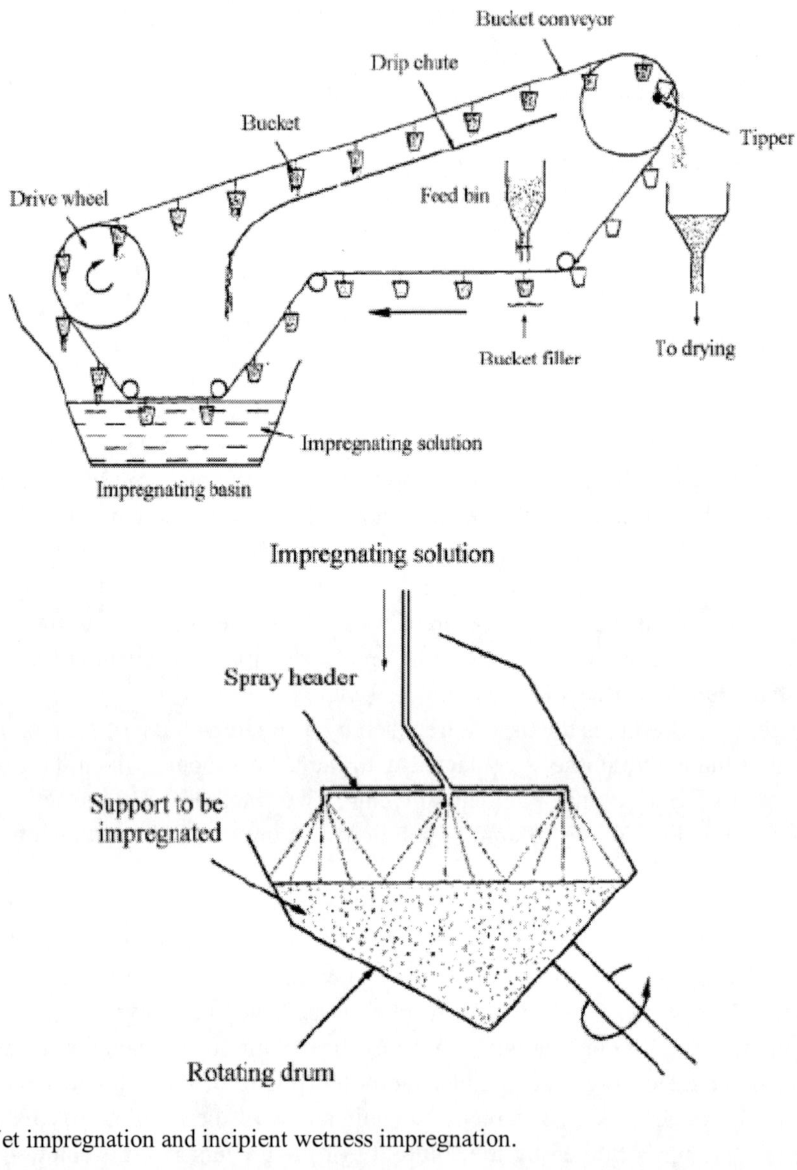

Figure 6. Wet impregnation and incipient wetness impregnation.

Adsorption [3,5,7,12] allows the controlled anchorage of a precursor (in an aqueous solution) on the support. The term adsorption is used to describe all processes where ionic species from aqueous solutions are attracted electrostatically by charged sites on a solid surface. Often consideration is not given to the difference between true ion exchange processes and electrostatic adsorption at the charged surface of oxides. Catalyst systems, which need charge compensating ions, are ideal materials for ion exchange (zeolites, cationic clays or layered double hydroxides). Instead most oxide supports, when placed in an aqueous solution, develop a pH-dependent surface charge. These oxides may show a tendency for adsorption of cations (SiO_2–Al_2O_3, SiO_2), or anions (ZnO, MgO) or both, cations in basic solutions and anions in acid solutions (TiO_2,Al_2O_3). The surface charge of an oxide depends on its *iso-electric* point alos known as point of zero charge (pzc) as well as on the pH and ionic strength of the solution.

In deposition–precipitation [3,5,8,9] two processes are involved: (1) precipitation from bulk solutions or from pore fluids; (2) interaction with the Support surface. Slurries are formed using powders or particles of the required salt in amounts sufficient to give the desired loading and then enough alkali solution is added to cause precipitation. However, precipitation in the bulk solution must be avoided, since it gives rise to deposition outside the pores of the support. A well-dispersed and homogeneous active phase is reached when the OH^- groups of the support (for example, the silanols of silica) interact directly with the ions present in the solution, thereby also determining the nature of the phase formed [3,12,14]. The nucleation rate must be higher at the surface than in the bulk solution and the homogeneity of the solution must be preserved. A method to obtain uniform precipitation is to use the hydrolysis of urea as a source of OH^- instead of conventional alkali. Urea dissolves in water and decomposes slowly at ca.90 °C, giving a uniform concentration of OH^- in both the bulk and pore solutions. Thus the precipitation occurs evenly over the support surface, making the use of urea the preferred method for amounts higher than 10–20%.

NEW METHODOLOGIES IN CATALYST PREPARATION

Catalyst Preparation Using Plasma

There are three main trends in the preparation of catalysts using plasma technologies: (1) plasma chemical synthesis of ultrafine particle catalysts; (2) plasma assisted deposition of catalytically active compounds on various carriers, especially plasma spraying for the preparation of supported catalysts; (3) plasma enhanced preparation or plasma modification of catalysts. Compared to conventional catalyst preparation, there are several advantages of using plasmas, including: (1) a highly distributed active species; (2) reduced energy requirements; (3) enhanced catalyst activation, selectivity, and lifetime; (4) shortened preparation time. These advantages are leading to many potential applications of plasma prepared catalysts. [15]

Aerosol Flame Synthesis of Catalyst

Various aerosol reactors and methods have been developed for the synthesis of a wide variety of metal and metal oxide particles. Compared to wet-chemical routes with various post-treatment steps, such as *filtration, washing, drying and calcination*, gas-phase processes allow the preparation of the desired material without any further post-processing. Flame processes for catalyst preparation also have the advantage that multi-component forms of catalysts can be readily prepared and their mixing can be controlled to obtain different distributions to prepare tailor-made nano composites.

Processes for aerosol flame synthesis (*AFS*) are classified first depending on the precursor state fed to the flame: vapor-fed AFS (*VAFS*) and liquid-fed AFS (*LAFS*). In the latter, if the liquid precursor solution drives the flame process (contributes more than 50% of the energy) it is called flame spray pyrolysis (*FSP*). If a non-combustible solution is fed into the flame then LAFS is called flame-assisted spray pyrolysis (*FASP*).

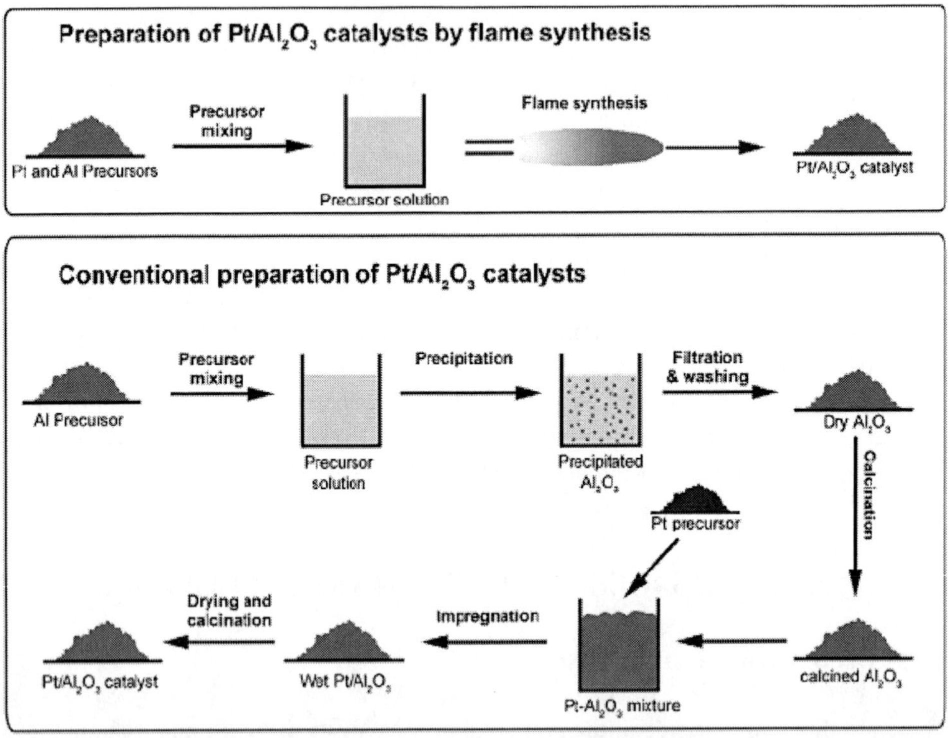

Figure 7. Comparison of conventional wet-phase and flame methods for the synthesis of catalysts as illustrated for Pt/Al_2O_3. Flame synthesis allows preparation of the catalyst in one step [17], whereas the conventional route by precipitation and impregnation is a multi-step process [16].

VAFS (Vapor-Fed Aerosol Flame Synthesis)

In this process volatile metal precursors (e.g. chlorides) are evaporated and fed into a flame that either supports the process (e.g.H_2/O_2 flame in fumed silica manufacture) or ignites the process. The metal precursor is converted into the metal oxide and starts to form particles by nucleation from the gas phase. The main *drawback* of this method is the need of *volatile precursors* at reasonable price. (Figure 8)

FASP (Flame-Assisted Spray Pyrolysis)

In FASP a non-combustible liquid precursor is dispersed into fine droplets that are evaporated and pyrolyzed by an external flame. Instead of an electrically heated tube as in conventional spray pyrolysis [18,19], an external hydrogen or hydrocarbon flame is used as the energy source during FASP. (Figure 8)

FSP: In FSP the metal precursor is a combustible liquid that is sprayed and ignited, Resulting in a product that is of nano size. The organic precursor solution is dispersed either ultrasonically [21, 24] or by gas convection through a nozzle [22,23] forming a fine spray which is ignited (Figure 8).

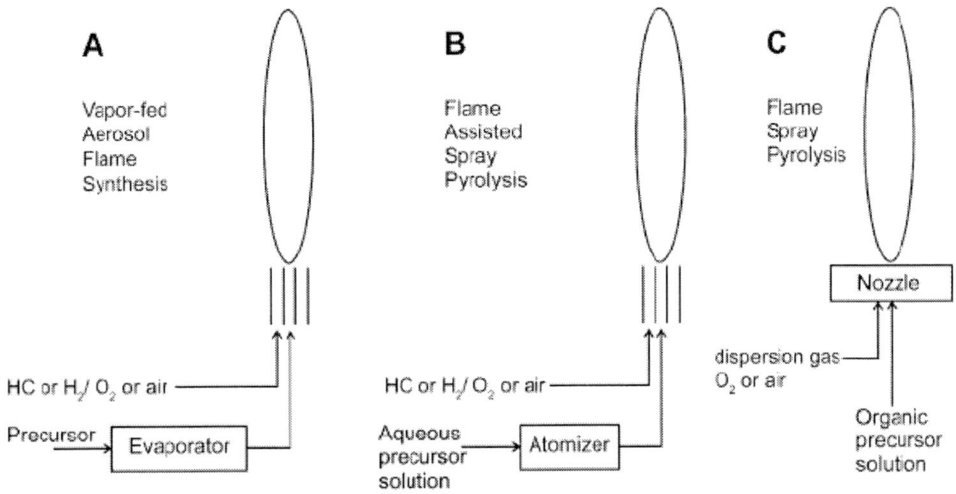

Figure 8. Schematic of selected flame configurations used for the synthesis of heterogeneous catalysts. (A) VAFS, (B) FASP and (C) FSP.

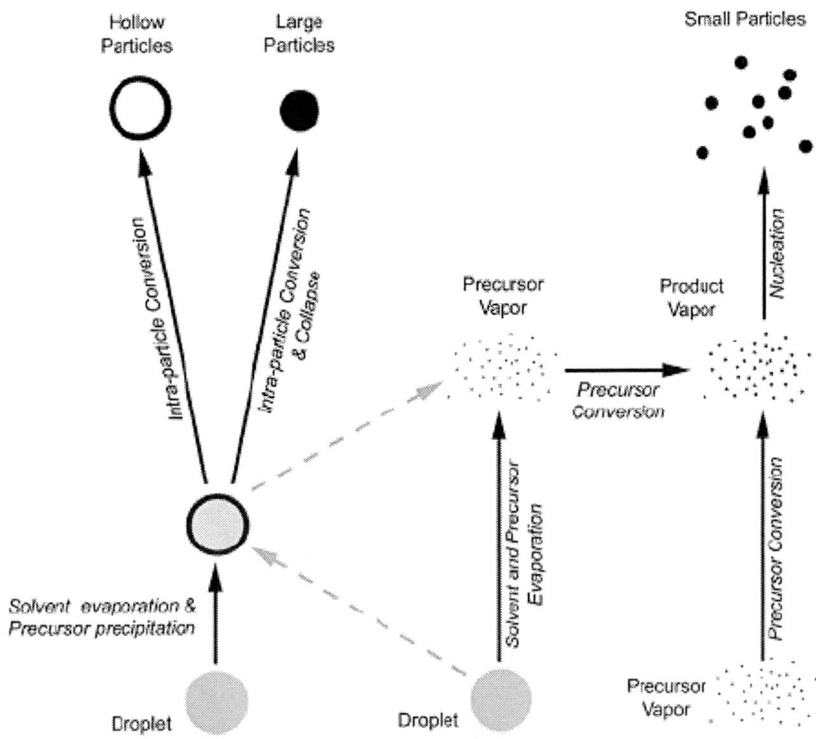

Figure 9. Schematic of the particle formation processes occurring during spray pyrolysis, FSP and VAFS. Two basic particle formation processes can be observed: particle formation in a droplet leading to large or hollow particles (spray pyrolysis and partly in FSP) and particle formation from the gas phase giving small nano-sized particles (VAFS, FSP and partly spray pyrolysis) [27].

In general, aqueous solutions of metal salts (i.e. nitrates) are sprayed in the external flame where the solvent evaporates from the droplets and metal precursors are converted in products. Supplying the energy by an external vapor flame allows much higher process temperatures and cooling rates compared to electrically heated hot-wall reactors where temperatures are usually relatively low (below 2000˚C) and residence times longer (up to a few seconds). Depending on process conditions, hollow, micron-or nano-sized particles are formed when the precursor reacts in the droplet or gas phase, respectively (Figure 9) [18,19,20].

The metal precursors evaporate in this spray flame and are combusted. Particles are then formed by nucleation from the gas phase (Figure 9). The process features short residence times (a few milliseconds) and high maximum process temperatures (upto 3000K) [25]. The main advantage of FSP over FASP is the formation of generally homogeneous, nano-sized particles through control of the precursor–solvent composition [26].

Catalyst Characterization

The characteristics of catalyst include:

- Chemical composition of the bulk and surface of the solid
- Bulk and solid structure, crystallite size and size distribution, crystallite morphology, porosity (micro,meso, and macro), and surface area.
- Surface chemical properties such as valence state, acidity, reactivity with different molecules, surface energy and surface electronic states.
- Aggregate properties such as particle size, magnetic properties, density, mechanical strength and attrition resistance.
- Catalytic properties, i.e. activity, selectivity, and activity stability [28].

Objectives of Catalyst Characterization

There are several important reasons for measuring the characteristics of solid catalysts:

- Understanding relationships among physical, chemical, and catalytic properties
- Elucidating causes of deactivation, designing procedures for regeneration
- Determining physical and chemical properties (e.g. composition, pore size, surface area, particle size and strength).
- Ensuring quality control in catalyst manufacturing

Most commonly used characterization techniques:

- *AAS* (atomic absorption spectroscopy): Quantitatively measures atomic composition based on photon absorption of a vaporized aqueous solution prepared from the starting material.
- *ADS* (adsorption): quantitatively measures amount of gas physically or chemically adsorbed on a solid; physical adsorption determines the total surface area and chemical adsorption determines the active site density.

- *AFM* (atomic force microscope) and STM (Scanning tunneling microscope): It measures the electronic structure of the surface with atomic resolution by mechanically scanning a probe over a surface without touching the surface.
- *AES* (Auger electron spectroscopy): Analysis of secondary electrons emitted from the surface upon electron bombardment yielding compositional information regarding the surface.
- *BET* (Brunaeur, Emmett and Teller method): determines the surface area of solids and/or the the distribution of mesopore sizes in these solids.
- *CAEM* (controlled-atmospheric electron microscopy): TEM system for studying samples in a controlled atmospheric environment over a fairly wide range of temperature.
- *CCSEM* (Computer controlled scanning electron microscopy): computer-automated SEM for conducting chemical and size imaging analysis of solid materials.
- *EDS or EDX* (energy dispersive X-ray spectroscopy): Low resolution, high efficiency X-ray diffraction of thin specimens in TEM.
- *EELS* (electron energy loss spectroscopy): Analysis of energy loss suffered by transmitted electrons in TEM for sample microchemical analysis.
- *EPMA* (Electron probe microanalysis): quantitatively measures composition of solids with a spatial resolution of the order of 1mm.
- *EPR* (electron paramagnetic resonance): It Studies about ionic impurities present by measuring the transition that occurs between electron spin energy levels of paramagnetic ions.
- *EXAFS* (extended X-ray absorption fine structure): Determines structural information about the surface on the atomic scale.
- *FEM* (field emission microscopy): Forms an image of the atomic arrangement at the surface of the tip of a material.
- *FTIR* (fourier transform infrared spectroscopy): Analyzes the structure of adsorbed molecules on a crystal surface under controlled atmospheric conditions. In addition, FTIR can also be used for kinetic studies because of its rapid scanning resonance.
- *ICP* (inductive coupled plasma): Performs chemical analysis of solid and liquid samples by vaporizing them in a plasma heater; it is more sensitive than AAS.
- *IR* (infrared spectroscopy): Analyzes the chemical structure of gas, liquid and solid phase molecules. It can also be used to study the structure and coverage of molecules on a catalyst surface during reaction.
- *Magnetic susceptibility measurements*: measures magnetic properties and particle size distributions of some metallic phases (especially cobalt, iron or nickel)
- *MAbS* (Mossbauer absorption spectroscopy): Measures oxidation states, atomic symmetry, lattice dynamic properties, and magnetic properties of selected elements, including Fe, Sn and Eu and their alloys.
- *MAS* (magic angle spinning): NMR method involving spinning of solid sample to determine its chemical structure.
- *MBS* (molecular beam scattering): Investigates the structure and reactivity of the surface and to study the angular and number distributions of scattered particles.
- *NMR* (nuclear magnetic resonance): Studies the chemical environment of nuclei having a magnetic moment due to splitting of the nuclear spins by external magnetic

field. It is useful for study of adsorbed or solid compounds of carbon, hydrogen and aluminum.

- *NS* (neutron scattering): Investigates various properties of solids, including crystal structure, vibrational transitions and crystallite or pore sizes
- *PIXE* (proton-induced X-ray emission): Quantitative compositional analysis to the ppm level of very small samples.
- *RBS* (Rutherford backscattering): determines species concentration in multi-surface layers of a solid
- *SEM* (scanning electron microscopy): determines the topography of a solid surface
- *SFG* (sum frequency generation): studies reaction intermediates during reaction
- *STEM* (scanning transmission electron microscopy): Determines nanotexture and nanostructure over a small area.
- *TEM* (transmission electron microscopy): Determines microtexture and microstructure of electron transparent samples.
- *TG or TGA* (thermogravimetric analysis): Measures weight loss (or gain) as a function of temperature in a controlled gaseous atmosphere; also studies reactions of solids with gases.
- *TPD* (temperature programmed desorption): Measures rate of desorption of adsorbed molecules as a function of temperature that allows the study of various adsorption sites, their binding energies, their surface concentration and their desorption kinetics.
- *TPR* (temperature programmed reduction): Measures rate of reduction as a function of temperature, which allows the study of oxidation states of the surface and bulk of a solid.
- *TPSR* (temperature programmed surface reaction): Measures rate of reaction of gaseous species with adsorbed species as a function of temperature to determine reactivity of different surface species, their reaction states and relative unsteady state reaction rates.
- *UPS* (ultraviolet photoelectron spectroscopy): Investigates about the electronic structure of the surface.
- *UV-VIS* (ultraviolet-visible spectroscopy): Measures absorption edge and band gap energies of surface species in solid materials
- *XRD* (X-ray diffraction): Characterizes bulk-crystal structure and chemical phase composition.
- *XPS* (X-ray photoelectron spectroscopy): Gives information regarding the surface chemical states (e.g. oxidation states)[28].

THE CHARACTERIZATION of a heterogeneous catalyst explains about the physical and chemical properties of a catalyst responsible for its performance in a given reaction. Measurements include composition, active sites, and the external and internal structure of the solid material through which reactants and products must be transported during catalysis.

Within the catalyst itself five fundamental processes must take place (in order of occurrence): (1) diffusion or transport of reactant's to active sites through the pore structure of the catalyst, (2) chemisorptions of reactant's on to active sites, (3) chemical reaction of chemisorbed species to produce product's, (4) desorption of product's from active sites, and (5) diffusion or transport of product's through the pore structure.

Figure 11 is an idealized picture of these five fundamental steps. The internal structure, comprising pores and surface area, is important for making the active catalytic sites accessible to the reactant molecules. The location of the active species is important for minimizing diffusional resistance [29].

Table. Catalyst Characterization methods and their Investigation

Characterstic	Method of Study
Physical Properties	
Surface area	Adsorption of N_2 or CO_2, CCSEM
Pore size, Size distribution	Extended N_2 and CO_2 adsorptions, Hg porosimetry, NMR spin relaxation of H_2O
Density, Skeletal (Solid)	He displacement or pycnometry
Density, Apparent (particle)	Hg displacement
Density, bulk (tap)	Tapping pack density
Catalyst particle size and size distribution	Sieving, electronic counting, laser light scattering, SEM,TEM, Optical imaging
Mechanical Properties	
Single pellet crush strength	Crushing tests in hydraulic press
Bulk Crushing Strength	Crushing test in hydraulic system
Attrition and abrasion resistance	Rotating drum, air jet, and ultrasonic jets
Chemical Properties	
Chemical State, bulk	TG, TPR, MAbS, FTIR, EPR, NS,NMR
Chemical State, Surface	XPS, TPSR, EXAFS, MAbS, FTIR, NMR
Chemical State , Surface additives	XPS, FTIR, RAMAN, NMR, EELS, HREELS, TPD, TPSR, MBS
Composition, bulk	Elemental analysis, AAS, ICP, XRF/XRD, SEM, TEM, MAS, TG, FTIR, Magn., NMR
Composition, Surface	AES, XPS, SIMS, EXAFS, ISS
Homogeneity	EPMA, IMP, LMMS, PIXE, SEM, TEM
Morphology, bulk	XRD, SEM,TEM, STEM, MAbS
Morphology, Surface	LEED, EXAFS, STM, TEM, STEM, FEM
Dispersion or Percentage exposed	Chemisorption, TEM, XRD
Surface Acidity	Adsorption and TPD of bases, IR, NMR
Surface Reactivity, active site concentration	Chemisorption, TPD, TPSR, transient kinetic methods, calorimetry

A common Scheme of various spectroscopic techniques is showed in Figure 10 giving information regarding the details that can be obtained from different spectroscopic techniques.

Method	Common scheme	Thickness analyzed	Quantitative measurements	Information obtained
UV–Vis	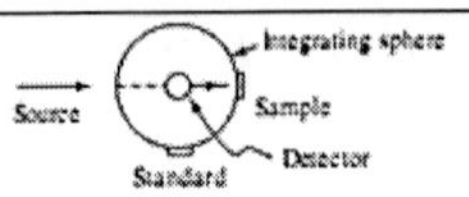	mm	Possible	Ion symmetry; adsorbing species; degree of oxidation
Transmission IR	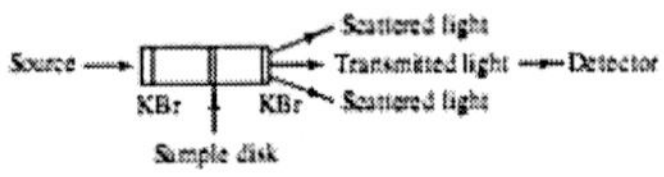	mm	Possible	Adsorbed species; surface active sites; functional groups
Reflection IR (RAIRS)		mm	Possible	Adsorbed species; surface active sites; functional groups
ATR	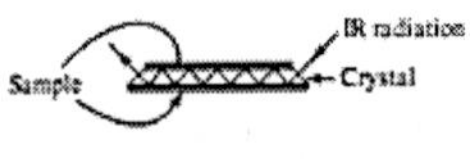	mm	Possible	Adsorbed species; surface active sites; functional groups
Diffuse reflection (DRIRS)	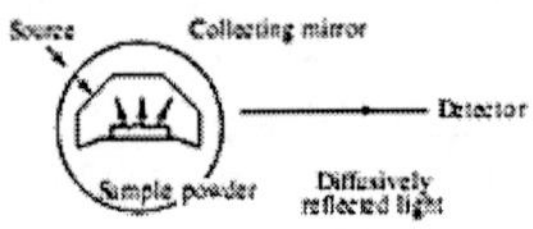	mm	Possible	Adsorbed species; surface active sites; functional groups
Emission IR	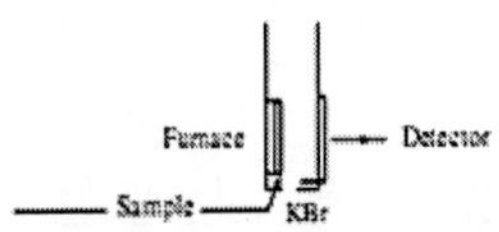	mm	Possible	Adsorbed species; surface active sites; functional groups
Raman	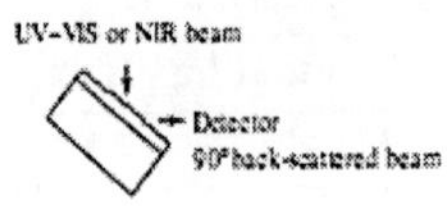	mm	Possible, but with difficulties	Functional groups; adsorbed sites; surface structure; bulk structure; structure of adsorbed species
EELS	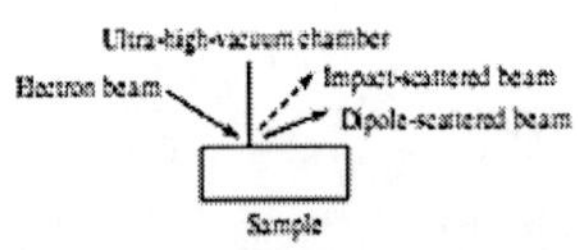	µm	-	Metal–ligand bonds; phase transitions; chemical structure
IETS	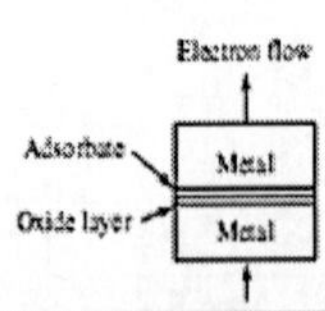	µm	Possible, but with difficulties	Vibrational spectra of minute quantities of materials

Figure 10. Basic Principles and applications of different spectroscopy methods used for surface chemistry analysis.

Aim of Catalyst Characterization

The catalytic properties of a surface are determined by its composition and structure on the atomic scale. Hence, it is not sufficient to know that a surface consists of a metal and a promoter perhaps iron and potassium – but it is essential to know the exact structure of the iron surface, including any defects and steps, as well as the exact location of the promoter atoms.

Thus, from a fundamental point of view, the ultimate goal of catalyst characterization should be to examine the surface atom by atom, and under reaction conditions. The well-defined surfaces of single crystals offer the best perspectives for atom-by-atom characterization, although occasionally atomic scale information can also be obtained from real catalysts under in-situ conditions. The many aspects that we need to study in order to properly understand supported catalysts on a fundamental level are shown schematically in Figure 12 [30].

The establishment of empirical relationships between the factors that govern catalyst composition, particle size, shape and pore dimensions on the one hand, and catalytic performance on the other hand, are extremely useful in the process of catalyst development. However, such relationships may not provide much fundamental insight into how the catalyst operates in molecular detail.

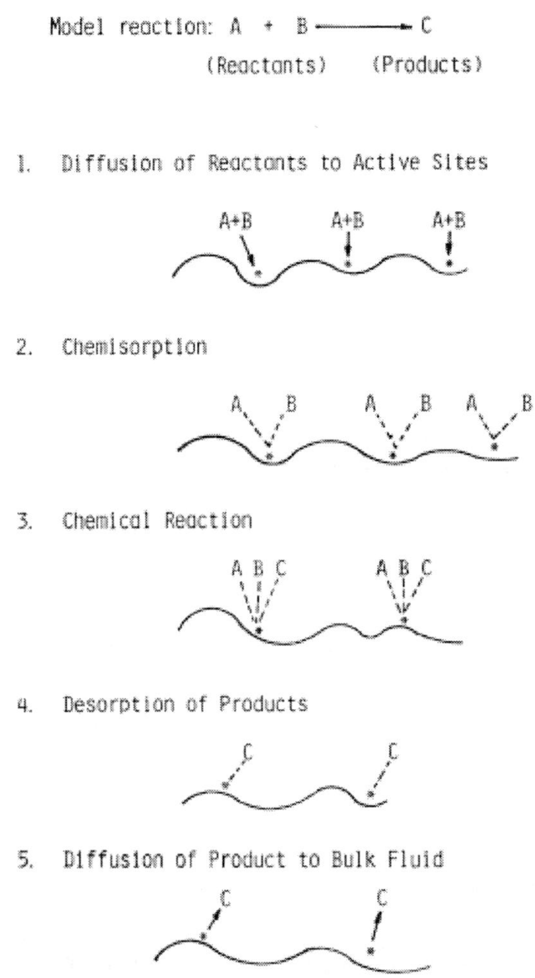

Figure 11. Schematic of the possible rate-controlling steps in a heterogeneously catalyzed reaction.

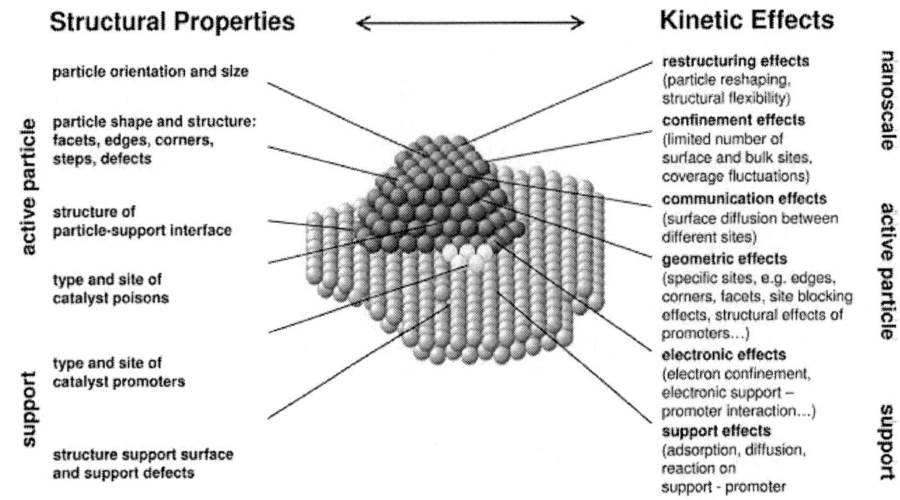

Figure 12. Scheme of the many aspects of supported catalysts that call for characterization on the molecular level.(Adapted from [30]).

Determining Physical Properties of Catalyst: Surface Area and Pore Size

Gas Adsorption: Surface Area

The most common procedure for determining the internal surface area of a porous material, with surface areas greater than 1 or 2 m^2 /g and upto~1200m^2 /g, is based on the adsorption and condensation of N_2 at liquid N_2 temperature. The partial pressure of N_2 above the sample is gradually increased; and N_2 molecules are physically adsorbed on to the surface, approaching monolayer coverage (first steep portion of isotherm shown in Figure 13A). Each adsorbed molecule occupies an area of the surface comparable to its cross-sectional area (~16.2 $Å^2$). By measuring the number of N_2 molecules adsorbed at monolayer coverage, one can calculate the internal surface area. In practice, coverage beyond a monolayer occurs, and at high relative N_2 partial pressures, condensation of liquid N_2 in the pores occurs. The Brunauer, Emmett, and Teller (BET) equation describes the relationship between volume adsorbed at a given partial pressure and the volume adsorbed at monolayer coverage:

$$\frac{P}{V(Po-P)} = \frac{1}{VmC} + \frac{(C-1)P}{VmCPo} \quad (3)$$

Here, P is the partial pressure of N_2, Po the saturation pressure at the experimental temperature, V the volume adsorbed at P, Vm the volume adsorbed at monolayer coverage, and C a constant.

This equation can be linearized by plotting P/V (Po − P) against P/P0, in which the slope is (C − 1)/VmC, whereas the intercept is equal to 1/VmC (Figure 13B). The sum of the slope and intercept yields the reciprocal of Vm. The most reliable results are obtained at relative pressures (P/P0) of between 0.05 and 0.3.

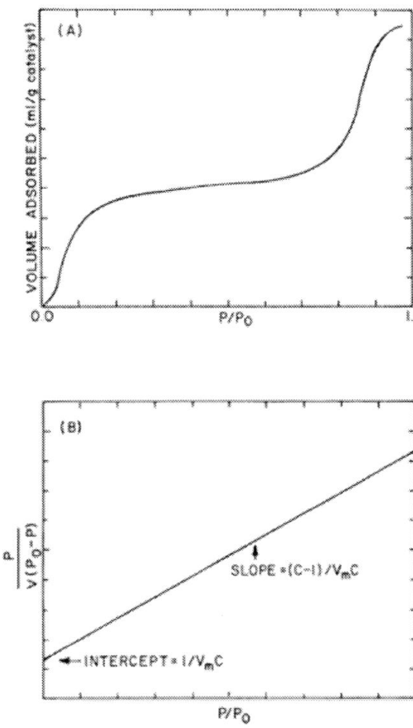

Figure 13. Measurement of surface area by the BET gas adsorption method. (A) Typical adsorption isotherm with a relatively flat curve in the region of monolayer adsorption. (B) Plot of the linear form of the BET equation between p/p0 = 0.05 and 0.3 used to calculate the monolayer coverage Vm.

Gas Adsorption: Pore Size

The same equipment as that for measuring surface area can be used to determine the pore size distribution of porous materials with diameters of 20 to 500 Å, except that high relative pressures are used to condense N_2 in the catalyst pores. The procedure involves measuring the volume adsorbed in either the ascending or the descending branch of the BET plot at relative pressures close to 1. Capillary condensation occurs in the pores in accordance with the Kelvin equation,

$$\ln\frac{P}{Po} = -\frac{2\sigma V\cos\theta}{rRT} \quad (4)$$

where σ is the surface tension of liquid nitrogen, θ the contact angle, V the molal volume of liquid nitrogen, r the radius of the pore, R the gas constant, T the absolute temperature, P the measured pressure, and Po the saturation pressure.

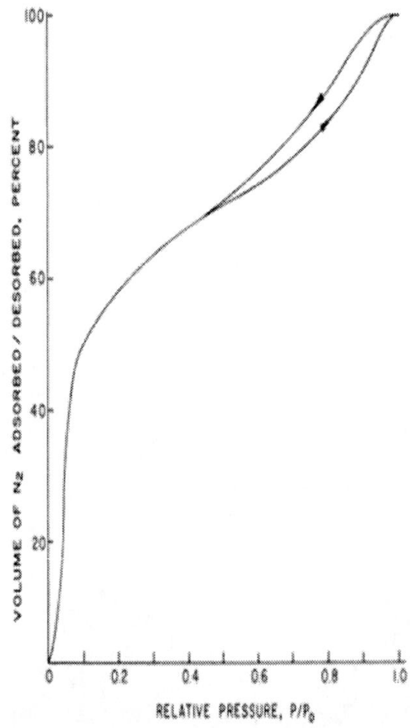

Figure 14. Nitrogen adsorption and desorption isotherms at 78K. Pore size distributions in the micropore range are calculated from the isotherms using the Kelvin equation.

Hysteresis in the adsorption–desorption isotherms (Figure 14) is a common observation for supports with a large fraction of small pores. It results from desorption from the meniscus at the end of a filled pore. The vapor pressure above the liquid at the pore mouth defines the pore radius in the Kelvin equation. Therefore, it is the desorption branch of the isotherm that is preferred in calculations of pore size distributions.

Morphology

Surface Texture via Electron Microscopy

Electron microscopic examination of catalyst materials, particularly those containing natural components, permits the identification of their origin.

Adhesion of Washcoats to Monoliths by Optical Microscopy

Refractory high surface area oxides are deposited from slurries on to the walls of the channels that make up monoliths in order to provide an adequate surface area to support the active catalytic species. Washcoats such as Al_2O_3 andTiO_2 are commonly used for pollution abatement applications (autoexhaust, stationary NOx abatement, etc.) where the monolith is usually a ceramic. Optical Microscopy enables one to see that the catalyzed washcoat follows the contour of the ceramic surface.

Location of Catalytic Species within a Support

Electron Microscopy

The basic electron optical system consists of an electron source, an array of magnetic lenses to collimate the electron beam, objective lenses, and a projector lens that focuses the images at the focal plane. The collimated electron beam passes through the specimen, is focused by the condenser lens on the back focal plane, and is magnified by intermediate lenses and finally the projector lens to form the final image. With the addition of deflector coils to the magnetic lens system, the electron beam can raster across a small area of the specimen. Defocusing of the objective lens increases the depth of field of the image in the transmission mode. Compared with that of optical microscopes the depth of field is much greater before any loss of resolution is observed. When the electrons are brought into focus in the back focal plane by the objective lens, an electron diffraction pattern is obtained. The diffraction pattern provides structural information of crystalline materials equivalent to X-ray diffraction patterns. The crystal structure and consequently identification of very small crystallites can be obtained by this technique, an important measurement in catalytic research.

Chemical Properties

Chemical Composition

Elemental Analysis:

Energy-dispersive analysis can be used to perform spot analysis within a catalyst. The electron beam can be focused on a particle or area whose analysis is desired, and the X-ray characteristics of the elements present are measured. This is a common method for poison analysis in a selected area of the catalyst. It should not be a substitute for a total analysis since only a small area is analyzed.

Analysis by X-Ray Diffraction:

Provided that a material is sufficiently crystalline to diffract X-rays and is present in an amount greater than ~1%, X-ray diffraction (XRD) can be used for qualitative and quantitative analysis. The principle of this technique is that crystal structures possess planes made by repetitive arrangements of atoms, which are capable of diffracting X-rays. The angles of diffraction differ for the various planes within the crystal, and thus every compound or element has its own somewhat unique diffraction pattern. The differences in these patterns, therefore, allow the differentiation of various structures within the catalyst. The compounds making up the catalyst sample can be clearly identified in the XRD pattern. Not only does the XRD pattern qualitatively identify the phases present in the catalyst, but the quantity of each phase can be determined by measuring the area under selected diffraction peaks relative to a standard.

Structural Analysis

Oxidation state:

A very convenient method for approximating the oxidation state of a component within a compound is to reduce or oxidize a sample in a controlled environment and measure the weight change by micro-balance techniques such as thermal gravimetric analysis. The catalyst weight (0.1–0.5g) is equilibrated at a particular temperature in a stream of inert gas. A reactive gas, such as H_2 for a reduction or O_2 for an oxidation, is introduced to the inert gas stream and the associated weight changes monitored. The weight change associated with reduction or oxidation can give information regarding the valence state of the catalytic component. By comparison with bulk chemical analysis, one can determine the extent of oxidized species present.

An alternative technique based on the same principle is called temperature programmed reduction. The TPR apparatus is similar to a gas chromatograph. A stream of reactive gas passes through a small bed of sample and the composition of the gas is monitored by a thermal conductivity detector. The temperature of the catalyst is raised at a linear rate and the detector measures the consumption of there active component of the gas as a function of temperature. Reduction and oxidation of the catalyst, desorption of chemisorbed gases, and catalyzed reactions can be studied quantitatively by this technique.

Dispersion of Catalytic Species

Chemisorption:

Selective gas adsorption, referred to as chemisorption, can be used to measure the accessible catalytic component on the surface indirectly by noting the amount of gas adsorbed per unit weight of catalyst. The stoichiometry of the chemisorptions process must be known in order to estimate the available catalytic surface area. One assumes that the catalytic surface area is proportional to the number of active sites and thus reaction rate. This technique has found use predominantly for supported metals. A gas that will selectively adsorb only onto the metal and not the support is used under predetermined conditions. Hydrogen and carbon monoxide are most commonly used as selective adsorbates for many supported metals. There are reports in the literature of instances in which gases such as NO and O_2 have been used to measure catalytic areas of metal oxides; however, due to difficulty in interpretation they are of limited use.

The measurements are usually carried out in a static vacuum system similar to that used for BET surface area measurements. The pressure of gas above the sample is increased and the amount adsorbed measured at equilibrium. When there is no further adsorption with increasing pressure (flat portion of Figure 15), it is assumed that the catalytic surface is saturated with a monolayer of adsorbate. Noting the amount of gas adsorbed and knowing its stoichiometry with the surface site, one can determine the number of catalytic sites. Approximating a value for the cross-sectional area of the catalytic component based on an assumed geometry, one can calculate its surface area and dispersion. It must be repeated that this technique measures only surface species capable of adsorbing the probing gas. The first two pulses are used for calibration and by pass the catalyst sample. The second set of pulses, passing through the catalyst are first diminished due to adsorption. Once saturation or monolayer coverage is reached, no further adsorption from the gas phase occurs. The amount adsorbed is found by the difference in areas under the peaks compared with those under the calibration pulses.

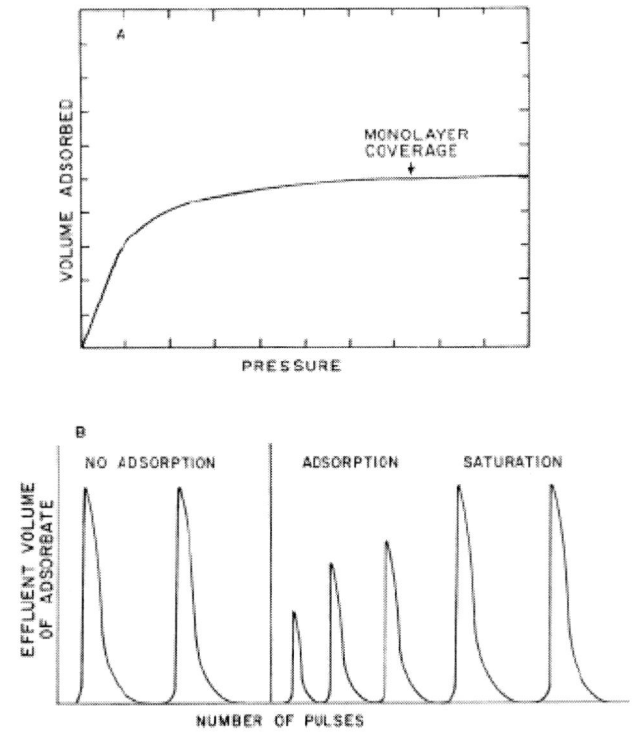

Figure 15. Chemisorption on a metal surface. A, Chemisorption isotherm showing approach to monolayer coverage; B, typical data from a pulsed chemisorptions technique.

Transmission Electron Microscopy

Transmission electron microscopy observes the transmission of electrons through a thin slice of material. Dense crystallites of metal or metal oxides prevent transmission and thus appear as dark spots on a photo micrograph. By image analysis a size distribution and average crystallite size can be calculated. By assuming a shape for the crystallites the dispersion, or ratio of surface atoms to total atoms in the crystallite, can be calculated. It should be understood that this technique measures only a small fraction of the catalyst, and hence obtaining data representative of the entire sample is quite difficult. Usually many different areas must be analyzed for proper statistics to be obtained.

Surface Acidity

Many reactions are catalyzed by acid sites on the surface of the catalyst. Isomerization, polymerization, aromatization, and cracking are catalyzed by Lewis and/or Bronsted acid sites.

Infrared Analysis

Infrared absorption can be used to estimate the relative amounts of Lewis and Bronsted acid sites on cracking catalysts. Bases complex with Lewis acid sites while proton transfer to the base occurs at Bronsted acid sites. Each has distinct, well-resolved infrared bands. For example, pyridine forms a complex with the Lewis acid site and produces an infrared absorption band at approximately $1450cm^{-1}$. Pyridinium ions form at Bronsted sites and produce an absorption band at approximately $1540cm^{-1.}$ The relative intensities of these two bands can be used to estimate the relative amounts of Lewis vs. Bronsted acid sites.

THE SURFACE OF CATALYST

A Surface Techniques

The tools available for surface composition characterization are electron spectroscopy for chemical analysis (ESCA), Auger spectroscopy (AES), ion scattering spectroscopy (ISS), and secondary ion mass spectroscopy (SIMS). ESCA spectroscopy is used more widely than the others for studying the surface composition and oxidation states of industrial catalysts.

Kinetic Catalyst Characterization

Catalysis is a kinetic phenomenon (i.e. acceleration of the reaction rate), and the kinetic characterization of catalysts, is an extremely important focus of heterogeneous catalysis research. Heterogeneous catalytic processes involve complex chemical transformations and transport phenomena. To obtain the kinetic dependence on reaction variables, the chemical phenomena must be distinguished from the transport phenomena. A common approach for extracting kinetic information is to measure the rate of chemical transformation from the rate of transport. Then, if the rate of transport is known it can be used as a "measuring stick" for determining the rate of transformation. Transport, however, is generally a complex process that involves both convection and diffusion, and it can differ in different parts of a catalytic system. As a result, transport processes can obscure the kinetic phenomena. Consequently, special experimental strategies, especially those based on the use of well-defined transport, are necessary to extract the kinetic characteristics from the kinetic-transport information. The process of kinetic characterization involves the collection and interpretation of reaction data using special experimental techniques and mathematical models. The key problem is determining the dependence of the chemical transformation rate (or its kinetic characteristics) on different reaction characteristics (e.g. gas composition, catalyst composition, temperature, etc.). In conjunction with other information, such as structural data related to the catalyst, kinetic information can provide guidance in the development of new or improved catalysts. The *main goal* of kinetic characterization is to extract the chemical transformation rates from the reaction-transport data and to determine their dependencies on reaction and structure characteristics.

Need for Determining Kinetics

Due to an increased pressure for short time-to-market development, lowered costs in chemical processing as well as a need to fulfill environmental regulations such as using cheaper and greener feedstock's, a significantly higher speed and efficiency in process optimization, and new materials discovery is required today. The combinatorial strategy based on high-throughput experimentation (HTE) fulfills these needs (i) by increasing the probability for new materials discovery, (ii) by increasing speed and efficiency for process development and optimization (higher and faster access to optimized operating conditions through the development of micro pilot technologies), and (iii) by improving safety and environmental aspects (reduced reactor volumes and integrated mass and heat transfer for a beneficial overall energy balance). Progress in the methods in catalysis research and development has reached an impressive rate. Companies implementing these techniques have recognized their power and importance in increasing productivity and adding value to processes and products. Companies providing technologies, instrumentation, and expertise are also expanding their effort, looking for a wider array of clients' needs. In parallel, among the academia groups having initiated an effective approach of combinatorial research under a more fundamental basis during the last 5 years and finding out new concepts trying to combine understanding and random approaches [33].

Strategies of Kinetic Testing of Industrial and Model Catalysts

CSTR and PFR:

Industrial catalysts are usually tested in a CSTR or a PFR system, at normal conditions. The experimental observables are the concentrations of gaseous substances, but not the concentrations of surface substances. The rate of chemical transformation is obtained by relating the amount of converted substance per unit of time to the flow rate of convective transport.

SSITKA (steady-state-isotopic-transient-kinetic-analysis)

The SSITKA approach is based on the determination of isotopic labels in the reactor effluent. The change in isotopic species versus time is measured following a step-wise change in the isotopic labeling of one of the reactants. SSITKA data can be used to determine the amount of certain surface substances (the number of "working active sites"), and to characterize the specific steady-state of the catalyst. Experimental data obtained from "CSTR-SSITKA" systems can contain significant experimental errors. Thus, the common procedure in SSITKA experiments is to use a steady-state PFR system rather than a CSTR system. To date, an unsteady-state version of the SSITKA experiment has not been developed and is likely to present significant experimental and modeling difficulties if developed in the future.

Surface science kinetic measurements

Surface science experiments are performed at ultra-high vacuum conditions, and convective transport does not play a role in the interaction of the gas phase with the solid. Kinetic data are obtained from thermal desorption or molecular-beam scattering experiments. A unique feature of the surface science approach is the possibility of obtaining information about the catalyst surface composition and structure. The composition of the locally supplied gaseous mixture (e.g., molecular beam) can also be monitored. On the other hand, surface science experiments are not well suited for investigating kinetics on industrial catalysts. One well- known difficulty arises when attempting to compare data from high-vacuum surface science studies with data from studies performed at the normal conditions. Another difficulty is due to the difference in materials used in surface science experiments from those used as industrial catalysts. Typically, surface science methods are applied to well-defined materials, such as single metal crystals, which are very different from technical multi-component catalysts, particularly mixed metal oxide catalysts such as those used in selective oxidation processes. Only some advances have been made to extend surface science experiments to include technical materials Surface science experiments can provide information about the rate dependence of both surface and gaseous concentrations, but typically have only been used to characterize catalytic surfaces that are very different from complex industrial ones.

TAP technique:

The TAP-2 (temporal analysis of products) reactor system is a relatively new research technique for investigating gas–solid reactions. It can be used to perform vacuum pulse-response experiments, commonly known as TAP experiments, temperature-programmed desorption/reaction experiments, and atmospheric pressure steady-state and transient experiments. A single catalyst sample can be tested over a wide domain of pressures (10^{3}–10^{-8} torr) and relaxation times (10^{3}–10^{-4}s) without exposing the sample to ambient conditions. The TAP system can be used to study model catalysts and technical catalysts including mixed metal oxide catalysts. The original TAP system was invented by Gleaves and Ebner in the 1980s and the TAP-2 system was developed by Gleaves and coworkers in the 1990s [31]. A detailed list of references related to TAP studies has been published in the theoretical paper [32]. A TAP pulse response experiment is a special type of transient response experiment that involves injecting a small gas pulse of very short duration in to an evacuated micro reactor containing a packed bed of particles. When the number of molecules in the pulse is sufficiently small ($\approx 10^{-10}$ mol) convective flow disappears, gas transport occurs by Knudsen diffusion, and the molecules move through the bed independently. The kinetic regime of a TAP Knudsen pulse response experiment is located at the boundary of the surface science regime. Inter-particle diffusion in a TAP Knudsen pulse response experiment acts as a "measuring stick" against which other transformation processes can be measured. TAP micro reactors can be loaded with particles in different ways. The simplest configuration is called a one-zone reactor, in which the reactor packing consists of a uniform bed of particles. A common reactor configuration used in many catalytic reaction studies is called a three-zone reactor, in which the catalyst bed is sandwiched between beds of inert particles. The main advantage of a three-zone reactor is that the catalyst zone can be easily maintained in an isothermal condition.

The thin-zone (TZ) reactor [34] is a new and very useful limiting case of a three-zone configuration. The thickness of the catalyst zone in the TZ reactor is very small compared to the length of the reactor. Having the catalyst zone very thin makes the change in the gas concentration across the catalyst zone small compared to its average value. A key advantage of the TZ configuration is that the catalyst bed can be changed uniformly by exposing the catalyst to a long series of small pulses. Another advantage of the TZ configuration is the simplicity of its mathematical model.

With testing and comparing of different catalytic compositions belonging to a given library (prepared or not by HT automated synthesis techniques), the combinatorial strategy is generally restricted to the evaluating as rapidly as possible simplified catalytic performances such as activity expressed by the degree of conversion of a key reactant, selectivity of the reaction to desired products, and therefore yield at simplified standard conditions. This is generally carried out by means of parallel reactors and permits only a primary ranking among various materials, which require neither, high accuracy in the process control nor sophisticated analytical techniques determining effluent compositions. However, this fast simplified testing procedure may eliminate good candidates which could have been revealed under other appropriate operating conditions. Accordingly, testing conditions should ideally be varied for each catalyst; looking out for maximum performance by probing several contact times/space velocities, temperatures, and partial pressures and eventually integrating time-on-stream deactivation. As such, basic kinetics derived from interpolation and extrapolation in the multi-parameter space could hint at optimum conditions, allowing a fair comparison of catalytic materials. However, such a mode of data acquisition is in general far too time consuming to accommodate the overall high throughput of the combinatorial loop. An elegant and efficient way to solve the problem is to perform fast kinetic measurements under transient conditions. As a matter of fact, in a study confined to steady state kinetics, only lumped kinetic parameters can be estimated rapidly whereas determining all relevant intrinsic parameters is possible but requires a large number of time-consuming experiments. By generating transient input signals, one observes signals out of the steady state and their relaxation may be related to the timescale on which the corresponding global reaction occurs at steady state. In turn, mathematical analysis of the relaxation curves may rapidly provide all the relevant intrinsic parameters of the reaction kinetics. In the steady-state-isotopic-transient-kinetic analysis (SSITKA), a reactant is stepwise substituted by its isotope equivalent and the isotopic concentration relaxation can be analyzed under overall steady-state conditions. In the temporal-analysis-of-product (TAP) reactor, concentration transients are generated by pulsing small and known amounts of reactants through a catalyst bed maintained under vacuum, and the pulse expansion and relaxation are analyzed at the reactor exit with a time resolution below milliseconds.. Numerous studies and reviews have demonstrated the unique power of these transient techniques for investigating key catalytic phenomena such as a non uniform surface and combination of kinetic, thermal, and hydrodynamic parameters [33]. How transient kinetic methods applied under HTE conditions may both accelerate the search for new materials and bring advanced fundamental insights into reaction mechanisms by providing kinetic parameters as descriptors of a given catalytic system.

Interaction of Inorganic Compounds with Oxides

Introduction: The interaction of different inorganic and organometallic complexes with oxide surfaces is new direction in the synthesis of catalysts containing catalytically active centers with particular properties. Physical methods allow the detection of all steps involved in the interaction between inorganic, organometallic or organic components and supports in the preparation of supported catalysts as well as stages of the formation of an active component. If the supported complex contains a fragment similar to the probe-molecule large piece of important information can be obtained.

The adsorption and catalytic properties of supported complexes can be significantly different from those of the initial complexes, due to interaction with a support. The reaction media and temperature can also affect the properties of supported complexes. Interaction between surface active sites and different reagents, including acids and alkalis, can also lead to modification of the surfaces.

Supported catalysts are synthesized by the deposition of the active metal precursor salt or other compound on to the surface of an oxide support (e.g. alumina, silica, titania, etc.). This can result in either isolated ions in different coordination (the latter is easily detected by both UV–Vis and ESR methods) in two-dimensional surface metal oxide overlayers or in three-dimensional metal oxide crystallites. Large three-dimensional crystallites can be detected by X-ray diffraction, in contrast to small three-dimensional crystallites (below 40 ° A), two-dimensional metal oxide overlayers and, in particular, isolated ions. Thus, the absence of an XRD peak may be caused by small crystallite size, their low concentrations or by the presence of a two-dimensional metal oxide overlayer. On the other hand, the presence of an XRD peak confirms that large crystallites are present and does not provide information about the presence or absence of a two-dimensional metal oxide overlayer.

Comparison between Raman, XRD and IR spectroscopy: For supported catalysts, Raman spectroscopy has a great advantage when compared with IR Spectroscopy, as the spectrum of a support is usually very weak in the Raman spectrum, and only the spectrum of the catalytically active component is observed. Unlike XRD, the Raman signal is not dependent on long range order and reveals only the vibrations of the M–O (metal oxide) bonds [35]. The different metal oxide structures that are present in the two-dimensional overlayers and in oxide crystallites produce different Raman and IR signals. This allows discrimination between these two structures. It is possible to estimate the degree of crystallinity of an oxide by using the half-widths of the Raman bands. Moreover, Raman spectroscopy enables the study of supported oxide layers down to monolayer coverage, whereas XRD studies require five layers as a minimum [36]. The Raman and XRD data obtained for thin films of RuO_2 on fused silica and alumina and showed that the bands characterizing RuO_2 are narrowed with an increasing of degree of crystallinity, as brought about by an increasing calcinations temperature.

Raman spectroscopy [37] and real-time surface-enhanced Raman spectroscopy (SERS) give quantitative information about thin films of oxides and the irrelevant amounts and the extent of oxidation [38]. Raman scattering is also very useful in studying ultra-fine-grained materials. For example, it has been established that grain size effects can produce large shifts within and also Broadening of the Raman spectrum [39].

IR and particularly FTIR spectroscopies can also be used successfully in a number of cases to evaluate the metal oxide structures, especially when the vibrations of an active component do not overlap with those of the support. When the vibrations of active component overlaps with the support the method of support-spectrum compensation [40], involving computerized subtraction of the support spectrum from the total spectrum, enables the detection of the spectrum of the active component.

Organometallic Complexes

Interaction of supported catalysts with organometallic compounds is mainly of the acid–base type. Coordinatively unsaturated cations and anions at the surface of the support provide Lewis acceptor and donor centers respectively. On adsorption, supported complexes are subjected to the electric potential over a large region on the solid surface and may involve interactions with more than one surface site. Characterization of the surface complexes is a primary requirement for the understanding of the mechanism of the interaction between organometallic compounds and supports, and study of the reactivity of the surface adducts. Spectral methods have found wide application in this field through observations of changes in the states of ligands and the character of their interactions with support surfaces.

METAL CARBONYLS AND NITROSYLS

Transition-metal carbonyls can interact with oxide surfaces in different ways. The main interactions can be divided, according to Zecchina and Arean [41], into ligand-centered and metal-centered types as follows:

The first class includes (i) weak physisorption through vanderWaals interactions, (ii) hydrogen-bonding to OH groups of partially hydroxyllated oxides, and (iii) oxygen bonding to a surface Lewis acid site, where as the second class comprises of (i) decarbonylation and clustering at strong Lewis acid sites, (ii) nucleophilic attack by Lewis basic sites, (iii) nucleophilic attack by Bronsted basic sites, (iv) Lewis-base- assisted redox dis-proportionation, and (v) protonation.

The primary interaction seems to be of the donor–acceptor type between CO ligands and Lewis acid sites at the support surface, forming an oxygen-bound surface adduct. Carbonyl ligands both linear and bridging surrounding the metal center have the oxygen atom pointing toward the exterior of the molecule, and form O-bound σ-adducts with Lewis acids. Formation of σ-adducts causes a perturbation of the CO vibrational spectrum. Low wave number shifts of 100–150cm^{-1} and 200–400cm^{-1} are typically observed for linear and bridging CO groups, respectively. Surface adducts formed at defective sites are more strongly held than those located on regular surface sites. Electron density withdrawal from the metal carbonyl is larger, and the perturbation of the CO vibrational spectrum is maximized in such cases. Electron withdrawal also facilitates transformation of bridging CO groups into linear ones.

As a rule, mononuclear carbonyls have high symmetry and correspondingly simple vibrational spectra. These spectra become more complicated after interaction of the complex with the surface, as new additional bands appear. IR spectroscopic studies of the $Ni(CO)_4$ interaction with various types of supports (covalent, SiO2; acid, γ-Al_2O_3; strongly acid, HY and HZSM-5 zeolites; basic, MgO) [42] showed significant differences in the spectrum observed. Upon interaction with a surface, a tetrahedral molecule of $Ni(CO)_4$, with Td symmetry in the gas phase, is reduced in symmetry.

The IR spectrum of $Ni(CO)_4$ on silicalite is very similar to that in hydrocarbon solution: Only the band at calcinations 2050cm^{-1} is observed, and physical adsorption through the vanderWaals force takes place. On highly dehydroxylated γ -Al_2O_3, $Ni(CO)_4$ forms oxygen-bound adducts. There are two low frequency bands at about 1950 and 1850cm^{-1}, corre-sponding to CO ligands σ-bound to surface Al^{3+} ions with different degrees of coordinative unsaturation. On some defective Lewis acid sites, the O-bound surface adducts are unstable and decompose with the elimination of CO ligands [41].

Sub carbonyl species are then formed, as well as $(Ni^{o})n$ clusters. The $Ni(CO)_4$ interaction with HY and HZSM-5 zeolites leads to the formation of hydrogen-bonded adducts. Dehydroxylation of both HY and HZSM-5 by heating under vacuum leads to a progressive elimination of Bronsted acidity and simultaneous creation of Lewis acidity. Upon the interaction of $Ni(CO)_4$ with fully dehydroxylated MgO, besides a number of bands inthe1800–2100cm^{-1} region, there are two absorption bands at about 1490 and 1050cm^{-1}, corresponding to the asymmetric and symmetric stretching modes of the CO_2^{2-} group in a structure commonly observed when metal carbonyls are dosed on basic oxides:

$$Ni(CO)_4 + Mg^{2+}O_2^{-} \rightarrow [(CO)_3NiCO_2]^{2-}Mg^{2+} \qquad (5)$$

In this reaction the CO ligand interacts with an ionic pair.

$$Mo(CO)_6 \leftrightarrow Mo(CO)_4 \underset{-CO}{\longleftrightarrow} Mo(CO)_3L \leftrightarrow Mo_2(CO)_6 \qquad (6)$$

UV–Vis spectroscopy has been applied to study the formation of the active catalyst component upon the interaction between $Mo(CO)_6$ and an alumina support [43]. These data establish that the activation of the $Mo(CO)_6/\gamma$ -Al_2O_3 catalyst is associated with the formation and rearrangement of sub-carbonyl structures on the surface. The formation of the $Mo_2(CO)_6$ active center may be represented as follows: where L is the ligand (possibly a surface hydroxyl group). The formation of $Mo_2(CO)_6$ is associated with a decrease in energy.

All of the above examples demonstrate the effect of support surface centers on the initial complex, which is thereby changed to give an active component. Another situation is observed when the supported complex decomposes, and oxide or metallic active components are formed from it. The properties of the new active component depend on the initial structure of the complex formed on the support, its transformation conditions, and also on the properties of the cluster formed on interaction with the support. As a rule, the use of such a system to prepare the catalyst allows the generation of a more dispersed and homogeneous distribution of the active component over the support in the cases of formation of oxide species in an oxidizing medium or metallic particles in a reductive atmosphere. IR and UV–Vis spectroscopy have become very useful as a method of controlling catalyst preparation. With this help, excellent low-temperature catalysts for waste gas combustion have been prepared from iron penta- and hexacyanide complexes ($M[Fe(CN)_5NO]$ and$M_2[Fe(CN)_6]$, where M= Mn , Fe, Co, Ni and Cu), synthesized on γ -Al_2O_3 supports [44].

Interactions with Simple Acids and Bases

One of the most widely used methods of modifying oxide surfaces is by interaction with acids or alkalis. Thus, the addition of Na^+ ions on alumina causes a decrease of the surface hydroxyl group liable for protic acidity, probably due to an exchange of H^+ of the hydroxyl group with Na^+ (LFB at around 3660cm^{-1} disappears at this) [45-47]. Support for this comes from the IR spectrum of NH_3 adsorbed on such a catalyst: there are no bands of NH_4^+ ions in this case [45,47]. IR data on the CO adsorption show that both the strongest Lewis acid sites (the band at 2235cm^{-1}), as well as the Bronsted acid sites, are fully depressed after incorporation of sodium, whereas the acceptor abilities of the weaker Lewis acid sites (the band at 2210cm^{-1} which shifts to 2205cm^{-1}) decreases [46]. In contrast, the treatment of alumina by HCl leads to the disappearance of two high-frequency bands of the OH groups due to the substitution of OH^- groups by Cl^- ions, accompanied by an increase in the strengths of both the protic sites and the LASs, probably because of the inductivity effect.

Impact of Drying on Supported Catalysts

During the impregnation step, the support is immersed in a solution of the inert precursor. In the case of capillary impregnation, the support is initially dry, while during diffusional impregnation the support is initially filled with the liquid solvent. In both cases air can be trapped in the pore volume of the particle, which may hinder impregnation. As the liquid penetrates the support due to capillary pressure, the entrapped air is compressed. The liquid penetration ends when the capillary pressure is equal to the pressure of the entrapped gas. For this reason the air is often evacuated before impregnation or the particles are exposed to a soluble gas (NH_3). If the pore radius is very small, the capillary pressure is much larger than the pressure of the entrapped air and air removal will not lead to any appreciable differences in the results. At the end of the impregnation step the distribution of the precursor in the support is typically a function of the metal/surface interaction (adsorption), the pore structure, and the diffusivity of the dissolved precursor in the impregnating solution.

Drying, which follows impregnation, causes the evaporation of the liquid solvent. Catalyst drying is usually carried out at temperatures between 50°C and 250°C. During drying, the liquid solution is transported by capillary flow and diffusion and the precursor may be redistributed by adsorption/desorption phenomena [48].

Drying is a three-stage process, consisting of a preheating period, a constant-rate period, and a falling-rate period. In the preheating period, the support is heated up by the drying medium. The rate of liquid vaporization at the support's surface and the rate of drying increase as the temperature increases. During the constant-rate period, vapor is removed from the support's saturated surface. Moisture transport inside the support is sufficient to keep the surface saturated. In this stage, the rate of drying depends on the rate of heat transfer to the evaporating surface. Therefore, the temperature of the saturated surface is constant and is called the wet-bulb temperature. During the first two stages, the liquid is transported toward the external surface by capillary flow and the ions dissolved in the liquid phase are transported by liquid convection and diffusion. The falling-rate period begins at critical moisture content, where the moisture transport inside the support is no longer sufficient to keep the surface saturated and the drying rate decrease. Dry patches appear near the surface and the temperature begins to rise. A drying front develops and vaporization takes place inside the solid. In this case convective flow of the vapor dominates the moisture transport and convective ion transport is negligible. Evaporation starts at the particle surface and a moving front, receding towards the particle center.

The four main categories of catalyst profiles are shown in Figure 16. The choice of the optimal catalyst profile in the support is determined by the required activity, selectivity, and by other characteristics of the chemical reaction (kinetics, mass transfer). Sometimes catalyst stability is an important factor. egg-yolk catalysts have the best activity when there is no mass transfer limitation under isothermal conditions.

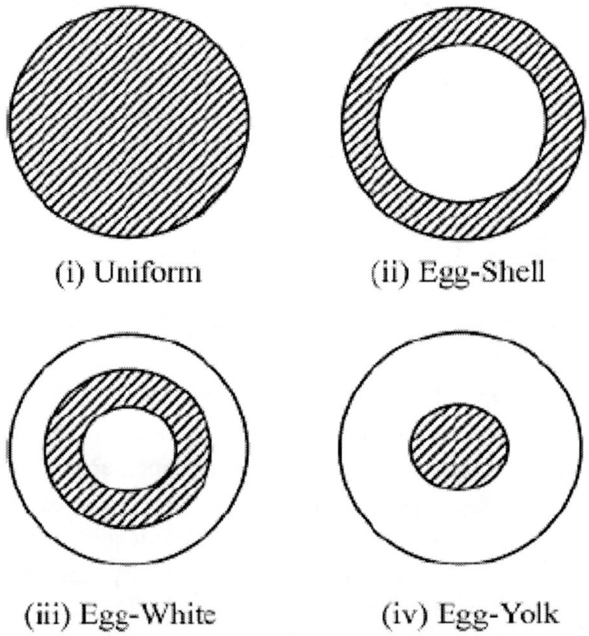

Figure 16. Main types of catalyst distribution.

Egg-yolk catalyst distribution is optimal when the reaction is kinetically controlled while an egg-white, or egg-shell catalyst gives optimal performance for diffusion control. Clearly, egg-shell catalysts are advantageous in the case of fast reactions with strong diffusional restrictions, because the active component is concentrated close to the external pellet surface. Additionally, egg-shell catalysts give higher selectivities for consecutive reactions, where the intermediate product is the desired one. When the external surface of the catalyst is exposed to poisoning or intense abrasion, as in fluidized or moving beds, the egg-white, and egg-yolk distributions give improved performance and have a longer life time because attrition does not impact the active layer.

How to Select Best Transition Metal for the Reaction

Modern density functional theory (DFT) calculations provide a method to resolve the electronic structure of even rather complicated model systems with adequate accuracy at a reasonable computational cost [49]. A class of systems for which DFT has been applied with considerable success involves chemical reactions taking place on solid surfaces [50]. Both in heterogeneous catalysis and in electrochemistry it is important to understand the chemical bonding of atoms and molecules to transition metal surfaces. The best transition metal catalyst for a given reaction will to a large extent be determined by the ability of the metal to bond the key reaction intermediates in just the right way [51]. When going through the periodic table, the adsorption properties of the pure metals vary enormously, and it is the changes in the electronic structure of the metallic surface which lead to these variations. An atomic-scale understanding of phenomena in both heterogeneous catalysis and in electrochemistry is therefore intimately tied to an understanding of the electronic structure of the catalyst or electrode surfaces. Only a small fraction of the research being carried out in the fields of electrochemistry and heterogeneous catalysis is related to electronic structure theory, and this seems to suggest that in both classes of systems large complexities are present which are necessary to take into account in order to actually understand specific reactions. This is perhaps true for most electrochemical reactions, where the complexity of the electrolyte–surface interface with varying ion concentrations and local field effects seems larger than for a gas phase heterogeneous reaction. Calculating the reaction rate for a particular catalyst is an enormous task, but if our target is to determine which of the elements in the periodic table will serve as the best catalyst, a much simpler approach is sufficient [52]. The reason for this is that the change in electronic structure from one element to the next in the periodic table implies large variations in adsorption and activation energies for the elementary steps of this reaction. In spite of changes in the specific adsorption sites for the intermediates, changes in relative coverage of active sites by various intermediates, the experimental trends are systematically reproduced by adding only the simplest level of micro-kinetic analysis to the results of the electronic structure calculations [52]. Recently simple models [53] have been introduced to treat electrochemical systems using DFT calculations, and it has become possible to start performing the same type of trend studies for electrochemical reactions as those that have been known in heterogeneous catalysis. Two cases where trends studies have been initiated are the oxygen reduction reaction [54] and the hydrogen evolution reaction. Whereas the ideas behind such trend studies have been known and used in the field of electrochemistry for half a century, the advent of DFT calculations have allowed for the development of systematic databases of adsorption energies. One benefit of this development is that DFT calculations can be used as the basis for the search after new electrode materials. Another benefit is that the entire established apparatus of electronic structure theory now can be applied to the surfaces of electrodes under electrochemical reaction conditions. In order to understand the variation in adsorption energies that form the basis for describing trends in surface reactivity in both heterogeneous catalysis and electrochemistry it is important to develop simple models or concepts being able to rationalize the data. A particularly useful model in heterogeneous catalysis, which is often used to relate changes in the electronic structure of transition metal surfaces to changes in chemical reactivity, is the d-band model [55].

The D-Band Model

In the d-band model variations in adsorption energies and activation barriers for a given reaction from one transition metal to the next are given, to a first approximation, by variations in the coupling between the adsorbates levels and the transition metal d-bands. The adsorbate–surface bond is viewed as consisting of two contributions:

$$\Delta E = \Delta E_0 + \Delta E_d \tag{7}$$

where ΔE_0 is the bond energy contribution from the coupling of the adsorbate states to the free-electron-like S-electrons and ΔE_d is the contribution from the extra interaction with the transition metal D-electrons. The two-step model is illustrated in Figure 17.

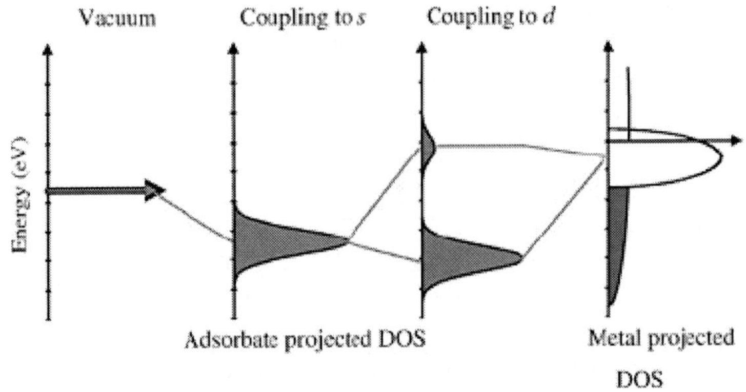

Figure 17. Schematic illustration of the formation of a chemical bond between an adsorbate valence level and the s-and d-states of a transition metal surface. From Ref. [50].

One of the basic assumptions of the d-band model is that ΔE_0 is independent of the metal. This is a quite reasonable first approximation since the S-bands of transition metals are broad and it is always half filled. It is not a rigorous approximation though. It will for instance fail when metal particles get small enough that the *sp* levels do not form a continuous (on the scale of the metal–adsorbate coupling strength) spectrum. It will also fail for metals where the *d*-states do not contribute to the bonding at all. The other basic assumption is that we can estimate the d-contribution as the non-self-consistent one-electron energy change as [50]:

$$\Delta E_d \cong \int \varepsilon(\Delta n'(\varepsilon) - \Delta n(\varepsilon))d\varepsilon \tag{8}$$

Here $\Delta n'(\varepsilon)$ and $\Delta n(\varepsilon)$ are the adsorbate-induced densities of states with and without the d-coupling included, respectively.

In general the coupling of the adsorbate states to the metal d-bands will depend on a number of parameters, the energy of the adsorbate state(s), ε_a, the d-density of states projected onto the metal atoms in direct contact to the adsorbate, $n_d(\varepsilon)$,and the coupling matrix element between the adsorbate and the surface states. In the following we will focus on situations where a given adsorbate couples to a specific kind of transition metal atom complex (a CO molecule bonding to a Pt atom, for instance) and study the effect of varying the surroundings (or ligands) to the relevant metal atoms (changing the nearest neighbor atoms of the Pt to which CO bonds, for instance). In these cases ε_a and the matrix elements are constant and only the $n_d(\varepsilon)$ will vary. We will focus on the first moment of the $n_d(\varepsilon)$, the center of the d-band, ε_d, as the simplest descriptor, and show that it describes variations in the adsorption energies quite well. It should be noted that apart from the d-band center, the interaction energy will depend on the shape and width of the projected d-density of states, $n_d(\varepsilon)$. These variations are, however often coupled to the d-band center variations, and can therefore be lumped into that dependence. To illustrate this point, consider a situation where the width (the second moment, W, of $n_d(\varepsilon)$ is decreased for some reason—it could be because the surface layer is strained so that the coupling, V_{dd}, of the metal d-states to the neighboring metal d-states is smaller ($W \sim |V_{dd}|$)or because the number of metal neighbors (the coordination number, NM)is decreased by creating a step or a curve on the surface ($W \sim N_M^{0.5}$). Changing W for a fixed $\varepsilon_d - \varepsilon_F$ would change the number of d-electrons. It is generally found that the number of d-electrons does not change for a given kind of metal, and the system compensates for this by shifting the d-states up in energy as illustrated in Figure 18. [50]

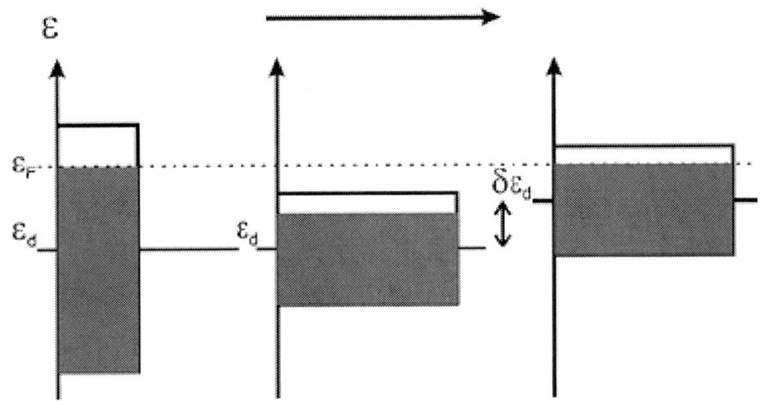

Figure 18. Schematic illustration of the coupling between bandwidth and d-band center for a band with a fixed number of d-electrons. When the bandwidth is decreasing the only way of maintaining the number of d-electrons is to shift up the center of the band.

Electric Field Effect in Heterogeneous Catalyst

Coadsorbed ions on the surface of metal catalysts often induce changes in the rate of catalytic reactions. The modification of the catalytic rate is connected to the change in the bond strength of the adsorbed species which take part into the reaction. For instance, an increase in the desorption energy of the final products will result in a considerable reduction of the turnover rate; a decrease in the bond strength of an adsorbed hydrogen or oxygen atom will facilitate the hydrogenation or oxidation of an organic adsorbate, and so on. The effect is well known as ‘ promotion’ or ‘ inhibition’ in catalysis and several experimental and theoretical studies have been dedicated to this important scientific and technological problem[56]. Various explanations have been proposed to rationalize the promotion effect, starting from the observation that electropositive (electronegative) adsorbates on metals form ions at low coverage [57]: (a) A direct interaction due to an overlap of the adsorbate and ion wavefunctions; (b) a change in the position of the donor-acceptor energy levels of an adsorbed molecule due to the coadsorbates [58]; (c) short- or long-range electrostatic interactions, etc. [59]. A generally accepted model assumes that the coadsorbed ions induce a polarization of the substrate metal with formation of an image charge of opposite sign below the surface [59]. The charge pair generates an inhomogeneous electric field at the surface which will affect the properties of the other adsorbates (atoms or molecules). The effect of electric fields on catalytic reactions has been studied experimentally; for instance Block and co-workers [60] have developed a field pulse technique in the field ion microscope that allows the investigation of the field effect on chemical reactions. The way the interaction between the non-uniform field due to the ions and the adsorbates takes place [61].

BEP (Brønsted–Evans–Polanyi) Relation and Volcano Curve in Heterogeneous Catalysis

One of the most fundamental concepts in heterogeneous catalysis is the volcano curve [1].It has been established empirically that a volcano-shaped curve is obtained when the activity of catalysts for a certain reaction is plotted as a function of a parameter relating to the ability of the catalyst surface to form chemical bonds to reactants, reaction intermediates, or products [63,64].

Such relations are interesting from a scientific point of view since they point to important aspects of the reaction, and they are also useful as guidelines in the search for new catalysts [65].

An important problem in connection with volcano curves is which fundamental parameters the catalytic activity depends on. Activities have been correlated with various electronic properties of the catalyst [62], and it seems natural to relate the activity to bond energies.

The question is which energy is the most relevant. In addition there is a problem of finding systematic databases of relevant surface thermochemical data. The fact that many volcano curves are plotted as a function of a bulk heat of formation is related to the latter problem—only bulk thermo-chemical data are widely available. It was decided that for dissociative chemisorptions of a number of molecules the activation energy depends linearly on the reaction energy1 [8]. Such a Brønsted–Evans–Polanyi(BEP) relation has often been assumed implicitly to hold for surface reactions. It is however, only with the extensive use of density-functional theory (DFT) calculations that it has been possible to establish such relations firmly for a number of systems. An unexpected result from these calculations was the findings that not only do many surface reactions follow BEP relationships, there are classes of similar reactions which follow the same "universal" relationship [66]. This is illustrated for N_2, CO, NO and O_2 dissociation in Figure 19. Subsequent calculations have confirmed these results and have shown similar universal BEP relations to hold also for other surface reactions [67].

It is clear that the linear BEP relationship in a number of cases leads directly to volcano curves where the fundamental parameter is the dissociative chemisorptions energy of the key reactant.

To understand how the volcano curve depends on the mechanism and on the number of possible rate-determining steps. An important outcome is that for a class of reactions involving key reactants that follows the same BEP relation, the maximum in the volcano curve is found generally to be in the same range of reaction energies, independent of the reaction, and that variations in the position of the maximum with reaction conditions should be understood.

It has been concluded that the dissociative chemisorptions energy is a good descriptor of the catalytic activity for a series of simple catalytic reactions. As an aid to understanding trends in catalytic activities for other reactions, we perform systematic DFT calculations of reaction energies for molecular and dissociative chemisorptions for a large number of molecules on stepped surfaces of the 13 transition metals most commonly used in heterogeneous catalysis.

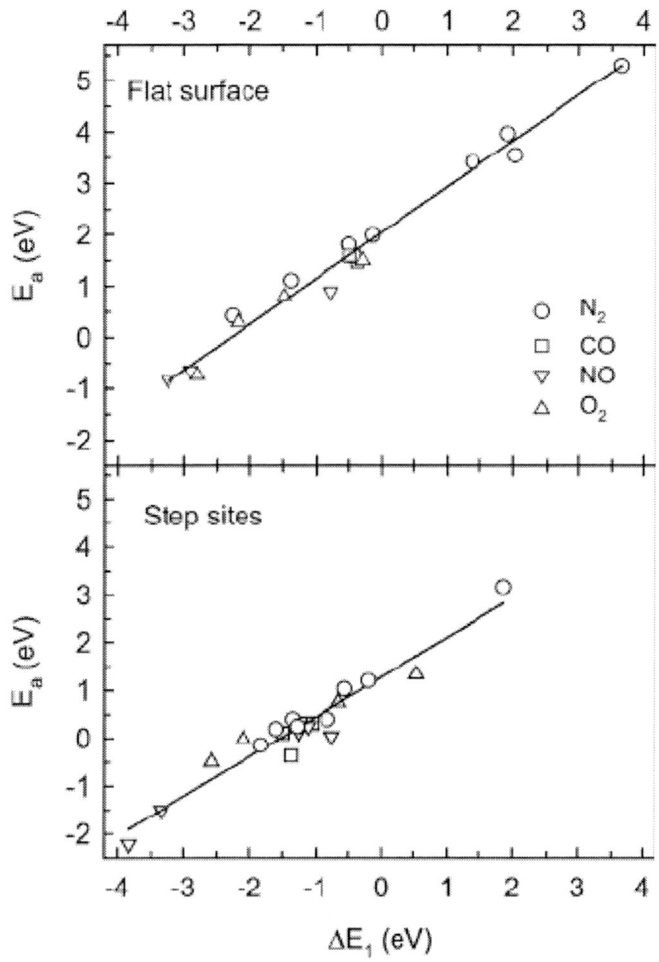

Figure 19. Calculated transition state energies (Ea) and dissociative chemisorptions energies (ΔE1) for N_2, CO, NO, and O_2 on a number of transition metal surfaces. Results for close packed as well as stepped surfaces are shown. Adapted from Ref. [8].

Spill Over

An important phenomenon sometimes observed in multi-phase catalysts is that active species formed on one of the phases move during the reaction to the surface of another phase it is usually called spill-over [68]. Of great importance in many catalytic reactions are the acid–base properties of solids. Important contributions to the interpretation of the relation between solid state properties and catalytic behavior have been obtained by computational modeling and simulation, including quantum mechanical calculations. Apart from the interpretation of catalytic behavior of catalysts in terms of solid state properties, attention is also given to new possibilities where ion or mixed conducting oxides are used as membranes separating reactants and reactant products. This gives exciting new possibilities for reactor configurations leading in particular to higher selectivities.

Point Defects

The defects under consideration here may be vacant lattice sites (called vacancies), ions placed at normally unoccupied sites (called interstitial ions), foreign ions present as impurity or dopant and ions with charges different from those expected from the overall stoichiometry. Electron defects may arise either in the form of ions present with charges deviating from the normal lattice ions, or as a consequence of the transition of electrons from normally filled energy levels, usually the valence band, to normally empty levels, the conduction band. In those cases where an electron is missing, i.e. when there is an electron deficiency, this is usually called an electron hole (often abbreviated to hole). Usually it is convenient to consider point defects, such as vacancies or electron holes, to be the moving entities in a lattice even though in reality of course the ions or electrons move through the lattice in the opposite direction.

In the absence of macroscopic electric fields ionic lattices must be electrically neutral everywhere. This requires that a charged defect must be compensated by the presence of a number of defects, together having a joint charge of opposite sign. This is called the electroneutrality condition. This means that these charged defects are always present as a combination of two (or more) types of defects. The charges of defects and of the regular lattice particles are defined with respect to the neutral, unperturbed (ideal) lattice and are called effective charges. These are indicated by a dot (•) for a positive excess charge, by a prime (') for a negative excess charge and by an x (^{x}) for effectively neutral defects (i.e. ions having their normal charges corresponding with the stoichiometry of the compound). The notation used for defects is from Kröger and Vink [69].

Protons in Oxides

The presence of defects, which contain hydrogen, sometimes has a great influence on chemical and catalytic properties of oxide materials. These types of properties are, the Brønsted acidity, proton conduction in the solid state [70].

For the Protonic defects in oxides the most important formation reaction is that between water molecules and oxygen vacancies

$$Oo^{x} + Vo^{\bullet\bullet} + H_2O(g) \rightarrow 2OHo^{\bullet} \quad (9)$$

In which two effectively positive hydroxyl-groups on regular oxygen positions are formed.

There are also some important defect reactions where Protonic defects are formed by reaction with hydrogen. The first is the reaction with electron holes according to

$$2h^{\bullet} + 2\,Oo^{x} + H_2 \rightarrow 2OHo^{\bullet} \quad (10)$$

For which the presence of excess holes is obviously necessary. An alternative possibility is the oxidation of hydrogen under formation of free electrons according to

$$2\,Oo^{x} + H_2 \rightarrow 2OHo^{\bullet} + 2e' \quad (11)$$

where the electrons are assumed to be donated to the conduction band in the case of band conductors.

There are two mechanisms for proton conduction. The first is that of proton hopping, also called the Grotthus-mechanism, in which the proton hops between adjacent oxygen ions. The other conduction mechanism is hydroxyl-ion migration, also called the "vehicle" mechanism. One possibility to distinguish these mechanisms is by measurement of the isotope effect, e.g., by comparing the magnitudes of H^+ and D^+ ion conduction. In the first mechanism a significant isotope effect is expected, whereas in the second, due to the large mass of the oxygen atom, this is almost absent.

Surface Phenomena and Defects

The defect concentrations at the surface also differ from those in the bulk. Depending on the sign of the energy difference this can either lead to an increased concentration, also called positive defect segregation to the surface, or the reverse which is called negative (or reverse) segregation. Due to the changes in band energies near the surface in semi-conducting oxides, the so-called band-bending or band curvature, the electron or electron hole concentrations in the conduction or valence bands, respectively, are different close to the surface from those in the bulk.

Even on "stoichiometric" surfaces there are non-idealities, which should not be called defects in the strict sense. These are the ledges shown in the cross-section of the (310) surface of CeO_2 in Figure This leads to the presence of sites of the same type of ion with different properties. At the interfaces between different oxides, such as in supported or mixed oxide catalysts, similar effects occur.

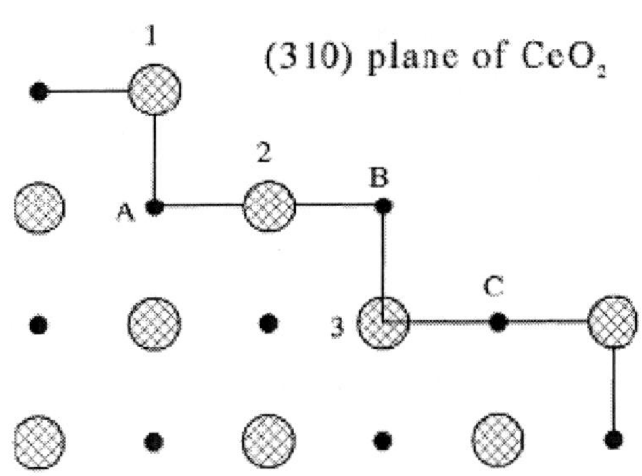

Figure 20. Cross-section through (310) plane of CeO_2. 1, 2, 3: three different surface oxygen sites; A, B, C: three different surface cerium sites.

Types of Adsorbed and Surface Oxygen

There are a number of oxygen species: O_2 (adsorbed molecule), O (adsorbed neutral atom), O_2^- (super oxide), O_2^{2-} (peroxide), O_3^- (ozonide), O^-, etc. which may be present on the catalyst surface. Bielanski and Haber[72] (also cited in [71]) presented a scheme of the different types of oxygen of importance in catalysis, shown in Figure.

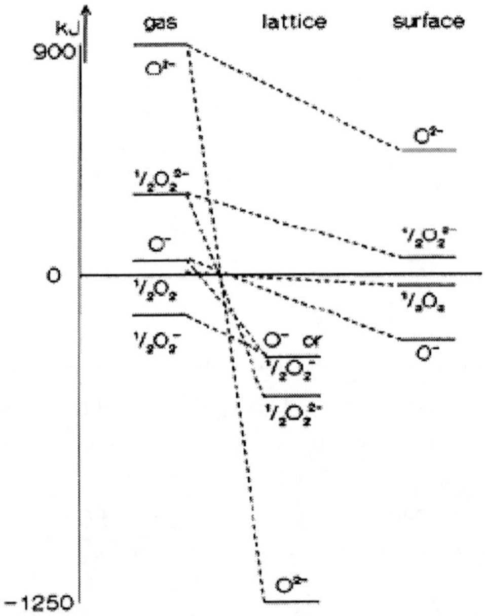

Figure 21.Survey of oxygen species.

Catalyst Deactivation

One of the major problems related to the operation of heterogeneous catalysis is the catalyst loss of activity with time-on-stream, i.e. "deactivation". This process is both of chemical and physical nature and occurs simultaneously with the main reaction. Deactivation is inevitable, but it can be slowed or prevented and some of its consequences can be avoided.

Chemical, Physical and Kinetic Aspects of Catalyst Deactivation

Deactivation can occur by a number of different mechanisms, both chemical and physical in nature. These are commonly divided into four classes, namely poisoning, coking or fouling, sintering and phase transformation. Other mechanisms of deactivation include masking and loss of the active elements via volatilization, erosion and attrition.

Poisoning

Chemical aspects of poisoning: Poisoning is the loss of activity due to the strong chemisorptions on the active sites of impurities present in the feed stream. The adsorption of a basic compound on to an acid catalyst (e.g. isomerization catalyst) is an example of poisoning. A poison may act simply by blocking an active site (geometric effect), or may alter the adsorptivity of other species essentially by an electronic effect. Poisons can also modify the chemical nature of the active sites or result in the formation of new compounds (reconstruction) so that the catalyst performance is definitively altered. Usually, a distinction is made between poisons and inhibitors [73]. Poisons are usually substances whose interaction with the active sites is very strong and irreversible, whereas inhibitors generally weakly and reversibly adsorb on the catalyst surface.

Upon poisoning the overall catalyst activity may be decreased without affecting the selectivity, but often the selectivity is affected, since some of the active sites are deactivated while others are practically unaffected. This is the case of "multi-functional" catalysts, which have active sites of different nature that promote, simultaneously, different chemical transformations.

Table 2 reports a list of the poisons typically encountered in some industrial catalytic processes. In some cases, due to the very strong interaction existing between poisons and the active sites, poisons are effectively accumulated on to the catalytic surface and the number of active sites may be rapidly reduced.

Table2. Examples of poisons of industrial catalysts

Catalyst	Process	Poison
Fe	Ammonia synthesis	CO, CO_2, H_2O, C_2H_2, S, Bi, Se, Te, P
Ni/Al_2O_3	Steam reforming	H_2S, As, HCl
Cu	Methanol synthesis, low-T CO shift	H_2S, AsH_3, PH_3, HCl
$SiO_2 \pm Al_2O_3$, zeolites	Catalytic cracking	Organic bases, NH_3, Na, heavy metals
Ni, Co, Fe	CO hydrogenation	H_2S, COS, As, HCl
V_2O_5	Oxidation	As
Pt, Pd	Automotive catalytic converters (oxidation of CO and HC, NO reduction)	Pb, P, Zn
Ag	Methanol oxidation to formaldehyde	Fe, Ni, carbonyls
Ag	Ethylene to ethylene oxide	C_2H_2
Transition metal oxides	Many	Pb, Hg, As, Zn

Poisoning of metal-based catalysts: metal catalysts of groups VIII B (Fe, Ru, Os, Co, Rh, Ir, Ni, Pd, Pt) and IB (Cu, Ag, Au), typical poisons are molecules containing elements of groups VA (N, P, As, Sb) and VIA (O, S, Se, Te). The surface metal atoms active in the catalytic reactions can be depicted as involved in the chemisorptions of the reactants (and of poisons as well) via their "dangling orbitals". Accordingly, any chemical species having a "proper electronic configuration" (e.g. unoccupied orbitals or unshared electron) or multiple bonds (e.g. CO, olefins, acetylenes, etc.) can be considered as potential poisons. Accordingly several molecules have been classified as having "shielded" or "unshielded" structures [74]: for example As in the form of arsine (AsH_3), having a lone pair, is a strong poison for catalysts such as Pt in hydrogenation reaction, whereas no effect on catalytic activity is observed on the decomposition of H_2O_2, possibly because As under oxidizing conditions it is present in the form of arsenate AsO_4^{3-}. Along similar lines the order of increasing poisoning activity for sulfur species, i.e. $H_2S>SO_2>SO_4^-$, can be explained.

Poisoning of metal oxide-based catalysts: Metal oxide-based catalysts are generally more resistant than metal catalysts to deactivation by poisoning. Acid catalysts (e.g. cracking catalysts) are poisoned by basic materials (alkali metals or basic N-compounds)[75]. Oxide catalysts other than acid catalysts are also poisoned by several compounds, and often by Pb, Hg, As, Cd. These compounds react with the catalyst active sites usually leading to a permanent transformation of the active sites which thus become inactive.

Preventing poisoning: Poisoned catalyst can hardly be regenerated, and therefore the best method to reduce poisoning is to decrease to acceptable levels the poison content of the feed. This is generally achieved by appropriate treatments of the feed, e.g. catalytic hydrodesulphurization followed by H_2S adsorption or absorption to remove S-compounds, methanation for the elimination of COx from the ammonia synthesis feed, adsorption over appropriate beds of solids to remove trace amounts of poisons (e.g. ZnO for H_2S, sulfured activated charcoal for Hg, alkalinized alumina for HCl). Another approach to prevent poisoning is to choose proper catalyst formulations an design. For example, both Cu-based methanol synthesis and low-temperature shift catalysts are strongly poisoned by S-compounds. In these catalysts significant amounts of ZnO are present that effectively trap sulfur leading to the formation of ZnS. The catalyst design (e.g. surface area, pore size distribution, pellet size) can also modify the poison resistance.

Coking

Chemical aspects of coking: For catalytic reactions involving hydrocarbons (or even carbon oxides) side reactions occur on the catalyst surface leading to the formation of carbonaceous residues (usually referred to as coke or carbon) which tend to physically cover the active surface. Coke deposits may amount to 15% or even 20% (w/w) of the catalyst and accordingly they may deactivate the catalyst either by covering of the active sites, and by pore blocking. Sometimes a distinction is made between coke and carbon. The difference is however somewhat arbitrary: usually carbon is considered the product of CO disproportionation ($2CO \rightarrow C + CO_2$), whereas coke is referred to the material originated by decomposition (cracking) or condensation of hydrocarbons.

The mechanism of coke formation on oxides and sulfides is rather complex but it can be roughly visualized as a kind of condensation – polymerization on the surface resulting in macromolecules having an empirical formula approaching CHx, in which x may vary between 0.5 and 1. It has been suggested that the pathway to coke, starting from olefins or aromatics, may involve: (a) dehydrogenation to olefins; (b) olefin polymerization, (c) olefin cyclization to form substituted benzenes, and (d) formation of polynuclear aromatics from benzene [76]. These mechanisms proceed via carbonium ions intermediates and accordingly they are catalyzed by Bronsted acid sites. The chemical nature of the carbonaceous deposits depends very much on how they are formed, the conditions of temperature and pressure, the age of the catalyst, the chemical nature of the feed and products formed.

Various analytical techniques have been used in order to characterize the nature, amount and distribution of coke deposits. The chemical identity of the carbonaceous deposits has been extensively investigated by IR [77]. Other techniques are well suited for this purpose, e.g. UV-Vis, EPR, ^{13}C-NMR. The amounts of coke deposited into the catalyst pores may be estimated by burning the coke with air and recording the weight changes via TG-DTA techniques and/or by monitoring the evolution of the combustion products CO_2 and H_2O.

Preventing coke deposition: The coke deposition may be controlled to a certain extent by using an optimal catalyst composition and an appropriate combination of process conditions. During the reaction an equilibrium is reached between the rate of coke production and the rate of coke removal by gasifying agents (e.g.H_2, H_2O andO_2 that remove coke as CH_4, CO and COx, respectively) so that steady-state conditions, corresponding to a certain level of coke present on the catalyst surface, are eventually reached. Otherwise, if the rate of coke deposition is higher than that of coke removal, a suitable regeneration procedure must be applied.

The catalyst composition does also affect significantly the coke deposition. Promoters or additives that enhance the rate of gasification of adsorbed carbon or coke precursors and/or depress the carbon-forming reactions minimize the content of carbon on the catalyst surface. For this reason alkali metal ions, e.g. potassium, are incorporated in several catalysts (e.g. Ni-based steam reforming catalysts, Fe_2O_3-Cr_2O_3 dehydrogenating catalysts, etc.). Potassium has several effects: it neutralizes acid sites which would catalyze coke deposition via the carbonium ion mechanism, and catalyzes the gasification of the adsorbed carbon deposits, thus providing an in-situ route for catalyst regeneration.

Sintering

Sintering usually refers to the loss of active surface via structural modification of the catalyst. This is generally a thermally activated process and is physical in nature. Sintering occurs both in supported metal catalysts and unsupported catalysts. In the former case, reduction of the active surface area is provoked via agglomeration and coalescence of small metal crystallites into larger ones with lower surface-to-volume ratios. Two different but quite general pictures have been proposed for sintering of supported metal catalysts, i.e. the atomic migration and the crystallite migration models. In the first case, sintering occurs via escape of metal atoms from a crystallite, transport of these atoms across the surface of the support (or in the gas-phase), and subsequent capture of the migrating atoms on collision with another metal crystallite. Since larger crystallites are more stable (the metal-metal bond energies are often greater than the metal-support interaction), small crystallites diminish in size and the larger ones increase. The second model visualizes sintering to occur via migration of the crystallites along the surface of the support, followed by collision and coalescence of two crystallites. As a matter of fact, sintering of supported metals involves complex physical and chemical phenomena that make the understanding of mechanistic aspects of the sintering a difficult task. The actual rate and the extent of sintering depends on many factors, including the metal oxide concerned, the initial crystallite size and the size distribution, the presence of additives that favor or promote sintering, the environment. The key variable is temperature, so that operation at low temperatures greatly reduces the sintering rate. Reaction atmosphere also affects sintering: water vapor, in particular, accelerates crystallization and structural change in oxide supports. Accordingly, over high-surface area catalysts it is desirable to minimize the water vapor concentration at high temperatures during both operation and activation procedures as well. The presence of specific additives is known to reduce the catalyst sintering. For example BaO, CeO_2, La_2O_3, SiO_2 and ZrO_2 improve the stability of γ-alumina towards sintering [78], whereas Na_2O enhances its sintering. In addition to a decrease in the surface area, sintering may also lead to a decrease in the pore openings, and eventually the pores close completely making the active species in accessible to the reactants.

Solid-State Transformation

Solid-state transformation is a process of deactivation that can be viewed as an extreme form of sintering occurring at high temperatures and leading to the transformation of one crystalline phase into a different one. These processes may involve both metal-supported catalysts and metal oxide catalysts as well. In the first case we can observe the incorporation of the metal into the support, e.g. incorporation of metallic Ni into the Al_2O_3 support (at temperatures near 1000˚C) with formation of inactive nickel aluminate, or reaction of Rh_2O_3 with alumina (in automotive exhaust catalysts) to form inactive $Rh_2Al_2O_4$ during high-temperature lean conditions. In the case of metal oxide catalysts or supports the transformation of one crystalline phase into a different one can occur, like the conversion of γ-into δ-Al_2O_3 with a step-wise decrease in the internal surface area from about $150m^2$ /g to less than $50m^2$/g. Several of these transformations are limited by the rate of nucleation. This process may occur due to the presence of some foreign compounds in the lattice or even on the surface. For example, V_2O_5 has been reported to favor the sintering of the TiO_2-anatase support as well as the anatase-to-rutile transformation in TiO_2-supported V_2O_5 catalysts. On the other hand, WO_3 effectively contrasts this phenomenon [79,80].

Other Mechanisms of Deactivation

Other mechanisms of deactivation include masking or pore blockage, caused e.g. by the physical deposit of substances on the outer surface of the catalyst thus hindering the active sites from reactants. Masking may occur during hydrotreating processes where metals (principally Ni and V) in the feedstock deposit on the catalyst external surface, or in the case of automotive exhaust converters by deposition of P(from lubricants) and Si compounds. Certain catalysts may also suffer from loss of active phase. This may occur via processes like volatilization, e.g. Cu in the presence of Cl with formation of volatile $CuCl_2$, or Ru under oxidizing atmosphere at elevated temperatures via the formation of volatile RuOx, or formation of volatile carbonyls by reaction of metals with CO.

Determination of Process Technologies from Catalyst Deactivation

Figure 22 shows the relation between possible reactor technologies and rapidity of catalyst deactivation for these cases. If the deactivation rate in a fixed-bed reactor is sufficiently low, no special on-site facilities for regeneration are required, and after an operating period spent catalyst can be discarded or regenerated elsewhere for reuse. This situation pertains if the life of the catalyst is about a year or longer.

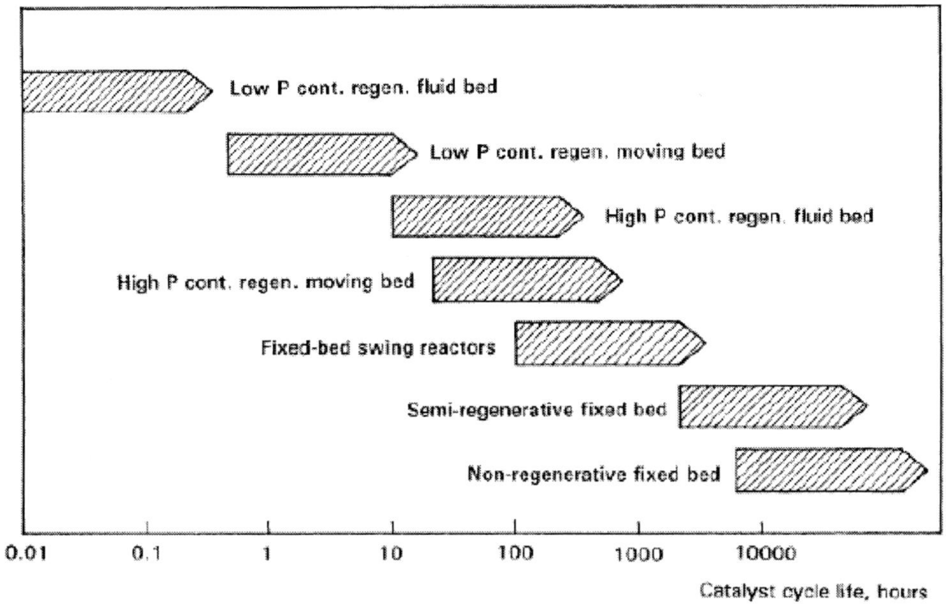

Figure 22. Viable process technologies as determined by rapidity of catalyst deactivation.

Assuming an average of 10 days downtime in 1year for a catalyst change in a process operating with oxidisable material under pressure (which change involves cooling down, depressurizing, and opening of the reactor, removal of spent catalyst, reloading with fresh catalyst, closing the reactor, repressurising and reheating), the availability of the unit is reduced by about 3%, which is generally considered acceptable. The loss of availability can be further reduced if the time of catalyst change can be made to coincide with the generally required periodic safety inspection of the unit. When catalyst life becomes shorter, e.g. about half a year, dedicated facilities for on-site regeneration become of interest, particularly when it concerns an expensive catalyst. In the so-called semi-regenerative mode of operation, the catalyst remains in the reactor, which is taken out of service during the regeneration campaign. The required facilities for carrying out the regeneration, such as compressors for inert gas circulation and dosing of air may be a permanent part of the unit or be installed on a temporary basis. With the relatively slow rate of catalyst deactivation in the above non-and semi-regenerative process modes, the effect of deactivation during a catalyst lifecycle is generally counteracted by raising the reactor temperature during operation so as to maintain the desired level of conversion or quality of the product [79].

Regeneration of Catalyst

Essentially, the three means of treating spent catalysts involve regeneration, possible use in another application and disposal. The choice depends on economic factors coupled to environmental legislation. Thus, for example, the disposal of vanadia in landfill even when encapsulated is not allowed on environmental grounds, and the excess cost involved in recycling the vanadium back to a fresh catalyst must be taken into the economic equation. Regeneration and reuse of catalysts are always the preferred option, but this is not always possible.

Catalyst Disposal

Although regeneration and re-use may be possible, it is certain that all catalysts will have to be replaced in the long term. Disposal of spent catalyst then depends both on economics and on local environmental laws. If the economic driving force is sufficiently large, recovery of catalyst components is justified. If not, the catalyst must be disposed of in a manner consistent with local legislation.

Requirements vary from place to place but, since catalysts are often used as landfill, the most relevant environmental legislation is usually based on leachability. Various methods are used to assess leaching, including ASTM-3987 and the USA Toxic Characteristic Leaching Procedure [81]. Acceptable limits may vary, but values for drinking water provide a good guide. These include less than 0.7 ppb Co, 7 ppb Mo, 13.4 ppb Ni, 7 ppb V and 0.3 mg/l Fe. If a catalyst cannot be discarded without exceeding these limits, it is necessary to recycle components or to encapsulate residues.

The procedure involved varies with the catalyst. Hydrotreating catalysts usually contain enough metals to warrant recovery and it is then necessary to dispose of residues. Cracking catalysts contain so little metal that recovery is not worthwhile despite the fact that sufficient metal may be deposited to cause environmental problems [81].

Component Recovery

In general, the recovery of metals from spent catalyst provides most economic advantage, but in some cases, environmental concerns may outweigh economics. Thus, for example, vanadia catalysts are so unwelcome in landfill that recovery at significant cost is essential [82].

Metal recovery from hydrotreating catalysts provides a convenient example of the necessary procedures of component recovery. The treatment has much in common with the processing of minerals and, in some cases, involves the some species. Spent catalysts often exist as metal sulphides, supported on alumina and heavily coated in carbonaceous materials [81]. Decoking may be necessary if only to destroy carcinogenic material and this can lead to the production of metal oxides. As a result, the treatment of sulphides and oxides needs to be considered. The easiest way to recover some components is by attrition. Metals can deposit throughout a catalyst, depending on the reactivity of the particular metal complex [83]. Vanadium and iron tend to deposit on the pellet exterior [83] and a simple abrasion process can produce a powder that is rich in these metals. However, the open pore structure of catalysts ensures that removal is far from quantitative. The most common approach is to recover totally metals in the catalyst. This is done by a combination of roasting and leaching.

Catalyst is first ground and roasted at 600°C to produce metal oxides and to remove coke. A second roast at 750°C in the presence of sodium carbonate produces soluble salts:

$$V_2O_5 + Na_2CO_3 \leftrightarrow 2NaVO_3 + CO_2 \quad (12)$$

$$MoO_3 + Na_2CO_3 \leftrightarrow NaMoO_4 + CO_2 \quad (13)$$

Leaching with water at 100°C is then followed by treatment with ammonium chloride or calcium chloride:

$$NaVO_3 + NH_4Cl \leftrightarrow NH_4VO_3 \downarrow + NaCl \quad (14)$$

$$2\,NH_4VO_3 \xrightarrow[500^\circ C]{} V_2O_5 + 2NH_3 + H_2O \quad (15)$$

$$NaMoO_4 + CaCl_2 \rightarrow CaMoO_4 + 2NaCl \quad (16)$$

Extension of the process to include roasting with caustic soda has claims to recover nickel. Nickel aluminate is produced, which may be separated by hot water digestion and converted to the sulphate by subsequent reaction with sulphuric acid. If roasting is unnecessary, it may be possible to consider less expensive two stage leaching, in which an alkali leach removes MoO_3 and V_2O_5 and a subsequent acid leach removes Ni and Co oxides. Pressure leaching may improve the efficiency of both operations. Total metals recovery involves the solubility of all metals, followed by the selective separation of individual components. The exact chemistry depends on the system. Thus, for example, better separation of vanadium from molybdenum species may be achieved by the use of a quaternary ammonium salt dissolved in an aromatic solvent, while preferential removal of nickel or cobalt requires treatment with an oxime [81]. It may be more economic to consider the selective leaching of individual metals in some cases. This demands either the selective formation of a given metal complex, or the formation of a range of complexes, only a few of which are soluble. The addition of suitable oxidizing agents to promote the formation of oxides from sulphides improves leachability. Other systems includes the use of iso-propanol to remove vanadium from oxidized catalyst, heteropolyacids such as molybdophosphoric acid or dilute mineral acids for the same purpose, Cl_2/H_2O to remove Ni, Co and V but not Mo and hot water alkaline leaching to remove Mo and V [81].

The general approach to any selective recovery is reasonably straight forward. The first question is whether a given reactant will form complexes with one or all of the catalyst components. The second question focuses on the solubility of the possible complexes. Finally, one must explore the possibility of recovering a desired metal salt from solution either by precipitation or by chemical reaction. The possibilities and the economic variations obviously depend on individual systems, with the recovery of small amounts of precious metal being probably more advantageous than large amounts of copper. Individual assessment is required. What is certain, however, is that residues from any leaching operation must be discarded with care. If low level removal of unacceptable impurities occurs on storage, encapsulation or fixation is required.

Final Disposal

Catalysts or catalyst residues often contain sufficient material to cause environmental problems and, as a result, must be encapsulated or stabilized before disposal. Encapsulation involves surrounding the waste with an impervious layer of sealant, such as bitumen, polyethylene or concrete. The encapsulant has to be stable over long periods of time, both with respect to mechanical shear and to weathering. Often, the encapsulant can add considerable bulk to the waste, thereby increasing the Volume (and cost) of disposal.

Stabilization involves a chemical reaction leading to a product which is less of an environmental problem. Stabilization of nickel by reaction with alumina to produce nickel aluminate is a good case in point. The bulk of the deposit is not greatly increased, and the leachability of the product is much reduced.

Chemical fixation and solidification (CFS) techniques have been developed that are directly applicable to waste catalysts. Again, the techniques focus on encapsulation in an inert organic matrix or by a variety of cementitious processes, including glassification and ceramics manufacture. Inclusion in Portland cement is the most cost efficient process, but pozzolanic concrete (in which lime is mixed with fine grained siliceous material) is known to be stable over long periods. Glassy materials, based on encapsulation/fixation within vitreous silicates or phosphates are also favored, but stress cracking and slow devitrification can occur [81].

Development of Catalyst Science

The classical picture of catalysis by a surface has been provided by Langmuir [84]. He proposed a surface with a finite number of equal vacant sites. Adsorbates are supposed not to interact. For the dependence of the surface coverage on gas pressure p be deduced the adsorption' isotherm:

$$\frac{\theta}{1-\theta} = kp \quad (17)$$

That has been named after him. θ is the site occupation and K the equilibrium constant. This expression is identical to the Michaelis-Menten expression applied in homogeneous catalysis for the fraction of reacted complexes.

Surface scientific investigation and recently especially STM-studies have convincingly demonstrated that the reality of the reactive surface is significantly more complex [85]. Generally, the interaction between adsorbates cannot be ignored. Lateral interactions may lead to non-ideal mixing behavior of adsorbed layers. Surface adsorbate layers are often present as ordered structures forming islands. This implies that the Langmuir adsorption term (13) has to be replaced by an expression as[86] :

$$\frac{\theta - x}{1-\theta - x} = kp \quad (18)$$

where x is a complex function of coverage and adsorbate lateral interaction energy.

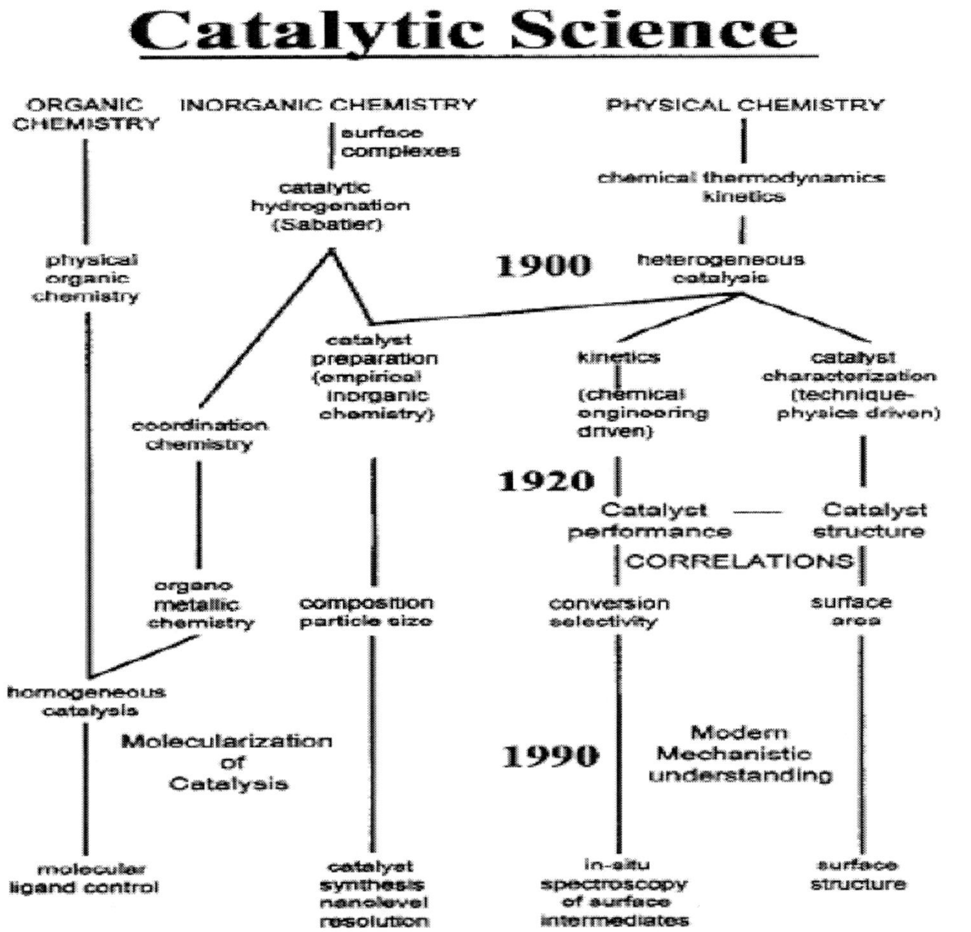

Figure 23. Scheme of Development of Catalytic Disciplines.

The ordering of adsorbates on a surface will strongly affect the rate of surface processes. For instance, the rate of dissociative adsorption for an ideal mixing surface layer will be ~$(1 - \theta)^2$, whereas when repulsive lateral interactions are strong the adsorption rate is ~$(1 - 2\theta)$. An adsorbed atom or molecule has an increased interaction when the coordination of surface atoms with neighboring surface atoms is decreased. This can provide a driving force to surface atom rearrangements with important consequences to catalysis.

At low surface coverage an adsorbed atom will usually not disrupt the surface lattice, but pull surface atoms outwards of the surface to form a surface complex embedded in the metal surface. Adsorbates tend to form Chemisorption complexes that are geometrically close to the analogous co-ordination complexes. The local surface structural change is counteracted by surface stress forces that respond to the local surface strain of the adsorption site. When more atoms absorb surface stress can be released by rearrangement of surface atoms [87], which will create sites that have a stronger interaction with the adsorbate. This may lead to the formation of ordered overlayers with transport of surface atoms. During the catalytic reaction cycle such sites may be formed and consecutively destroyed. Facetting of surfaces to give surface phases in reaction may occur after an initial period. The relative stabilities of surface phases are a strong function of surface coverage and composition that can be strongly reaction- condition-dependent. Under particular conditions the changes in surface phases can be made to oscillate which leads to fascinating spatial-temporal patterns.

The microscopic picture that evolves is a dynamic one. The time scale of a catalytic event is typically of microsecond or larger. However, surface atoms will locally rearrange on a picosecond time scale, the same time scale at which adsorbate molecules will equilibrate to the surface. On a time scale varying between a nanosecond and a microsecond, surface diffusion and surface reaction events can take place. The presence of lateral interactions between atoms adsorbed in the surface layer can also have an important consequence in the selectivity of a reaction. The chemical reactivity of adsorbed atoms may change as a function of surface concentration and composition.

For instance, the selectivity of the epoxidation reaction of ethylene by oxygen appears to be strongly dependent on the composition of the surface oxygen layer present on the silver catalyst [88]. At high oxygen surface coverage selective conversion of ethylene to the epoxide occurs. The surface layer then has a stoichiometric composition of AgO, with part of the oxygen atoms located in a subsurface layer and other oxygen atoms adsorbed in the external surface layer. The nature of the adsorbed oxygen atoms is electrophilic and favorable for insertion into the n-bond of ethylene. At low surface coverage the adsorbed oxygen atoms are Lewis basic in nature and assist CH bond cleavage. The change in reactivity of adsorbed atoms with coverage is a direct consequence of lateral interactions mediated by the changed valency of silver with surface coverage. Steric and geometric rearrangements in the surface layer that vary with surface layer composition may also affect the selectivity of a reaction.

Dissociation of an adsorbed molecule produces molecule fragments adsorbed to the surface. Adsorption of these fragments to the surface is necessary in order to make the dissociation reaction thermodynamically possible.

A surface ensemble of several atoms is usually required for dissociation to occur. Reduction paths for dissociation are such that optimum activation of adsorbed molecules occurs with large orbital overlap.

Pre-exponents of surface reactions often show low activation entropies in line with the low mobility of the transition state due to the strong interaction of the dissociating molecule with the surface. When other molecules become co-adsorbed the surface concentration of vacant sites will decrease and hence the rate of dissociation will be decreased. Also the adsorption mode of a molecule (and hence its interaction energy) may be strongly affected by co-adsorbates.

Even the presence of weakly co-adsorbed species may affect the course of the reaction. In water gas shift reaction selectivity of CO with H_20 depends strongly on the number of H_20 molecules co-adsorbed at the zinc reaction center of a model catalyst. A similar phenomenon appears in the decomposition of formic acid on TiO_2 (110). Co-adsorption of several formic acid molecules changes the selectivity compared to the case of formic acid adsorbed and isolated from the others [85]. The influence of a water layer can dramatically change dissociative Chemisorption of a molecule. In an ultra-high vacuum on platinum, methanol will dissociatively adsorb as a methoxy species and hydrogen, and hence form a metal-oxygen bond.

From the water phase dissociative adsorption of methanol will primarily occur by CH cleavage and formation of a metal-carbon bond [85]. The dissociative adsorption on a transition metal surface requires a strong interaction with an ensemble of surface atoms. The mechanism of solid acid catalyzed reactions also illustrates the importance of close contact between reactant and catalyst surface atom.

The overall activation energy now also depends on the heat of adsorption of the molecule. It appears that differences in heat of adsorption may determine to a significant extent the overall rate of acid catalyzed reactions.

The four important general concepts to catalysis are the following:

a. Self regenerating catalytic cycle: The rate of a reaction is maximum for an optimum interaction strength between reactant and catalyst; dissociative adsorption and desorption compete as rate limiting steps.
b. Chemisorption: Causes a local electronic and geometric disturbance; lateral interactions may lead to non-ideal mixing and site reconstruction.
c. Site specificity: Dissociation usually requires an ensemble of atoms, selectivity depends on surface topology, surface atom coordination and surface composition. Adsorption complexes have a close resemblance with coordination complexes. Stereo and shape selectivity can be tuned by micropore size, ligand shape and promoters.
d. Site preparation, phase stability: The catalyst structure adapts to reaction conditions. Catalytic site formation and destruction can be a part of the reaction cycle. Catalyst deactivation may occur by the formation of less-active, thermodynamically stable phases. Deactivation can also occur by poisoning due to co-product formation or the presence of impurities in the feed [89].

Selectivity and Activity

The selectivity of catalytic processes, like their overall activity, is defined by the kinetics of the reactions involved.

However, the parameters that control selectivity are typically more sensitive and difficult to adjust than those defining the total rate [90].This is illustrated by the energy diagram provided in Figure 24.

There, it is seen that the high initial barrier $\Delta G^{\ddagger}_{rls}$ for the conversion of the reactant to the intermediate limits the turnover frequency of the overall process. On the other hand, it is the difference in activation barriers from the intermediate to each of the two possible products, $\Delta G^{\ddagger}_1 - \Delta G^{\ddagger}_2$, that defines selectivity. The criteria for lowering $\Delta G^{\ddagger}_1$ with respect to $\Delta G^{\ddagger}_2$ are not likely to be the same as those required to modify $\Delta G^{\ddagger}_{rls}$.

Controlling the relative heights of activation energy barriers in order to tune catalyst selectivity is a subtle and difficult task. Fortunately, many parameters can be varied to improve catalytic performance.

From a practical point of view, changes can be exerted on the catalyst itself, on the reactants, and/or on the reaction conditions. However, in order to be able to predict selectivity in catalysis, a better understanding of the mechanisms of the underlying chemical reactions is needed [91].

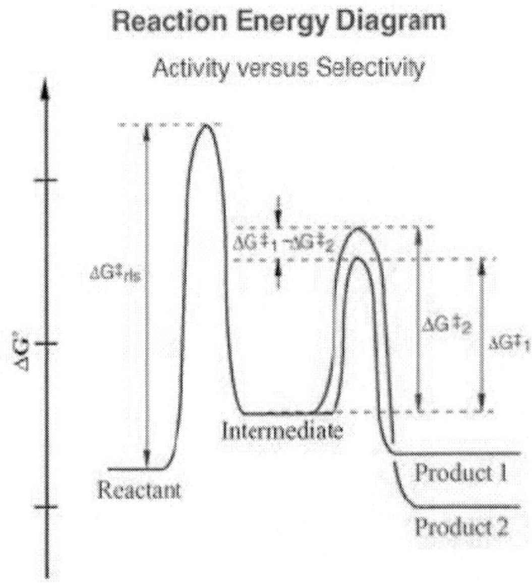

Figure 24. Simplified energy diagram for catalytic processes illustrating the different criteria needed to optimize total reactivity vs. selectivity. It can be seen that while the overall rate is controlled by the energy barrier of the initial conversion of the reactant into the intermediate, $\Delta G^{\ddagger}_{rls}$,the selectivity, defined as the yield ratio of product 1 to product 2, is determined by the difference in barrier heights $\Delta G^{\ddagger}_{1} - \Delta G^{\ddagger}_{2}$ for the two subsequent reactions available to the intermediate. Notice that a relative variation of only 10% in the absolute value of either $\Delta G^{\ddagger}_{1}$ or $\Delta G^{\ddagger}_{2}$ is sufficient to switch from the exclusive formation of one product to the other [90].

Metal-Support Interaction

Supported metals represent systems of usually well dispersed metal particles in intimate contact with insulator or semiconductor oxides. They must contain sites where the distance between metal atoms and support molecules is in the range of atomic dimensions and one can expect that these composite 'adlineation' sites may have peculiar propensities. It may, however, be hard to distinguish between structural and bonding effects. The unique properties of metal-support borderline sites were first emphasized by Schwab [92]. Boudart [93] termed the case of a support, modifying the electron density of small metal clusters by charge transfer or polarization, the 'Schwab effect of the second kind'. He regarded this effect to be of importance for catalysis. The 'Schwab effect of the first kind', i.e. the metal affects its semiconductor support, may have less practical significance from the catalytic point of view [94].

The idea that metals might behave differently when placed on different supports had been largely viewed as an electronic effect (band theory of delocalized electrons as opposed to shared electrons in a molecular orbital) [95] not involving chemical bond formation between the zero-valent metal and the support. The perception, the new concept that is illustrated post-SMSI, is that one or more elements that comprise the support can chemically react with metal particles. It is this reaction that provides a thermodynamic driving force for moving these elements onto the surface of the metal particles (the so-called decoration model where a sub oxide of the support covers or “decorates” the metal surface) and alters its catalytic activity. It would be fair to say that now there is no such thing as an inert support. Instead each support has a varying interaction, depending on the reducing/oxidizing treatment and the temperature of this treatment. Non-Faradaic electrochemical modification of catalytic activity (NEMCA) would not at first appear to have anything to do with SMSI, the NEMCA originates from an electrical current or potential between the catalyst and a second electrode, both deposited on the same solid electrolyte (support). However, we have come to understand that the result of this applied potential is to move oxide species (in the case of oxygen-conducting electrolytes such as YSZ) onto the catalytic metal electrode decorating the metal surface in a way parallel to SMSI. Moreover, we now know that for those reactions involving charge transfer (all reactions to some degree), NEMCA and metal–support interactions in general, can be interpreted in terms of a work function change that accompanies the decoration of the metal by cations, anions, and/or ion pairs. Thus, we find that the concept of SMSI illuminates NEMCA, that NEMCA in turn teaches us more about the phenomena of SMSI, and that both concepts are more fully realized when viewed as a whole.

Interaction at the Metal Oxide Interface

SMSI (Strong metal Support Interaction): In retrospect, perhaps the place to begin thinking about the metal–oxide support interface, even in anticipation of chemical reaction between the metal and oxide, is classical thermodynamics and this is the approach taken by Knözinger and Taglauer [96]. Of course, the thermodynamics of spreading and wetting only provide a framework for thinking about the energetic of the metal–oxide interface because very small clusters of metal on an oxide surface are not in thermodynamic equilibrium (they are meta-stable with respect to sintering) and a sub-oxide of molecular dimensions (as has been observed for SMSI [97]) does not exhibit the bulk phase that is presumed by classical thermodynamics. While the tendency to undergo SMSI can be correlated with oxide reducibility [98], it can also be correlated with surface free energies; i.e., TiO_2, V_2O_5, Nb_2O_5, and Ta_2O_5 have smaller surface free energies than SiO_2, Al_2O_3, ZrO_2, and MgO [96] and the former are known to more easily undergo SMSI. The surface free energies of reducible oxides increase if they become oxygen deficient. Thus, we can deduce that it is probably the concentration gradient and the molecular bonding between the noble metal and the sub-oxide that are the driving forces for spreading, and that migration probably precedes reduction to the final meta-stable state, and reduction decreases the mobility of the sub-oxide limiting the decoration to a thickness on the order of a monolayer. These predictions pretty much concur with surface science findings from model systems, but there exists an inherent contradiction that reduction may be necessary for migration and immobilization of the migrated species. Perhaps there is some intermediate reduced state that migrates. Consider the case of Pt clusters on TiO_2 (110) studied [97] using low-energy ion scattering (LEIS), which is mainly sensitive to the top most layer, while angular-dependent (electron detection 10–70° off crystal normal) X-ray photo electron spectroscopy (XPS) averages over several layers. When the LEIS of Pt is very significantly attenuated (see Figure 23), the $4f_{3/2}$ and $4f_{5/2}$ Pt XPS peaks are hardlychanged,even at the70° take off angle(see Figure 25), suggesting a covering of Pt by a layer of TiOx on the order of a monolayer. It was found that x is less than or equal to one and depends on time and temperature, but no evidence for metallic Ti was found (in the limited temperature range).

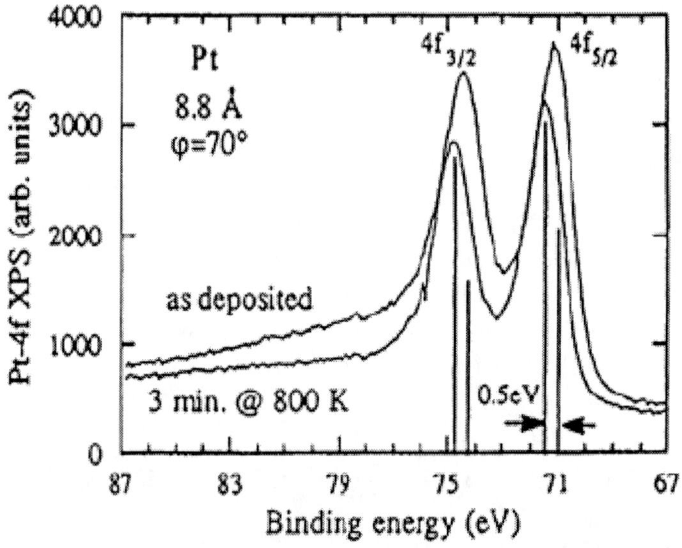

Figure 25. XPS of the Pt 4f doublet, plotted as a function of binding energy, for an 8. 8-Å Pt layer. (Top spectrum) As deposited (Pt evaporation with sample at room temperature); (bottom spectrum) layer heated at 800K for 3 min (grazing emission).

In a more recent scanning tunneling microscope (STM) study of Pt on TiO_2 (110) proposed an ordered atomic model that resembles a slightly oxygen-rich, oxygen-terminated $TiO_{1.1}$ double layer. The film exhibits a dipole moment, caused by the uneven distribution of negative and positive charges between the O-rich first and Ti-rich second layer. Assuming that a similar (dipolar) structure prevails locally when the Pt is not completely covered, these local dipoles would result in a work function change that would affect the adsorption and reaction

The comparison of metals (interaction with TiO_2) as follows: "the less reactive the deposited metal towards the oxide, the less effective the wetting, and the more reduced the suboxide migrating on the top of the islands." The primary comparison was made between Pt and Fe. On performing comparison between Pt and Rh it was found that Pt enters into the SMSI state more easily (at a lower temperature or in a shorter time) than does Rh and that the SMSI state for Pt/TiO_2 is more difficult to reverse than Rh/TiO_2 is, and that the (non stoichiometric) $M_{sur}TiOx$ surface complex is more oxygen-rich on Rh than on Pt [99]. These two generalization appear to be mutually consistent but may just be different ways of saying that the overall metal–support interaction is driven by the competition between M–Ti and M–O bond formation, which would then order the three metals Fe < Rh < Pt in terms of their tendency to enter the SMSI state.

The consequences of SMSI depend on the metal, the oxide, and the reaction and so generalizations are not simple. To the extent that decoration occurs, it is likely to slowdown all reactions and be particularly detrimental to the more structure sensitive ones, Of course,steady-state reaction in the SMSI state more or less rules out oxidation reaction, because both the oxygen and the water produced tends to reverse SMSI.

Non-Faradaic Electrochemical Modification of Catalytic Activity (NEMCA)

The NEMCA effect, is used to describe the increase in catalytic activity (up to 90-fold) and selectivity of a gas exposed electrode on a solid electrolyte cell upon application of a potential. This phenomena is well documented and has been observed on various surfaces (Ni, Au, Pt and Pd) supported by O^{2-}, Na^{+} and proton conducting solid electrolytes.

Metals on YSZ, particularly Pt, the ion-carrying current in the solid electrolyte will be O_2^- and it is the chemistry of this species at the three-phase boundary (tpb; between the gas phase, solid electrolyte, and metal) and the species that it evolved to as it migrates over the metal surface that is of interest. The essential facts that have been established experimentally are that an oxygen anion, perhaps O_2^-, migrates across the three-phase boundary and is more or less uniformly distributed over the metal surface. This species can be distinguished from the oxygen species adsorbed from the gas phase (in a given temperature range) by temperature-programmed desorption (TPD) and XPS. The electrochemically pumped oxygen species is more strongly bonded to Pt than oxygen adsorbed from the gas phase (as determined by TPD; it desorbs at higher temperature), but the latter is promoted (becomes more weakly bonded) by the former. The electrochemically formed oxygen has an O 1s binding energy of about 528.8eV (compared to 530.2eV for normally chemisorbed atomic oxygen) and is therefore more negatively charged and, perhaps for this reason it is much less reactive with gas-phase H_2 and CO in the UHV background [22] or with organic reductants. The strongly bonded ionic state (formed electrochemically) acts as a sacrificial promoter of the weakly bonded state (formed from gas-phase adsorption). The desorption activation energy of the weakly bonded state is linearly correlated with the catalyst potential. It has been argued thermodynamically that the catalyst surface potential can be equal to the applied over potential only if (i) equilibrium exits between oxygen ions in the electrolyte support and the catalyst surface, (ii) oxygen-ion coverages are intermediate, and (iii) there are no lateral interactions between surface ions. This confirms that the catalyst potential and work function have a pronounced effect on the binding energy of atomic oxygen and therefore can be expected to affect the rate of oxidation reactions involving adsorbed atomic oxygen. [99]

Role of Triplet States in Catalysis

Before consideration of spin effects in catalysis one has to consider some general features of chemical reactions between stable diamagnetic substances. The nature of activation barriers in bimolecular reactions is determined by exchange repulsion between closed shell molecules. It can be explained by higher weight of the inter-molecular triplet spin pairing (Figure 26).Spins are singlet-coupled inside the ground state molecules AB and CD, but intermolecular pairing is arbitrary: all possible pairing schemes are equally probable. Just from statistics there are three triplets and one singlet state for pair of non -correlated electrons. In a framework of VB method the intermolecular interaction between two closed-shell systems (Figure 26) can be described by exchange integrals for two singlet-paired states and six triplet-paired states. The singlet pairs are stabilized by exchange integrals, the triplets are much stronger destabilized. The total balance of the exchange intermolecular interaction is repulsion, which produces a barrier.

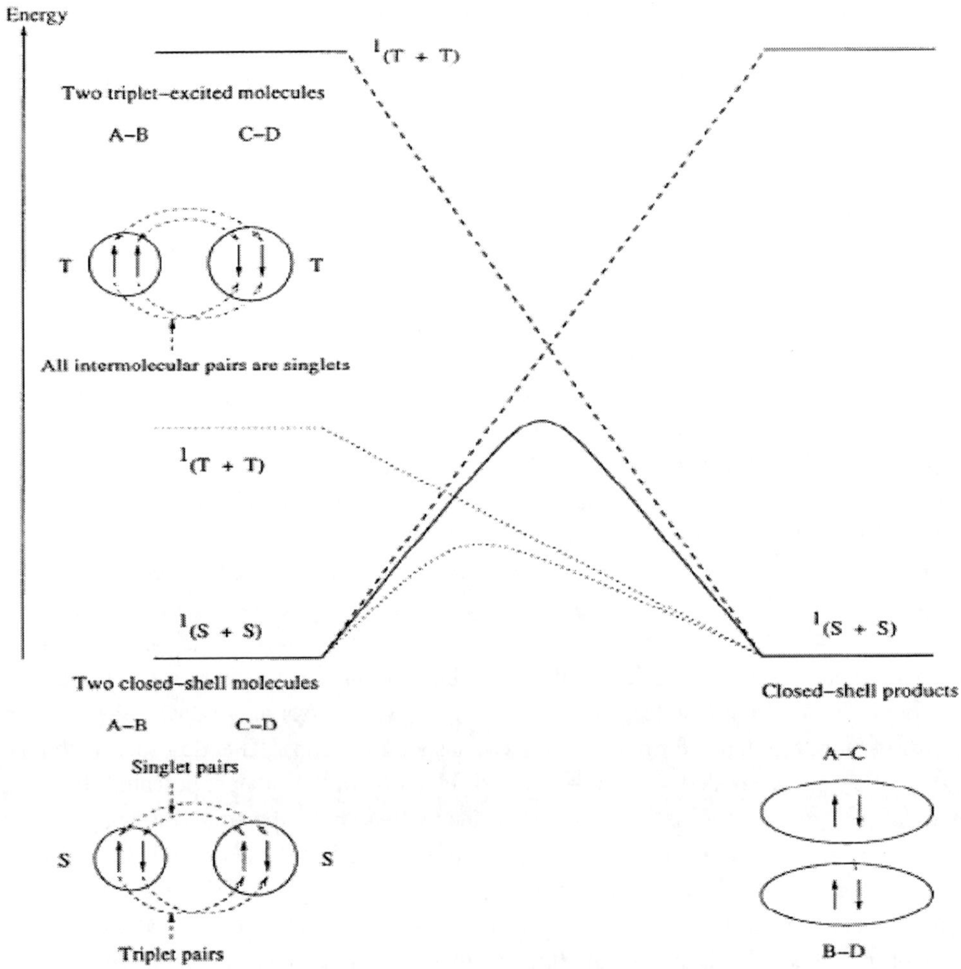

Figure 26. Exchange repulsion between two closed-shell molecules (triplet pairing prevails) and dependence of the activation barrier on the "double-triplet" excited state coupled to the singlet. The lower the "double-triplet" state the lower activation barrier (dashed lines).

In order to solve a question "how to diminish the barrier?" one has to note that the repulsion will be changed by strong intermolecular attraction when the triplet excited covalent states in each molecule are involved. In addition to intermolecular attraction a cleavage of each bond occurs and new cyclization (or insertion) products AC + BD will be formed. It does not matter what kind of products are considered (it should be also cyclization or insertion); exchange repulsion between closed shell reagents is determined by the fact, that new intermolecular overlaps growing along the reaction path, trends to destabilize spin-pairing inside molecules [101]. Two excited triplet states are coupled into the total singlet state; the avoided crossing between two singlet states produces the activation barrier. These are the ground singlet state and the "double-triplet" singlet excited state; their diabatic behavior is shown by dashed lines in Figure 26.

The height of the barrier, the reaction path and the heat of reaction are determined mostly by the properties of the triplet states. The thermal chemical reactivity is often coded by the triplet excited states of the diamagnetic reagents [100]. Thus, the triplet excited states involvement is a useful way of understanding of the nature of barrier and chemical reactivity.

It does not mean that the triplet excited states participate in thermal chemical reaction as kinetic intermediates.

This is only a correlation diagram of VB method, which indicates that at the transition state region there is a strong mixing of two singlet states with different spin-coupling patterns. The avoided crossing realizes the spin-uncoupling process during reaction. If the energy of one (or both) triplet states in the reagents is lowered in some way, the activation barrier would be diminished (dotted lines in Figure 26). This is what a catalyst really does in chemical activation.

An important advantage of the VB approach to catalysis is that it relates the excited triplet state and spin-uncoupling scheme to the traditional way in which chemists rationalize chemical structure and reactivity [100]. In order to construct diabatic electronic states representing specific resonance structures of a catalytic system one has to include the triplet excited state of activated chemical bond. A simple three-electron VB scheme and its generalization to many-electron catalytic systems illustrate the crucial importance of the triplet excited state of activated molecule. In the VB method an account of this state includes some excited states of the catalyst. It has been concluded that only the low-spin state of the catalyst is reactive and the small energy gap between high-spin and low-spin states is the main electronic factor of *catalytic activity.* It is known that the ionic contributions are very important in the transition state region and ionic structures are considered as important factor for catalytic activity.

Optimization

Optimization of a catalytic process usually means minimization of expenses at preservation of basic process parameters or improvement of the process parameters at the same expenses. The following two problems are to be resolved due to optimization of the active component distribution:

- To decrease the total active component loading in the bed at a given outlet reactant conversion.
- To increase conversion at a given active component amount.

The reaction rate is affected by diffusion limitations, reactant concentration, catalyst surface temperature. In the kinetic regime under isothermal conditions, the reaction rate is only determined by the total loading of the active component. Otherwise, the redistribution of the active component concentration between zones with different temperatures, reactant concentrations and different intensity of mass transfer can result in variations in the outlet reactant conversion. In order to demonstrate that, let us formulate mathematical setting of the problem of optimization of the active component distribution along the catalyst bed length [102].

Future of Catalysts

Homogeneous Catalysts

New phases which have been tested as media for homogeneous catalysis include expanded liquids and induced melts. Expanded liquids are organic liquids into which large quantities of gas have been dissolved. As a result, the volume of the liquid phase increases greatly (sometimes well over10-fold) and other physical properties change as well, including viscosity, surface tension, diffusion rates, solubility of reagent gases, solubility of catalysts and substrates, density, polarity, and melting point. Rate constants and equilibrium constants for reactions dissolved in the organic liquid also change as a result of the expansion of the liquid. The gas is typically CO_2 but could be any gas relatively near its critical temperature. While most organic liquids undergo this expansion when exposed to CO_2 (over 30 bar), a few such as diols, triols, liquid polymers and ionic liquids do not expand significantly. Nevertheless, the physical properties of the liquid polymers and ionic liquids do change with increasing CO_2 pressure. Compared to super critical fluids, expanded liquids have many advantages (high diffusion rates for example) without the cost of the higher pressure required by pure SCFs. Homogeneously catalyzed reactions in expanded liquids [103] and even expanded ionic liquids [104] have been reported. CO_2 dissolved into an alcohol in this way can act as a catalyst precursor for acid-catalyzed reactions.

Induced melts are liquids or softened-solid phases prepared by the addition of CO_2 pressure over an organic solid. The pressure of CO_2 (or any other gas near its critical pressure) causes the melting point of the solid to decrease such that it melts or atleast softens. Solvent less reactions of solids, which are typically very slow for obvious mass-transfer reasons, can be accelerated by the addition of pressurized CO_2 in this manner, without the pressure being high enough for the CO_2 to act as a compressed gas solvent [105]. The principal disadvantage of this technique is that the melting point depression is usually limited to a 20 or 30 K drop, although for some compounds the melting point depression can be significantly greater. Classes of homogeneous catalysis that have received less attention in SCF studies include phase transfer catalysis, photo-catalysis, and electro-catalysis. Homogeneous electro-catalysis in SCFs seems to be entirely unexplored. In the case of an insoluble polymer, the SCF can serve to swell the polymer and transport the SCF-soluble catalyst into the swollen or porous solid polymer. This technique had been used earlier with free-radical initiated polymerization in $scCO_2$ inside a polymer. Homogeneous catalysis for the modification of SCF-soluble polymers is also possible [106].

Mechanochemical Activation as a Tool for Increasing Catalytic Activity

The development of mechanochemical activation (MCA) of solids has shown its considerable promise for the preparation of catalysts with novel properties and improvement of the catalyst activity and selectivity. During MCA a solid (a catalyst in our case) accumulates an excess potential energy as elastic and plastic deformations and great variety of defects accompanied by an increase in its reactivity. This excess energy can considerably affect both the reactivity and activity of catalysts. The system reactivity is universal in the sense that it appears as a universal tendency for the decrease of its thermodynamic parameters. This tendency can involve different relaxation channels, including catalytic reactions yielding the desired products. MCA allows creating defects of a fixed type, to form a surface by faces and edges not typical of catalysts in the usual state. All this extends a range of bond energies and properties of the intermediate surface species and is the reason of change in activity and selectivity [107].

Kinetics and Rational Thermodynamics

In chemistry, the *law of mass action* is a mathematical model that explains and predicts behaviors of solutions in dynamic equilibrium. It can be described with two aspects: 1) the equilibrium aspect, concerning the composition of a reaction mixture at equilibrium and 2) the kinetic aspect concerning the rate equations for elementary reactions. Both aspects stem from the research by Guldberg and Waage (1864-1879) in which equilibrium constants were derived by using kinetic data and the rate equation which they had proposed. Guldberg and Waage also recognized that chemical equilibrium is a dynamic process in which rates of reaction for the forward and backward reactions must be equal. It has been proved that the reaction rate is a function of concentration and temperature, which can be approximated by *polynomial.*

Many irreversible thermodynamic approaches fail to derive the mass-action kinetic law directly, particularly in its traditional form, i.e., expressing reaction rate through products of rate constant and concentrations. *Samohýl's* rational thermodynamic approach is probably the only one successful in this regard. The methodology and resulting (rate) equations possess new and inspiring points summarized as follows. Rate equations obtained by this method may be interpreted within the framework of traditional mass-action kinetics as including terms corresponds to individual steps of reaction mechanism and their forward and reverse rates. It is also seen that the rate of any step may be affected also by species, which do not directly take part in this step or, in other words, by the rate of (some) other steps. Similarly, the effect of an inert on a reaction rate can be naturally included.

This interpretation enables to propose reaction mechanism significant for description of reaction kinetics. Each term in the resulting (polynomial) rate equation, i.e. each of its difference term is considered to represent the difference between forward and reverse rate of one step of reaction mechanism. Complexity of the mechanism can be changed by the number of components, which consequently determines the number of independent reactions, or by the degree of approximating polynomial. Resulting rate equations contain only whole-number reaction orders with respect to any component. Rate equation for, e.g. product formation, automatically contains concentrations of all species present in the reaction mixture and not only of those taking part directly in the reaction step(s) in which the product is formed. This form of rate equation is not unknown to the traditional mass-action kinetics although its practical use is rare. Rational thermodynamic procedure operates with independent reactions only, i.e., starts with the set of independent reaction steps. However, the resulting rate equations, in the sense of above interpretation, may contain also other reactions relevant for the description of reaction kinetics. Resulting "kinetic" mechanism may be thus different from and more complex than initial, let us say "thermodynamic" one. The procedure does not preclude ("thermodynamically") dependent reactions to be kinetically important reaction steps, which are not among independent one's may occur in the rate equation. Rational thermodynamics naturally differentiates thermodynamic from kinetic independence [108].

Moreover, rates of all dependent reactions can be unambiguously expressed from the rates of selected independent reactions by simple operations from linear algebra. Kinetic equilibrium criteria are fulfilled with more general equation than usual Guldberg–Waage, considering only true thermodynamic equilibrium constants, with no need for any "kinetic" equilibrium constant, given as a ratio of rate constants in forward and reverse reaction directions. Moreover, the method does not need "backward" rate constants and enables the direct usage of thermodynamic equilibrium constants, which can be rigorously computed from thermodynamic data, in rate equations.

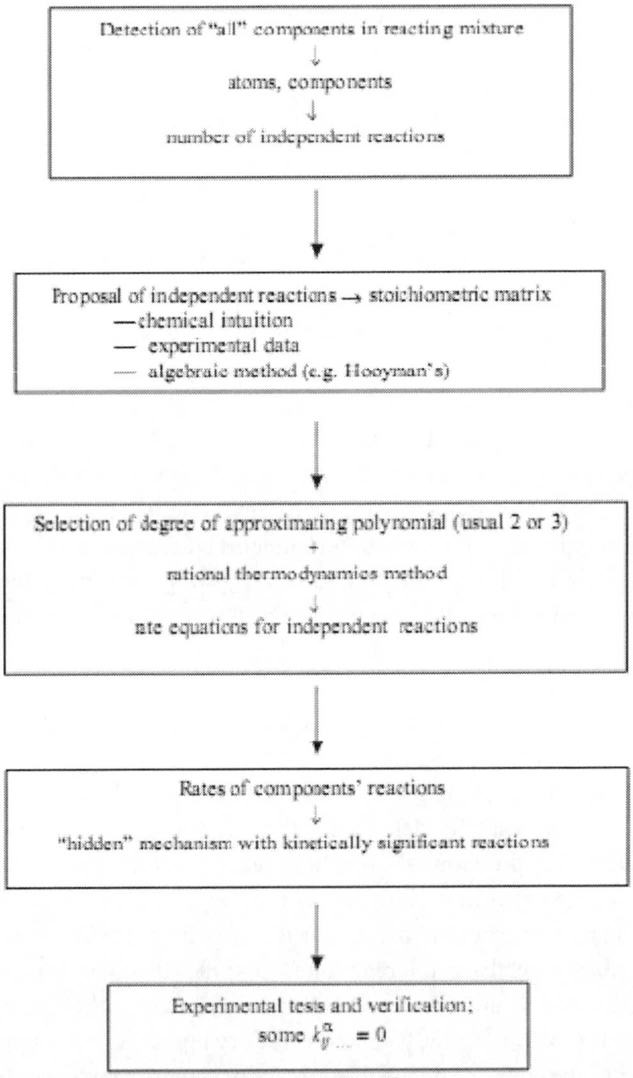

Figure 27. Instructions of rational thermodynamic to kinetics.

Equilibrium conditions are not introduced by a priori stating that rate is a difference between forward and reverse rates as usual but differences like this naturally follows applying requirement of general validity of vanishing rate in equilibrium. Reaction mechanism for a given set of reacting species (and selected degree of approximating polynomial) directly appears in the rate equations. Therefore, this method can be used to design suitable mechanism for reaction under study, which fully and consistently with (irreversible) thermodynamics describes reaction kinetics. Some arbitrariness in selecting independent reactions or modifications of approximating polynomial can be viewed as a way of proposing and testing various mechanisms for a particular reaction. It should be noted, that in this procedure mechanism steps with "single simple" intermediate, i.e. intermediate which is sole product in one reaction (step) and sole reactant in another step, are not supported. Similarly, the same component cannot stay as a sole reactant in two (or more) reaction steps. To summarize, Figure 27 gives "rational thermodynamic guide" for studying chemical kinetics. The basis consists in determining relevant reacting species to be able to construct respective stoichiometric matrix. Concentration profiles of all relevant constituents are the only necessary experimental inputs in practical applications. No preliminary deductions on mechanism are needed. In most real situations, the degree of approximating polynomial should be two or three as this corresponds to reaction steps of second or third order, resp., in reaction mechanism. For simple reacting mixtures, first or second order can usually only recover conventional mass-action (Guldberg–Waage) rate equation. Resulting rate equations and mechanism "hidden" in them are to be confronted with experimental data to eliminate some terms with negligible coefficients (rate constants) and to put experimental or computed equilibrium constant into the rate equation(s). Final equation(s) and kinetically relevant mechanism follow [108].

REFERENCES

[1] J. F. Le Page, in: G. Ertl, H. Knözinger, J. Weitkamp (Eds.), *Handbook of Heterogeneous Catalysis*, vol.1; Wiley/VCH: New York/ Weinheim, 1997, p.49.

[2] R. Montarnal, J. F. La Page, *La Catalyse au Laboratoire et Dans L'industrie*, Masson: Paris, 1967, p.231.

[3] C. Perego, P. L. Villa, Catalyst preparation methods, *Catal. Today*, 1997, *34*, 281-305.

[4] M. Campanati, G. Fornasari, A. Vaccari, Fundamentals in the preparation of heterogeneous catalysts, *Catal. Today*, 2003, *77*, 299-314.

[5] M. V. Twigg (Ed.), *Catalyst Handbook*, 2nd ed.; Wolfe: London, 1989.

[6] F. Schüth, K. Unger, in: G. Ertl, H. Knözinger, J. Weitkamp (Eds.), *Handbook of Heterogeneous Catalysis*, vol.1; Wiley/VCH: New York / Weinheim, 1997, p.72.

[7] J. -F. Le Page, J. Cosyns, P. Courty, E. Freund, J. -P. Franck, Y. Jacquin, B. Juguin, C. Marcilly, G. Martino, J. Miquel, R. Montarnal, A. Sugier, H. Van Landeghem, *Applied Heterogeneous Catalysis: Design, Manufacture, Use of Solid Catalysts*; Technip: Paris, 1987.

[8] J. W. Geus, in: G. Poncelet, P. Grange, P. A. Jacobs (Eds.), *Preparation of Catalysts III, Studies in Surface Science and Catalysis*; vol.16; Elsevier: Amsterdam, 1983, p.1.

[9] P. T. Cardew, R. J. Davey, P. Ellitt, A. W. Nienow, J. P. Winterbottom, in: B. Delmon, P. Grange, P. A. Jacobs, G. Poncelet (Eds.), *Preparation of Catalysts IV, Studies in Surface Science and Catalysis*; vol.31; Elsevier: Amsterdam, 1987, p.15.

[10] J. T. Richardson, *Principles of Catalyst Development*; Plenum Press: New York, 1989.

[11] M. Che, O. Clause, Ch. Marcilly, in: G. Ertl, H. Knözinger, J. Weitkamp (Eds.), *Handbook of Heterogeneous Catalysis*, vol.1;Wiley/VCH: New York/Weinheim, 1997, p.191.

[12] J. W. Geus, J. Van Dillen, in: G. Ertl, H. Knözinger, J. Weitkamp (Eds.), *Handbook of Heterogeneous Catalysis*, vol.1; Wiley/VCH: New York/Weinheim, 1997, p.240.

[13] S. Bhatia, *Zeolite Catalysts: Principles and Applications*; CRC Press: Boca Raton, FL, 1990.

[14] L. A. M. Hermans, J. W. Geus, in: B. Delmon, P. Grange, P. A. Jacobs, G. Poncelet (Eds.), *Preparation of Catalysts II, Studies in Surface Science and Catalysis*; vol.3; Elsevier: Amsterdam, 1979, p.113.

[15] Chang-jun Liu ,Gheorghi P.Vissokov ,Ben W.-L. Jang, catalyst preparation using plasma technologies, catal. today 2002, *72,*173-184.

[16] G. Ertl, H. Knözinger and J. Weitkamp (Eds), *Handbook of Heterogeneous Catalysis*, Vol.1. Wiley-VCH: Weinheim, 1997.

[17] R. Strobel, W. J. Stark, L. Mädler, S. E. Pratsinis and A. Baiker, Flame-made platinum/alumina: structural properties and catalytic behavior in enantioselective hydrogenation, *J. Catal.,*2003 *213*, 296–304.

[18] G. L. Messing, S. C. Zhang and G. V. Jayanthi, Ceramic powder synthesis by spray-pyrolysis, *J. Am. Ceram. Soc.* 1993, *76*, 2707–2726.

[19] K. Okuyama and I. W. Lenggoro, Preparation of nanoparticles via spray route, *Chem. Eng. Sci.* 2003, *58*, 537–547.

[20] L. Mädler, Liquid-fed aerosol reactors for one-step synthesis of nano-structured particles, *KONA* 2004, *22*, 107–120.

[21] M. Sokolowksi, A.Sokolowska, A. Michalski and B.Gokieli, The 'in-flame-reaction' method forAl_2O_3 aerosol formation, *J. Aerosol. Sci.* 1977, *8*, 219–229.

[22] L. Mädler, H. K. Kammler, R. Mueller and S. E. Pratsinis, Controlled synthesis of nano structured particles by flame spray pyrolysis, *J. Aerosol. Sci.* 2002, *33*, 369–389.

[23] C. R. Bickmore, K. E. Waldner, D. R. Treadwell and R. M. Laine, Ultrafine spinel powders by flame spray pyrolysis of a magnesium aluminum double alkoxide, *J. Am. Ceram. Soc.* 1996, *79*, 1419–1423.

[24] Kilian and T. F. Morse, A novel aerosol combustion process for the high rate formation of nanoscale oxide particles, *Aerosol Sci. Technol.* 2001, *34*, 227–235.

[25] R. Jossen, Controlled synthesis of mixed oxide particles in flame spray pyrolysis, *Dissertation* 16401, ETH, Zurich 2006.

[26] W. J. Stark, L. Mädler, M. Maciejewski, S. E. Pratsinis and A. Baiker, Flame synthesis of nano crystalline ceria-zirconia: effect of carrier liquid, *Chem. Commun.*, 2003, 588–589.

[27] R. Strobel, A. Baiker, S. E. Pratsinis, Invited review paper Aerosol flame synthesis of catalysts, *Advanced pow. Technol.* 2006, *17*, 457-480.

[28] C. H. Bartholomew; Robert J. Farrauto, *Fundamentals of Industrial Catalytic Processes*,2nd ; Wiley–VCH: Weinheim, 2005; pp 118-196.

[29] Catalyst Characterization. *Encyclopedia of Physical Science and Technology,*3rd , Academic Press: Liverpool, UK, 2004; Chemical Engineering, pp 501-526.

[30] J. W. Niemantsverdriet, *Spectroscopy in Catalysis*, 3rd; Wiley–VCH: Weinheim, 2007; pp 1-9.

[31] Gleaves,J. T., Yablonskii, G. S., Phanawadee, Ph., and Schuurman, Y., TAP2: An interrogative kinetics approach, *Appl. Catal. A: Gen.*, 1997, *160*, 55–87.

[32] Constales D., Yablonsky G. S., Marin G.B., and Gleaves,J .T. Multi-zone TAP-reactor theory and application: I. The global transfer matrix equation, *Chem. Engg. Sci.*, 2001, *56*,133–149.

[33] C. van Veen, D. Farrusseng, M. Rebeilleau, T. Decamp, A. Holzwarth, Y. Schuurman, and C. Mirodatos, Acceleration in catalyst development by fast transient kinetic investigation, *J. Catal.*, 2003, *216*, 135-143.

[34] Shekhtman S. O., Yablonsky G. S., Gleaves J. T. , and Chen, S. Thin-zone TAP-reactor theory and application, *Chem. Engg. Sci.*, 1999, *54*, 4371–4378.

[35] Jeziorowski, H. and Knozinger, H.(1979),'Raman and ultraviolet spectroscopic characterization of molybdenum on alumina catalysts', J. Phys. Chem., 83,1166–1173.

[36] Mar, S.Y., Chen, C.S., Huang, Y.S. and Tiong, K.K. (1995), 'Characterization of RuO_2 thin films by Raman spectroscopy', Appl. Surf. Sci., 90, 497–504.

[37] Graham, G.W., Potter, T.J. and Weber, W.H. (1989),' Formation of platinum oxide during the thermal oxidation of nobel-metal alloys', J. Vac. Sci. Technol., 7,1694–1696

[38] Tolia, A.A., Smiley, R.J., Deglass, W.N., Takoudis, C.G. and Weaver, M.J. (1994) 'Surface oxidation Of rhodium at ambient pressures as probed by surface-enhanced Raman and X-ray protoelectron spectroscopy', J. Catal., 150, 56–70.

[39] Zuo,J. and Xu, C.(1996),'Raman spectra of nanophase Cr_2O_3', J. Raman Spectrosc., 26,921–923.

[40] Davydov, A.A. and Goncharova, O.I. (1989) 'Use of the compensation method to study the IR spectra of Mo-containing systems deposited on oxide carriers', Zh. Prikl. Spektrosk. (in Russian); J. Appl. Spectrosc. (English translation), 50, 76–80.

[41] Zecchina, A. and Arean, C. (1993),' Structure and reactivity of surface species obtained by interaction of Organometallic compounds with oxidic surfaces: IR studies', Catal. Rev. Sci. Eng., 35,261–317.

[42] Guglielminotti, E., Zecchina, A., Boccuzzi, F. and Borello, E. (1980),'Growth and properties of metal clusters', in Bourdon ,J.(Ed.) Proceedings of the 32nd International Meeting of the Societe de Chimie Physique, Villerbane, September 24–28, 1979 Elsevier, Amsterdam, pp. 165–210.

[43] Matyshak, V.A., Kadushin, A.A. and Edreva-Kardzhieva, R.M. (1983), 'Effect of temperature of the Support pre-treatment on activation of $Mo(CO)_6/Al_2O_3$ catalysts for C_3H_6 methatesis', Kinet. Katal. (in Russian); Kinet. Catal. (English translation), 24, 1255–1258.

[44] Ivasenko, V.L., Sayapin, E.V., Davydov, A.A. and Egorova, G.M. (1986), 'Metal-complex heterogeneous catalysts for purification of waste gases from nitric acid production', Z.Prikl.Khim. (inRussian); Russ.J.Appl.Chem. (Englishtranslation), 61,431–434.

[45] Shekochkhin, Yu.M. and Davydov, A. A. (1976),' Study of reactions on Al_2O_3 by means of IR spectroscopy', (in Russian), Adsorptsiya Adsorbenty, 4,75–80.

[46] Gordymova, T.A. and Davydov, A.A.(1979) 'Spectroscopic investigation of propylene complexes on the surface of γ –aluminum oxide and mechanisms of isomerization', Dokl. ANSSSR (in Russian); Proc. AN USSR (English translation), 245,635–639.

[47] Davydov, A.A. and Shekochihin, Yu.M.(1969), 'IR spectra of NH_3 adsorbedonAl_2O_3', Kinet. Katal. (in Russian); Kinet. Catal. (English translation), 10, 163–171.

[48] Lekhal, B. J. Glasser, J. G. Khinast, Impact of drying on the catalyst profile in supported impregnation catalysts, *Chem. Engg. Sci.*, 2001, *56*, 4473-4487.

[49] W. Kohn, Rev. Mod. Phys., 1999, 71, 1253.

[50] T. Bligaard, J. K. Nørskov, Ligand effects in heterogeneous catalysis and electrochemistry, *Electrochim. Acta,,* 2007, *52*, 5512-5516.

[51] T. Bligaard, J. K. Nørskov, S. Dahl, J. Matthiesen, C. H. Christensen, J. Sehested, The Brønsted–Evans–Polanyi relation and the volcano curve in heterogeneous catalysis, *J. Catal.*, 2004, *224*, 206-217.

[52] Logadottir, T. H. Rod, J. K. Nørskov, B. Hammer, S. Dahl, C. J. H. Jacobsen, The Brønsted–Evans–Polanyi Relation and the Volcano Plot for Ammonia Synthesis over Transition Metal Catalysts, *J. Catal.*, 2001, *197*, 229-231.

[53] B. Anderson, Y. Cai, The Reversible Hydrogen Electrode: Potential-Dependent Activation Energies over Platinum from Quantum Theory, *J. Phys. Chem. B*, 2004, *108*, 9829-9833.

[54] J. K. Nørskov, J. Rossmeisl, A. Logadottir, L. Lindqvist, J. R. Kitchin, T. Bligaard, H. Jonsson, Origin of the Overpotential for Oxygen Reduction at a Fuel-Cell Cathode, *J. Phys. Chem. B*, 2004, *108*, 17886-17892.

[55] Hammer, J. K. Nørskov, Theoretical surface science and catalysis—calculations and concepts, *Adv.Catal.*, 2000, *45*, 71-129.

[56] M. Campbell, *Catalysis at Surfaces*, Chapman and Hall: London, 1988.

[57] P. S. Bagus, C.R. Brundle, G. Pacchioni, F. Parmigiani, Mechanisms responsible for the shifts of core-level binding energies between surface and bulk atoms of metals, *Surf. Sci.*, 1993, *19*, 265-283.

[58] H. J. Kreuzer, in: D. J. Dwyer and F. M. Hoffmann (Eds.), *Surface Science and Catalysis. In Situ probes and Reaction Kinetics*, Vol. 482, ACS Symp. Series: Washington, 1992, p. 269.

[59] J. K. Norskov, *Phys. Rev. B,* 1982, *26* , 2875.

[60] J.H. Block, H.J. Kreuzer, L.C. Wang, Electrostatic field effects in surface reactivity: Adsorption, dissociation and catalytic reaction of nitric oxide, *Surf. Sci.* 1991, *246*, 125-134.

[61] G. Pacchioni, J. R. Lomas, F. Illas, Electric field effects in heterogeneous catalysis, *J. Mol. Catal. A: Chem.* , 1997, *119*, 263-273.

[62] M. Boudart, in: G. Ertl, H. Knözinger, J. Weitkamp (Eds.), *Handbook of Heterogeneous Catalysis*, Wiley–VCH, Weinheim, 1997.

[63] J. A. Dumesic, D. F. Rudd, L. M. Aparicio, J. E. Rekoske, A. A. Treviño, *The Micro kinetics of Heterogeneous Catalysis*, Am. Chem. Soc., Washington DC, 1993.

[64] M. Boudart, G. Djéga-Mariadassou, *Kinetics of Heterogeneous Catalytic Reactions*, Princeton Univ. Press, Princeton, NJ, 1984.

[65] E. Benazzi, L. Leite, N. Marchal-George, H. Toulhoat, P. Raybaud, New insights into parameters controlling the selectivity in hydrocracking reactions, *J. Catal.*, 2003, 217, 376-387.

[66] J. K. Nørskov, T. Bligaard, A. Logadottir, S. Bahn, M. Bollinger, L. B. Hansen, H. Bengaard, B. Hammer, Z. Sljivancanin, M. Mavrikakis, Y. Xu, S. Dahl, C. J .H. Jacobsen, Universality in Heterogeneous Catalysis, *J. Catal.*, 2002, *209*, 275-278.

[67] Michaelides, Z.-P. Liu, C. J. Zhang, A. Alavi, D. A. King, P. Hu, Identification of General Linear Relationships between Activation Energies and Enthalpy Changes for Dissociation Reactions at Surfaces J. Am. Chem. Soc., 2003, *125*, 3704-3705.

[68] Delmon, Influence of species diffusing at surfaces on reactions of solids and catalytic processes, *Solid State Ionics*, 1997, *101*–103, 655–660.

[69] F. A. Kröger, H. J. Vink, Relations between the Concentrations of Imperfections in Crystalline Solids, *Solid State Phys.* 1956, *3*, 307-435.

[70] H. Iwahara, T. Yajima, H. Suzuki, T. Yogo, Protonic conduction in $SrZrO_3$-based oxides, *Solid State Ionics* 1992, *51*, 101-107.

[71] P. J. Gellings, H. J. M. Bouwmeester, Ion and mixed conducting oxides as catalysts, *Catal.Today*, 1992, *12*, 1–105.

[72] Bielanski, M. Najbar, Adsorption species of oxygen on the surfaces of transition metal oxides, *J. Catal.*, 1972, *25*, 398-406.

[73] J. Haber, J. H. Block, B. Delmon, Manual of methods and procedures for catalyst characterization, *PureAppl.Chem.* 1995, *67(8)(9)*, 1257.

[74] E. B. Maxted, The Poisoning of Metallic Catalysts, *Adv. Catal.,* 1951, *3*, 129-178.

[75] P. O'Connor, A. C. Pouwels, in: B. Delmon, G. F. Froment (Eds.), *Catalyst Deactivation* 1994, Elsevier: Amsterdam, 1994, p.129.

[76] C. Gates, J. R. Katzer, G. C. A. Schuit: *Chemistry of Catalytic Processes*, McGraw-Hill: New York, 1979.

[77] P. E. Eberly, C. N. Kimberlin, W. H. Miller, H. V. Drushel, Coke Formation on Silica-Alumina Cracking Catalysts, *Ind. Eng. Chem. Proc. Des. Dev.,* 1966, *5,* 193-198.

[78] P. Burtin, J.P. Brunelle, M. Pijolat, M. Soustelle, Influence of surface area and additives on the thermal stability of transition alumina catalyst supports. I: Kinetic data, *Appl.Catal.*, 1987, *34*, 225-238.

[79] S. T. Sie, Consequences of catalyst deactivation for process design and operation, Appl. Catal. A: Gen. 2001, *212*, 129-151.

[80] P. Forzatti, L. Lietti, Catalyst Deactivation, *Catal. Today*, 1999, *52*, 165-181.

[81] L. Trimm, The regeneration or disposal of deactivated heterogeneous catalysts, *Appl. Catal. A: Gen.*, 2001, *212*, 153-160.

[82] J. R. Donovan, R. D. Stolk, M. I. Unland, in: B. E. Leach (Ed.), *Applied Industrial Catalysis*, Vol.2, Academic Press: New York, 1983.

[83] P. W. Tamm, H. F. Harnsberfer, A. G. Bridge, Effects of feed metals on catalyst aging in hydroprocessing residuum, *Ind. Eng. Chem. Proc. Des. Dev.*, 1981, *20*, 262-273.

[84] Langmuir I., The constitution and fundamental properties of solids and liquids, *J. Am. Chem. Soc.*, 1916, *38*, 2221.

[85] R. A. Van Santen, catalysis: closing the molecular and Macroscopic sciences gap, *Chem. Engg. Sci.*, 1995, *50*, 4027-4044.

[86] Kang, H.Ch. Weinberg, W.H., Kinetic modeling of surfaces rate processes, *Surf. Sci.*, 1994, *299*, 755.

[87] Nerskov, J., Chemisorption on metal surfaces, *Rep. Prog. Phys.*, 1990, *53*, 1253.

[88] van Santen, R.A. and Kuipers, H.P.C.E., The mechanism of ethylene epoxidation, *Adv. Catal.*, 1987, *35*, 265.

[89] P. G. Menon, Diagnosis of industrial catalyst deactivation by surface characterization techniques, *Chem. Rev.*, 1994, *94*, 1021.

[90] Zaera, Outstanding Mechanistic Questions in Heterogeneous Catalysis, *J. Phys. Chem. B*, 2002, *106*, 4043-4052.

[91] Zaera, Tuning selectivity in hydrocarbon conversion catalysis, *J. Mol. Catal. A: Chem.*, 2005, *228*, 21-26.

[92] G.M. Schwab, Electronics of Supported Catalysts, *Adv. Catalysis*, 1978, *27*, 1-22.

[93] M. Boudart, G. Djega-Mariadassou, *Kinetics of Heterogeneous Catalytic Reactions*, Princeton University Press: Princeton, 1984.

[94] K. Hayek, R. Kramer, Z. Paal, Review Metal-support boundary sites in catalysis, *Appl. Catal. A: Gen.*, 1997, *162*, 1-15.

[95] D. Murzin, On the kinetic interpretation of metal-support interaction, *J. React. Kin. Catal. Let.*, 1995, *55*, 275-281.

[96] Knözinger, E. Taglauer, in: G. Ertl, H. Knözinger, J. Weitkamp (Eds.), *Handbook of Heterogeneous Catalysis*, VCH: Wenheim, 1997; pp 216.

[97] F. Pesty, H. -P. Steinrück, T. E. Madey, Thermal stability of Pt films on TiO_2 (110): evidence for encapsulation, *Surf. Sci.*, 1995, *339*, 83-95.

[98] S. J. Tauster, S. C. Fung, Strong metal-support interactions: Occurrence among the binary oxides of groups IIA–VB, *J. Catal.*, 1978, *55*, 29-35.

[99] L. Haller, New catalytic concepts from new materials: understanding catalysis from a fundamental perspective, past, present, and future, *J. Catal.*, 2003, *216*, 12-22

[100] S. S. Shaik, What happens to molecules as they react? A valence bond approach to reactivity, *J. Am. Chem. Soc.*, 1981, *103*, 3692-3701.

[101] B. F. Minaev, Spin effects in activation of hydrocarbons The role of triplet states in catalysis, *J. Mol. Catal. A: Chem.*, 2001, *171*, 53-72.

[102] V. M. Khanaev, E. S. Borisova, A. S. Noskov, Optimization of the active component distribution through the catalyst bed, *Chem. Engg. Sci.*, 2004, *59*, 1213-1220.

[103] G. Musie, M. Wei, B. Subramaniam, D. H. Busch, Catalytic oxidations in carbon dioxide-based reaction media, including novel CO_2-expanded phases, *Coord. Chem. Rev.*, 2001, *219*, 789–820.

[104] P. G. Jessop, R. Stanley, R. A. Brown, C. A. Eckert, C. L. Liotta, T. T. Ngo, P. Pollet, Comparing neoteric solvents for asymmetric hydrogenation: supercriticalfluids, ionicliquids, and expanded ionic liquids, *Green Chem.*, 2003, *5*, 123–128.

[105] P. G. Jessop, S. DeHaai, D. C. Wynne, D. Nakawatase, Carbon dioxide gas accelerates solvent less synthesis, *Chem. Commun.,* 2000, 693–694.

[106] P. G. Jessop, Homogeneous catalysis using supercritical fluids: Recent trends and systems studied, *J. Supercritical fluids*, 2006, *38*, 211-231.

[107] R. A. Buyanov, V. V. Molchanov, V. V. Boldyrev, Mechanochemical activation as a tool of increasing catalytic activity, *Catal. Today*, 2009, *144*, 212-218.

[108] M. Pekar, What can kinetics learn from rational thermodynamics, *Chem. Engg. Sci.*, 2004, *59*, 4103-4112.

In: Recent Progress in Chemistry...
Editor: A. K. Haghi, pp. 173-203
ISBN: 978-1-61668-501-0

Chapter 11

PHOTOCATALYSIS: AN EMERGING TECHNOLOGY

Suresh C. Ameta[1], P. B. Punjabi[1], Rakshit Ameta[2] and Indu Bhati[1]
[1]Department of Chemistry, M.L. Sukhadia University,
UDAIPUR-313001 (Raj.) INDIA
[2]Department of Pure and Applied Chemistry, University of Kota,
KOTA-342005 (Raj.) INDIA

INTRODUCTION

The most important challenge for chemists of 21st century is to combat against ever increasing environmental pollution. To have a neat, clean, healthy and green environment, there is an urgent desire to search such an approach, which may be applicable at room temperature, safe in handling economic, ecofriendly, and above all, the main requirement it itself is not harmful any more to environment in any manner. In the last two-three decades, photocatalysis has attracted the attention of scientific community all over the world due to its multifarious applications in various fields like energy conversion, environment, technology, waste water treatment, self cleaning process, decomposition of offensive atmospheric odors as well as toxins, the fixation of CO_2 and the decomposition of NO_X and chlorofluorocarbons on global scale [1-4].

The term "photocatalysis" is composed of two individual parts – photo, which indicates light and the second part is catalysis. Catalysis is the process, where a catalyst/substance participates in modifying/altering the rate of a chemical transformation without any permanent chemical change in itself or its consumption in the end of the reaction. Such substance is known as a catalyst. Generally, catalyst increases the rate of a reaction by reducing the activation energy, but many catalysts are well recognized for retarding effect on the rate of a reaction. Although the term "photocatalysis" has been debated for a long since 1983, at IUPAC level, but it is still reserved for such reactions only, which utilizes light to activate a substance (particularly a semiconductor), which modifies the rate of a chemical reaction without being involved itself. The IUPAC definition of photocatalysis [5] is: "a catalytic reaction involving light absorption by a substrate" seems broad enough to encompass all the preceding contexts. Chlorophyll in plants is a typical example of a natural photocatalyst. The chlorophyll (photocatalyst) is quite different than man-made photocatalyst, as chlorophyll captures sunlight to turn water and CO_2 into O_2 and glucose, but on contrary, photocatalyst creates strong oxidizing agent, hole (h^+) and an electron (e^-) to reduce the substrate.

The field of photocatalysis has been excellently reviewed by Gratzel [6], Pellizetti and Serpone [7], Somorjai [8], Schiavello [9] and Ameta et al [10,11].

MISTAKEN CONCEPT OF PHOTOCATALYSIS

In a photochemical reaction, light always acts as a reactant and never as a catalyst. Light is absorbed by any substrate or catalyst or semiconductor. The energy of light/photon is used to create/generate electronically excited molecules, which never comes out in the same form after the reaction and further, it cannot be used in photoinduced or photoassisted reactions. As per definition of a true catalyst, the rate of reaction is enhanced by reducing the energy of activation (E_a), where the light absorbed cannot alter the energy levels of the reactant or product, or the free energy change of the overall reaction. So the concept of 'photocatalysed reactions' seems to be fundamentally improper. The expressions 'photoinduced' or 'photoactivated' or 'photoassisted' reactions are more accurate. These expressions reflect that light is used as a reactant and not as catalyst. However, the term photocatalysis is still used quite widely.

CLASSIFICATION OF PHOTOCATALYTIC TRANSFORMATIONS

On the basis of simplest experimental evidence the photocatalytic reactions may be gathered into two major classes: (a) First, when the source of light is switched off, even than the reaction goes on for a while, it establishes that one photon enables transformation of an infinite number of molecules i.e., quantum yield > 1 and (b) But in second case, the absence of light immediately checks the reaction. [12] It means, here one photon is required for each molecular transformation, i.e., quantum yield ≤ 1. These are called photoassisted catalysed reactions.

1. Photoinduced catalytic reactions

   a. Photogenerated catalyst

$$\text{Cat.} \xrightarrow{h\nu} \text{Cat.}^* \ (\text{Cat.}^* = \text{Excited state of catalyst})$$

$$\text{Cat.}^* \longrightarrow \text{Cat.}' \ (\text{Cat.}' = \text{Active state of catalyst})$$

$$\text{Cat.}' + \text{S} \longrightarrow \text{P} + \text{Cat.}$$

   b. Photoinduced chain reactions

$$I \xrightarrow{h\nu} I^* \qquad (I^* = \text{Photoinitiator})$$

$$I^* + S \longrightarrow S' + I$$

$$S' + S \longrightarrow S' + P$$

1. Photoassisted catalytic reactions : These reactions requires continuous irradiation.

   a. Photoassisted reaction with a photoassistor

$$Cat \xrightarrow{h\nu} Cat^*$$

$$Cat^* + S \longrightarrow Cat + P$$

   b. Photoassisted reaction with a short lived intermediate in ground state as catalyst

$$Cat \xrightarrow{h\nu} Cat^*$$

$$Cat^* \longrightarrow Cat'$$

$$Cat' + S \longrightarrow Cat + P$$

   c. Photosensitized reactions

$$Cat \xrightarrow{h\nu} Cat^*$$

$$Cat^* + S \longrightarrow Cat + S^*$$

$$S^* \longrightarrow P$$

The photocatalytic reactions can be further classified into two categories on the basis of physical state/appearance of reactants.

(i) Homogeneous Photocatalysis

If catalyst and reactant both are in the same phases in a photocatalytic reaction i.e., gas, solid or liquid, then such reactions are named as homogeneous photocatalytic reactions.

Practical interest in soluble catalyst arises from the desire to develop such systems that are more active, per metal atom and more selective their heterogeneous counterparts. Due to the higher activity, the catalyst may perform its function under milder conditions.

Different dyes / organic substances and coloured coordination compounds are best example of homogeneous photocatalysts. By absorption of radiation energy (vis / UV), one or more components of the system (in excited state) posses geometries, metal-ligand bond orders, acid-base strengths and redox potentials sufficiently different from the corresponding properties of their counterparts in ground state [13,14].

(ii) Heterogeneous Photocatalysis

If the catalyst and reactant both are in different phases in a photocatalytic reaction, then such reactions are termed as heterogeneous photocatalytic reactions. The common case of this kind is a solid photocatalyst in contact with either liquid or a gas phase. Combination of bulk and surface properties reflect the activity and selectivity of this solid photcatalyst.

For homogeneous processes, the quantity 'specific rate' can be used, as all the measurements in determining the rate of reaction are used in normality, molatiry, etc. while for heterogeneous processes other all quantities are appropriate (like volume, mass and/or area). So in such a case, it is much better to determine rate related to active sites of catalyst. As the determination of active sites is not an easy task and therefore, rate in terms of active sites is considered as a rough approximation.

Generally rates are often referred to as turnover number, turnover frequencies or turnover rates. Another commonly used parameter in photocatalysis is quantum yield (ϕ). According to IUPAC – 'Quantum yield (ϕ) is the number of defined events, which occur per photon absorbed by the system. For a photochemical reaction, the integral quantum yield is:

$$\phi = \frac{\text{Amount of reactant consumed or product formed}}{\text{Amount of photon absorbed}}$$

Thus, the quantum yield states that how many molecules (moles) are transformed per number (einstein) of absorbed photons? For homogeneous systems, the determination of the molecules (moles) transformed and the number of absorbed photons is relatively simple and hence, calculation of quantum yield is relatively easy. But such determination of absorbed photons in heterogeneous systems does not exist, because in those systems only some attacking photons are being absorbed, whereas some of the photons are scattered in various ways depending on the particle size. [15] With the help of this feature (the particle size), a method was proposed for the determination of reflected fraction of the total impinging photons. [16] Using this fraction, the actual absorbed photons numbers can be evaluated and then quantum yield of heterogeneous system can be calculated.

SEMICONDUCTOR PHOTOCATALYSTS

A photocatalyst is basically a semiconductor. Semiconductor is a substance, where the energy gap between conduction band (lowest unoccupied molecular orbital LUMO) and valence band (highest occupied molecular orbit HOMO) ranges from 1.5 to 3.0 eV. The energy difference between the valance band and the conduction band is known as the 'band gap' (E_g). On the basis of this band gap, the materials are classified in three categories (i) E_g < 1.0 eV, Metal or conductor, (ii) E_g > 5.0 eV, Insulator or non-conductor and (iii) $E_g \sim 1.5$ to 3.0 eV, Semiconductor. These are given in Figure 1.

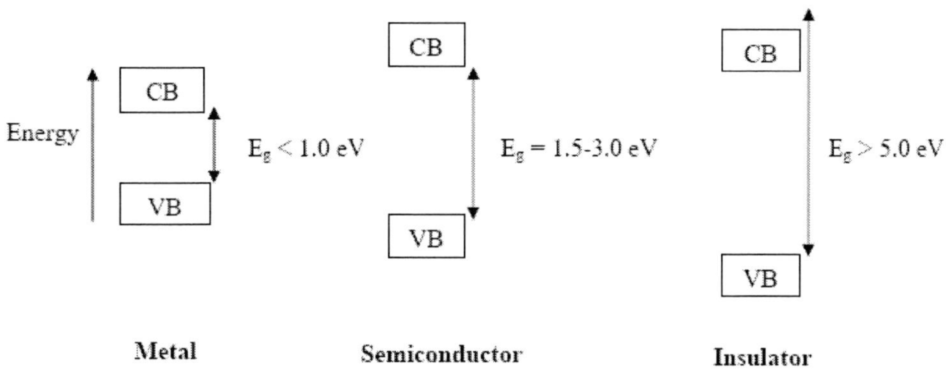

Figure 1. Band Gap and Nature of Material.

The band gaps of various semiconductors are given in Table 1.

Table 1. Band gaps of different semiconductors

Semiconductor	Band gap (eV at 300 K)
ZnS (Wurtzite)	3.91
ZnS (Zinc blende)	3.54
TiO_2	3.20
ZnO	3.03
WO_3	2.60
CdS	2.42
CdSe	1.70

If sufficient energy (equal to or more than band gap) is given to a semiconductor, e^- from the valence bond jump to the conduction band leaving behind a hold in valance band. Therefore, semiconductor may conduct some electric current on exposure to light containing energy equal to or more than corresponding to the band gap of semiconductor. Hence, semiconductors are capable conducting electricity even at room temperature in presence of light.

Mechanism of Photocatalyst

When the photocatalyst is illuminated by light, the energy of photons is utilized by the ev of valence band promoting it to the conduction band while a hole (h^+) is created in valance band by removal of this electron. This process creates 'photo-excitation state'. Thus, a photocatalyst absorbs appropriate radiation (VIS/UV) from sunlight or illuminated light source (fluorescent lamps), producing a pair of electron (e^-) and hole (h^+). This excited e^- can be used for reducing an acceptor substrate whereas the h^+ may be utilized for oxidation of donor molecules.

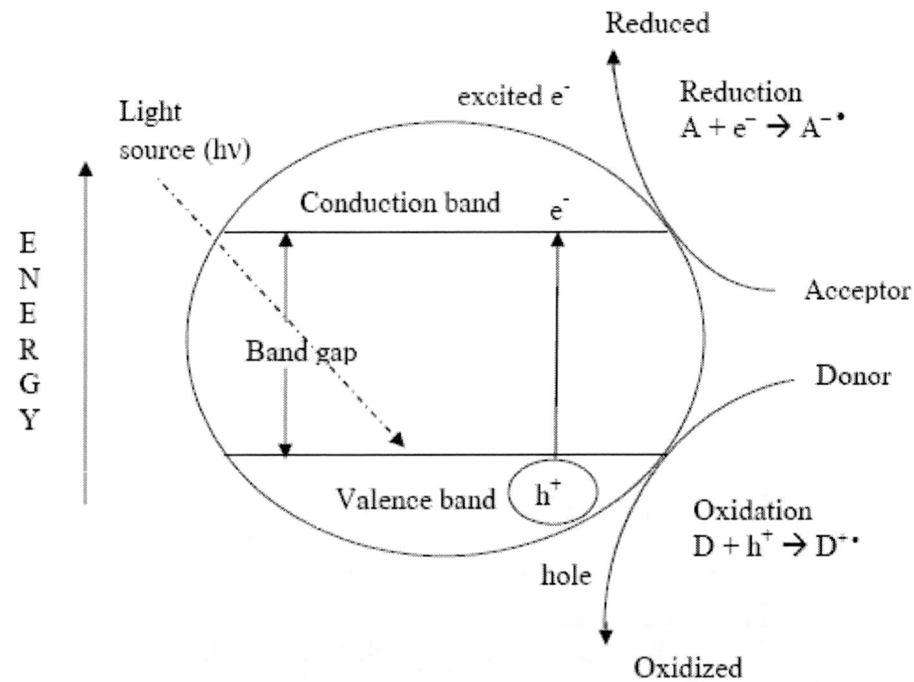

Figure 2. Photocatalysis ($e^- - h^+$ pair generation).

In photogenerated catalysis, the photocatalytic activity (PCA) depends on the ability of the catalyst to create e^- - h^+ pairs. But in absence of any other influence or process, the excited sate may diffuse down the concentration gradient of e^--h^+ pair, as there is an equal possibility for e^- to loose the energy and get back its position in valance band, where hole was present.

(i) Covalent Semiconductors

Si, Ge and III-V compounds have cubic diamond structure where each atom is surrounded by four isoelectronic neighbours. In these types of semiconductors, the valence and conduction bands are formed by overlapping of 4 identical hybrid sp^3 orbitals. Band widths are in the range of 15-20 eV, while band gaps are between 0.2-2.0 eV.

(ii) Influence of the type of overlapping orbitals

The band structure in some cases can be explained on their chemical stoichiometry. Two oxides SnO_2 and TiO_2 (rutile) have same crystallographic structure and band gaps are also of the same order of magnitude: SnO_2 = 3.59 eV, TiO_2 = 3.03 eV.

The difference in thae electronic bands in SnO_2 is due to overlapping of s and p orbitals. Valence band is mainly composed of O_{2p} orbitals while the conduction band consists mainly of Sn_{5s} and Sn_{sp} orbitals. Hence large band size is due to strong s and p character, whereas in TiO_2, conduction band and to some extent valence band have, essentially d-character and therefore, much narrower as compared to in SnO_2 [17].

(iii) Lattice distortion and deviation from stoichiometry

Lattice distortion and deviation from stoichiometry also influence the bands and as a consequence, the band gap is also affected. It can be better explain by the example of SiO_2, where bands are formed by four identical hybrid sp^3 orbitals of Si overlapping with p_x and p_y orbitals of O. Therefore, the width and shape of the band, depends on the Si-O-Si angle. [18] When O ions are missing in the lattice then unbound atomic Si states are introduced at an energy level corresponding to the free atom. When more and more O ions are missing, then these Si levels can overlap to form a new Si-type band.

(iv) Influence of crystallographic structure – rutile v/s anastase

TiO_2 exists in two crystallographic structures i.e., rutile and anastase. In both these forms, an octahedron if O contains a titanium ion at the centre and each O has three coplanar neighbour titanium cations. In rutile form, O ions form a slightly distorted hexagonal compact lattice (hcp) with Ti-O-Ti angle roughly equal to 120° whereas in anastase form, O ions form a cubic face centered (cfc) lattice in which one Ti-O-Ti angle is about 180° while the two others are close to 90°. It is responsible for the difference in their band gaps. The band gap of rutile phase is 3.03 eV whereas anastose phase has 3.2 eV as the band gap.

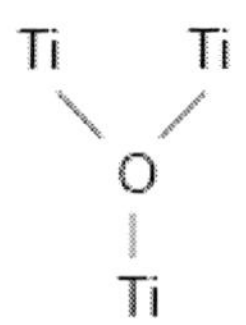

O hcp (Rutile)

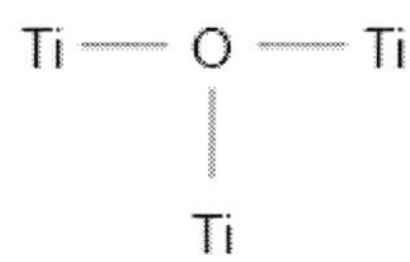

O cfc (Anatase)

SEMICONDUCTORS USED AS A PHOTOCATALYST

(1) Zinc Oxide (ZnO)

Zinc oxide is an inorganic compound with the formula ZnO. It is often called a II-IV semiconductor because Zn and O belong to the 2nd (II) and 6th (VI) group of periodic table, respectively. This semiconductor has favourable several properties: good transparency, high electron mobility, wide direct band gap of ~ 3.3 eV at room temperature [19], luminescence etc. It has n-type character, even in the absence of intentional doping. It may be due to nonstoichiometry but still the subject is controversial. [20] It is inexpensive, relatively abundant, chemically stable, non-toxic and easy to prepare. Because of these advantages, it has strong market opportunities and attracted the attention of a large number of chemists working in the field of semiconductor or photocatalysis. Mathews [21] used it in photocatalytic oxidation reactions of aromatic hydrocarbons containing phosphorus, sulphur, N and halogen. The final products were confirmed as CO_2, H_2O, PO_4^{3-}, SO_4^{2-}, NO_3^- and X^-. Cui [22] and Lu [23] explained the mechanism of photocatalytic oxidation by ZnO. By getting sufficient energy, valance band electrons absorb the energy and jump to the conduction band and this shift produces holes in the valence band. Electrons and holes can move to surface of semiconductor and these can react with other substances. In addition, they can react with O_2 and water to produce strong oxidizing agent like hydroxyl radicals, which oxidize the organic pollutants completely into CO_2, H_2O and N_2. Basic reactions of these process are as follows:

$$ZnO + h\nu \longrightarrow h^+ (V.B.) + e^- (C.B.)$$

$$h^+ + H_2O \longrightarrow H^+ + {}^{\bullet}OH$$

$$e^- + O_2 \longrightarrow O_2^-$$

$$O_2^{-\bullet} + H^+ \longrightarrow HO_2^{\bullet}$$

$$2\,HO_2^{\bullet} \longrightarrow O_2^{\bullet} + H_2O_2$$

$$H_2O_2 + O_2^{-\bullet} \longrightarrow {}^{\bullet}OH + {}^{-}OH + O_2$$

$$H_2O_2 + h\nu \longrightarrow 2\,{}^{\bullet}OH$$

$$\text{Organic molecules} + {}^{\bullet}OH + O_2 \longrightarrow CO_2 + H_2O + \text{other products}$$

Many studies have been carried out on photocatalytic oxidation of various substrates by ZnO. Chun [24] used ZnO as a photocatalyst and xenon lamp as a light source for the photocatalytic oxidation of aniline. Hema et al. described the mechanism for photooxidation of aniline by molecular oxygen where the final product was identified as azobenzene.

$$ZnO \xrightarrow{h\nu} ZnO^* \xrightarrow{\text{Aniline}} ZnO : \text{Aniline}$$

An e^- is transferred from aniline to ZnO resulting in the formation of a cation radicals simultaneously at some other site, where oxygen is adsorbed, the e^- may be transferred to oxygen resulting in the formation of superoxide anion ($O_2^{-\bullet}$). These cation and anion radicals will give rise to the final product azobenzene as follows:

$$C_6H_5\overset{+\bullet}{N}H_2 + O_2^{-\bullet} \longrightarrow C_6H_5\overset{\bullet}{N}H + HO_2^{\bullet}$$

$$C_6H_5\overset{\bullet}{N}H + H\overset{\bullet}{N}C_6H_5 \longrightarrow C_6H_5NH - NHC_6H_5$$

$$C_6H_5NH - NHC_6H_5 \xrightarrow{O_2} C_6H_5N = NC_6H_5$$

Photocatalytic oxidation by ZnO has been used in the treatment of paper mill waste water [25], of 4-nitrochlorobenzene [26], of effluents from pharmacy (discharged from the cephalosporin and penicillin producing industries) [27], of ladnfill leachate [28], etc. The leachate of landfill is an organic waste water with complex compositions. It has many organic pollutants and about 10 kinds of metal ions. In addition to being applied as a photocatalyst, ZnO has also been used to mimic the functions of chloroplasts in the near UV according to the e^- theory of catalysis of photosynthesis [29]. It was also found that Gram –ve Escherichia coli and Gram +ve Lactobacillus helveticus [30] can be inactivated by ZnO exposed to light Kraft E1 effluent was also completely removed by Ag doped ZnO photocatalyst [31].

Photocatalyst is useful not only to remove or decompose organic materials (pollutants), but it can be used for collection of high purity heavy/novel metals i.e., gold and silver from waste water.

Surprising, the silver ions in the rest used photo film developing solution were completely removed in just 15 min using ZnO nanopowder. The photocatalytic elimination of Hg (II) ions from aqueous solution using ZnO powder has been investigated by Domenech and Andres [32]. It was reported that the oxygen was produced under U.V. light from aq. suspension of ZnO containing Hg (II), Ag (I) and CrO_4^{2-} ions as conduction band e^- acceptors [33]. It has been shown that the main source of the oxygen was oxidation of water by these ions.

$$2\,Hg_2^{2+} + 2\,H_2O \longrightarrow O_2 + 4\,H^+ + 2\,Hg$$

$$4\,Ag^+ + 2\,H_2O \longrightarrow O_2 + 4\,H^+ + 4\,Ag$$

It was also observed in this study that the rate of oxygen evolution decreases in the order Hg (II) > Ag (I) > CrO_4^{2-} and it depends on the efficiency of these e^- acceptors to scavenge the photogenerated e^- in the conduction band of the semiconductor.

This concept may be used in decreasing COD and in increasing levels of BOD/O_2.

As we are all aware that CO is dangerous to human life after a particular limit and it is very difficult to decompose or remove it. Steinbach and Harbarth [34] performed the oxidation of CO and desorption of oxygen irradiating of ZnO single crystals by UV radiation under ultra high vacuum conditions.

ZnO semiconductor can also be used in photocatalytic reduction of CO_2 generating formic acid, formaldehyde and methyl alcohol [35].

$$ZnO + h\nu \longrightarrow e^-(CB) + h^+ ($$

$$H_2O + 2\,h^+(VB) \longrightarrow \tfrac{1}{2}\,O_2 + 2\,H^+$$

$$CO_2 + 2\,H^+ + 2\,e^-(CB) \longrightarrow HCOOH$$

$$HCOOH + 2\,H^+ + 2\,e^-(CB) \longrightarrow HCHO + H_2O$$

$$HCHO + 2\,H^+ + 2\,e^-(CB) \longrightarrow CH_3OH$$

$$CH_3OH + 2\,H^+ + 2\,e^-(CB) \longrightarrow CH_4 + H_2O$$

The photooxidation of organic substances like isopropanol [36], phenol [37], acetamides [38] etc. over ZnO has been studied. A cellulose bleaching effluent can also be degraded by photocatalysis using TiO_2 and ZnO supported on glass Rasching rings [39]. The effluent was completely decolorized and the total phenol contents was reduced by 85% on 120 min treatment with both these catalysts. The photocatalytic activity of commercial ZnO powder has also been compared with that of TiO_2 [40] and it was found that the photodegradation rates were high for ZnO, suggesting that it absorbs larger fraction of the solar spectrum as compared to TiO_2.

(2) Titanium Dioxide (TiO_2)

Titanium dioxide is also known as titanium (iv) oxide or titania. IT is the naturally occurring oxide of titanium with chemical formula TiO_2. TiO_2 occurs in nature as rutile, anatase and brookite, but is also found additionally as two high pressure forms (i) a monoclinic baddeleyite – like form and (ii) an orthorhombic α-PbO_2- like form. Both found recently at the Ries crater in Bavaria [41,42] The most commonly available rutile form is the most stable form [37] of TiO_2. Particularly, anatase form of TiO_2 is used as a photocatalyst under ultraviolet light. Recently, it has been investigated that if TiO_2 is spiked with nitrogen ions or doped with metal oxide, then it may act as photocatalyst under visible as well as in UV light. The photocatalytic activity of TiO_2 was discovered by Fujishima in 1972 [43]. This process operating on the surface of the titanium dioxide was called the Honda-Fujishima effect [44] in their ....

Fujishima and his group discovered one more unique characteristic of TiO_2, i.e., superhydrophilicity [45], which is a phenomenon, in which a surface repel the water. Thus, the surface doesn't allow the moisture to form beads on it, which is a center of attraction for dust particles and pollutants. With the help of this property of TiO_2, he developed a new era of self cleaning glass and anti-fogging coatings. This phenomenon has been discussed later on in this chapter. When TiO_2 is incorporated into exterior building materials, such as ceramic tiles, paving stones or in paints, it reduces the concentrations of airborne pollutants i.e., VOC (Volatile organic compounds) and NO_x as well as SO_x [46] in presence of sunlight. Itlacementi Group developed a photocatalytic cement using TiO_2 and it was included in Time's Top 50 Inventions of 2008 [47]. Although TiO_2 can oxidize organic materials directly but the strong oxidative potential is due to its positive holes in valence band. It is also used as a hydrolysis catalyst. In presence of light, TiO_2 also cleave water into hydrogen and oxygen.

$$H_2O + n\,h\nu \xrightarrow{TiO_2} H_2 + \tfrac{1}{2} O_2$$

It gives a totally newer concept of energy production. The hydrogen may be collected and used as a fuel as it has been advocated very strongly as the fuel of future. Efficiency of this process can be greatly improved by doping the titanium oxide with carbon [48]. To the increase catalytic activity of rutile samples platinization is another good technique [49], as effective optical band steadily decreases with platinization. In phenol degradation, acidic pretreatment of TiO_2 enhances its photocatalytic properties [50].

TiO2 has a great potential in industrial technology for detoxification, degradation or remediation of waste water due to several factors like –

i. TiO_2 has a large band gap of 3.20 eV.
ii. This photocatalyst is inexpensive and has a high turnover.
iii. TiO_2 can be supported on suitable reactor substrates.
iv. Use of TiO_2 avoids direct photolysis as it forms photocyclized intermediate products i.e., $^{\bullet}OH$, hydroxyl and H_2O_2.

$$TiO_2 \xrightarrow{h\nu} h^+(VB) + e^-(CB)$$

$$2\,h^+ + H_2 \longrightarrow \tfrac{1}{2}\,O_2 + 2\,H^+$$

$$O_2 + 2\,e^- + 2\,H^+ \longrightarrow H_2O_2$$

$$H_2O_2 + Fe^{2+} \longrightarrow {}^{-}OH + {}^{\bullet}OH + Fe^{3+}$$

i. Irradiation of TiO_2 surface can produce a highly amphiphilic surface [51-53] even under weak light irradiation. Such amphiphilic surfaces have anti-fogging and self-cleaning properties and,

ii. The process occurs under ambient conditions i.e., very slowly, however, the direct UV light exposure increases the rate of reaction.

iii. Complete oxidation of some proteins by TiO_2 photocatalyst to CO_2 has been studied by Muszkat et al. [54]. De Heredia et al. [55] made a clear difference ion the oxidation of p–hydroxybenzoic acid by UV radiation alone and a combination of TiO_2 / UV radiations. The rate constants obtained for the TiO_2 / UV system are clearly greater (Between 220 and 435%) than those obtained in the direct UV photooxidation alone. Furthermore, it has been shown that the addition of water, fill the hydroxyl ion reservoir on the TiO_2 surface and hence, the presence of water in gas phase strongly affects the performance of titanium photocatalyst [56].

(3) Cadmium Sulphide (CdS)

CdS is a direct band-gap semiconductor with a band-gap of 2.42 eV at 300 K. The conductivity increases, when irradiated with light [57]. When CdS is added to sulphide solutions and irradiated with the light hydrogen gas is generated [58]. It may be very useful in energy conversion and storage. The photocatalytic reaction involves generation of e^- - h^+, which is pairs followed by these reacting with water and sulphide, respectively

$$H_2S \xrightarrow{CdS} H_2 + S \qquad \Delta H_f = +9.4\ \text{kCal Mol}^{-1}$$

Production of an e^- - h^+ pair

$$CdS + h\nu \longrightarrow e^- + h^+$$

Reduction of water

$$2\,e^- + 2\,H_2O \longrightarrow H_2 + 2\,OH$$

Oxidatation of sulphide

$$2\,h^+ + S^{2-} \longrightarrow S$$

Borgarello [59] observed the oxidation of H_2S (a byproduct of coal and petroleum industry) to sulphur at CdS semiconductor surface.

CdS carried by carbon nano-tubes (CdS / CNTs) is a new candidate for photocatalytic degradation of organic dyes [60]. Obviously the capability for degradating organic molecules by CdS / CNTs is higher than that of separated CdS alone. If CdS is mixed with TiO_2 suspensions, then the effect of mixed semiconductors on the photodegradation of chlorinated organics i.e., 2-chlorophenol increases drastically [61].

Due to small band gap in CdS (yellow color), it is able to undergo oxidation were in the visible region, but it is usually unstable and photodegrade with time [62] and therefore, it is generally used as a component of mixed semiconductor. It has been found that CdS / TiO_2 semiconductor composite gives better result than CdS or TiO_2 alone [63]. Heat treatment of CdS with KCl [64] and KBr [65] enhances its photocatalytic activity. Rhodium loaded CdS particulate system is a very good photo catalyst for the photo catalytic transformation of CN^- to SCN^-. [66] Wang et al. [67] carried out the photo catalytic reduction of environmental pollutant Cr (VI) over CdS powder under visible light illumination. Photocatalytic oxidation of phenol [68], to sulphate [69], indoles [70], sulphite benzhydrol [71] and photoreduction of methyl orange [72], {$[Pt\,Cl_6]^{-2}$} [73], molecular nitrogen [74], photo bleaching of aberchrome 540 [75], anionic and cationic dyes [76], photo hydrodimerization of 2, 5-dihydrofuran [77], etc. has been studied well in presence of combination of CdS with different components i.e., ZnS, TiO_2, zeolite, Ru (II)-dinitrogen complex [Ru (HEDTA) (N_2)] and somwtimes Pt / RuO_2, and with CdS alone.

(4) Zinc Sulphide (ZnS)

It is typically encountered in 3 forms. The first more stable cubic form sphalerite [78], second hexagonal form wurtzite [78] and third one is its tetragonal form, which is a very rare mineral polhemusite (Zn, Hg)S. Both sphalerite and wurtzite are intrinsic and wide band gap semiconductors. The sphalerite cubic form has a band gap of 3.54 eV at 300 K whereas the wurtzite hexagonal form has a band gap of 3.91 eV. It can be doped as both n-type semiconductor and p-type semiconductor, which is ordinarily unusual for the II-VI semiconductors. The photo catalytic oxidation with efficient reductive H_2 evolution of organic substances like triethylamine, tetrahydrofuran and methanol by colloidal ZnS [79] has been reported. In this process, the dimeric products 2,3-bis (diethylamine) butane 2,2'-bistetrahydrofuryl or ethylene glycol were obtained, respectively. In this reaction, ZnS controls one e^- transfer from organic substance giving free carbon radicals, which leads to carbon-carbon bond formation [80] or their addition reaction with carboxyl compounds [81]. Nano-crystalline ZnS is also used for the photo oxidation of organic compounds [82].

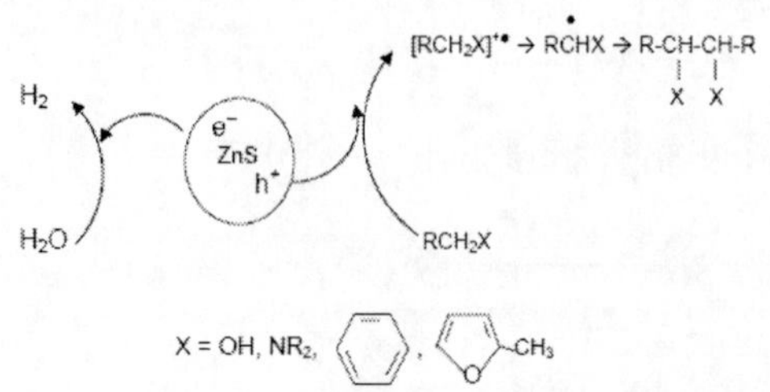

Photoformed e^- on ZnS nanocrystalline reduces water to hydrogen, while the photoformed hole oxidizes aliphatic amines, alcohols and some alkylated aromatic compounds leading to coupling reaction. The surface of ZnS and PbS were modified by interfacing PbS on ZnS and ZnS on PbS nanoparticles by Anil and Anshuman [83]. They found better results in photocatalytic activity of the semiconductors after treatment. CdS-ZnS has been used as a photocatalyst also for photodehyrodimerisation of 2,5- dihydrofuran [84]. Ren et al. [85] reported the effect of surface charge on reductionrate of methyl orange photocatalyzed by zinc sulphide. Clavdia et al. [86] demonstrated the efficient photocatalytic degradation of p-nitrophenol (p-NP) and acid orange 7 (AO7) using ZnS nanocrystals (~3 to 5 nm diameter) with > 50% yield. The photocatalytic properties of the prepared ZnS was also evaluated by decomposing methyl orange (MO) [87]. Photo active ZnS / TiO_2 nano-composites [88] enhanced visible-light photocatalytic degeneration of the aqueous parathion. In this context, efforts have been made for enhancement the photocatalytic performance of ZnS by fabricating $ZnS/Zn_3(PO_4)_2.4H_2O$ heterostructured hemispheres taking rhodamine-B as a voided system [89]. The enhancement in photocatalytic performance may be attributed to the high specific surface area, larger size and the hetero structured architectures. ZnS-CdS has also been used for the removal of crystal violet from industrial effluents [90]. It has been found that complete degradation of phenol is possible in a reasonable time (i.e., less than 5 hrs) when concentration of phenol is $\leq$ 100 ppm [91]. ZnS has also been used for photocatalytic reduction of nitrite and nitrate to ammonia. The photocatalytic activity of hydrophilic ZnS nano-crystals with narrow size distribution was also investigated for the liquid phase photocatalytic degradation of basic violet SBN (BVS) dye under UV irradiation [92].

(5) Ternary and Quaternary Compounds as Photocatalyst

The active ingredients currently being used as catalyst in the catalytic converters in developed countries are noble metals such as Pt, Pb, Rh etc. The alternative catalysts for these noble metals and binary chalcognides may be in the form of mixed metal oxides, which has been paid moderate attention. The main drawback associated with binary semiconductors are that majority of them are alter white or off white in colour. Therefore, they absorb only U.V. light and as a result, can not utilize a major part of solar insolation. Since most of ternary oxides are coloured in nature and hence, they absorb UV as well as visible light.

Zirconium and germanium phosphates are well investigated in heterogeneous photocatalytic oxidation of naphthalenes to phthalic anhydride. [93] Monaci et al. [94] reported photooxidation of some organic compounds like 1, 2- dimethylbenzene, 1,4-dimethylbenzene or 1, 2, 4, 5-tetramethylbenene in a polar solvent using zirconium phosphote as a photocatalyst. Synergistic dehydration of rhodamine-B (Rb-B) by combining electrooxidation and photocatalysis has been performed using of $ZnWO_4$ [95]. Photocatalysis using ZnO-Fe_2O_3 coupled catalyst can remove partially the organic matter dissolved in Kraft–Black liquor obtained from pinus wood cooking [96]. Sharma et al. [97] examined the catalytic activity of transition metal oxides and mixed oxides. They reported the decomposition of H_2O_2 at higher temperatures from 313 K to 343 K over transition metal oxides, Fe_2O_3, CoO and TiO_2 as well as in the spinal system like $FeCo_{1.5}$ $Ge_{0.5}$ and $Ga_{1-X}Fe_XNiCoO_4$. Some examples of ternary metal oxides are $NdCoO_3$, $MnFe_2O_4$, Bi_2WO_6,

$SrTiO_3$, $LaMeO_3$, $CeFeO_3$ etc. Jose [98] applied $LaCrO_4$ for the treatment of coloured waste water contaminated by azure-B, methylene blue and alizarin red. $LaFeO_3$ was also investigated as photocatalyst against brilliant green, methylene blue and alizain red dyes. Sin et al. [99] studied photocatalytic degradation and kinetics of orange-A using nanosized Sn (IV) WO_2 as a photocatalyst. Photocatalytic bleaching of azure-B dye was observed over lanthanum chromium oxide $LaCrO_3$ catalyst, which has been synthesized using coprecipitation method and specific heating cycles [100]. Photocatalytic degradation of benzene in the gas phase was achieved for the first time with a nanocrystalline $CaSb_2O_5(OH)_2$ photocatalyst [101]. SiO_2 / TiO_2 binary and boron SiO_2 / TiO_2 ternary oxides are very effective photocatalysts in the oxidative decomposition of trichloroethylene [102]. In NbO_4 nanoparticles in the range of 10-30 nm [103] offers a high photocatalytic activity under illumination with visible light. Although ternary and mixed metal oxides have been given very little attention but it is a very important area of researches in photochemistry. There are many mixed oxides, which have been synthesized using various methods, characterized using different techniques and their photocatalytic activity has been studied by numerous researchers. some examples of these oxides are beta – $ZrMo_2O_8$ [104], orthorhombic perovskites $LnVO_3$ and $Ln_{1-X}Ti_XVO_3$ (Ln = Ce, Pr and Na) [105], $LnVO_4$ and $LnMo_{0.15}V_{0.85}O_4$ (Ln = Ce, Pr and Nd) [106], nano–$GdCoO_3$ [107], nano–$Ce_{1-x-y}Ti_xPdyO_{2-3}$ and nano–$Ce_{1-x}PdxO_{2-3}$ [108], Ba $Bi_2Mo_{4-x}W_xO_{16}$ (0.25 < X < 1) [109], $BiNbO_4$ and $BiTaO_4$ [110], Li $Bi_4M_3O_{14}$ (M = Nb, Ta) [111], $Ti_{1-x}M_xO_{2\pm d}$ (M = W, V, Ce, Zr, Fe and Cu) [112], $Pb_3Sn_2O_6$ nanocrystals [113], monoclinic nanosized $BiVO_4$ [114], $SiW_{12}O_4$ [115] etc.

APPLICATIONS OF PHOTOCATALYSIS

Photoctalyst has been found useful not in only oxidation and degradation of some organic molecules, but also in the synthesis of organic compounds, isomerization / transformation, waste water treatment, self cleaning, sterilization, etc. Some of the important applications are described here.

(1) Waste Water Treatment

Anthropogenic activities include rapidly growing industrialization, series of new constructions, many fold increase in transportation, aerospace movements, developmental and enhancement in technologies i.e. nuclear power, pharmaceutical, pesticides, herbicides, agriculture etc. These all are most desirable activities for human development and welfare, but sidewise also lead to generation and release of objectionable materials into the environment. Thus, it polluted the whole environment making our life quite miserable on this beautiful earth. The situation, if not controlled timely, would become a malignant problem for the survival of mankind on the earth. Many rivers are being polluted by effluent water from industries and domestic sectors. This creates a problem for aquatic life by turning water of no use. So it is utmost necessary to solve this problem of water pollution.

Commonly used waste water treatment processes include adsorption, oxidation, biological, chemical and physical methods. These methods suffer from the one or other drawbacks. However, photocatalysis provides more cleaner and environment healthy / benign method for waste water treatment. [116] Some examples for waste water treatment are given below:

(a) Dyes

Around 10^9 kg and more than 10,000 different synthetic dyes and pigments are produced annually world wide and used extensively in dyeing and printing industries. It is estimated that about 10% dyes are lost in industrial waste water and pollute the water resources. Generally, most of dyes are toxic, carcinogenic, highly soluble and non-degradable under the typical aerobic conditions found in conventional biological treatment systems. This colored water cannot be used for any useful purpose like drinking, cleaning, cooling, irrigation etc. Each method of waste water treatment has its own limitations. In this context, photocatalytic treatment of waste water seems to be a promising technique. Photocatalytic transformation converts the colored water into colorless water, which can atleast be used for cleaning, washing, cooling, steam based projects (I.e. power plants) and irrigation etc. Since 1970, till now, a great evolution can be seen in the range of semiconductor photocatalyst. Different photocatalysts are used in decolorization of coloured water. These are –

TiO_2 : Brown and Darwent [117] carried out photoreduction of methyl orange dye using TiO_2 as a photocatalyst. Mu et al. [118] investigated TiO_2 mediated photocatalytic degradation of orange-II in presence of Mn^{2+} in solution. Papic et al. [119] suggested the advanced oxidation process (AOP) for the treatment of waste water containing azo dyes. The photocatalytic degradation of ethyl violet in aqueous TiO_2 suspension was studied by Chen et al. [120]. It has been shown in many cases that photocatalytic degradation of water contaminants can be enhanced by decreasing particle size as it increases its surface area and hence, it increases the number of active sites available on the surface of photocatalyst. The photocatlytic degradation of congo red catalysed by nano-sized TiO_2 was observed by Wahi et al. [121]. Photodegradation, decolorization and demineralization of malachite green in aqueous suspension of TiO_2 nonaparticles under aerated conditions has been reported by Kominami et al [122].

Vinodgopal et al. [123] made a photosensitization approach for the degradation of textile 920 dye and acid orange. Liu et al. [124] investigated the mechanism of photoxidation of alizarin red in TiO_2 dispersion under visible light while Mrowtez and Selli [125] observed the effect of iron species in photocatalytic degradation of a azo dye in TiO_2 suspension. Similarly Rao et al. [126] measured the influence of metallic species of TiO_2 for the photodegradation of dye and dye intermediates. Zhiyong et al. [127] synthesized and observed the stabilization of TiO_2 on thin polyethylene films (LDPE). It was reported that LDPE-TiO_2 film was able to mediate the complete decolorization of orange-II and it is seven times more active catalytically than TiO_2 suspension only. The photodegradation of eosin Y, an anionic xanthene fluorescent dye, has also been investigated in aqueous heterogeneous solution containing TiO_2 [128]. It was found that the disappearance of the organic molecule follows approximately a pseudo-first order kinetic according to the Langmuir-Hinshelwood model. When eosin Y was treated with TiO_2, its toxicity decreases up to 85% [128] due to photo oxidation. The use of aqueous TiO_2 suspension may improve the efficiency of detoxification of waste water containing hazardous chemicals in presence of solar radiations [129,130]. The presence of TiO_2 (as photocatalyst) showed several advantages, such as reduction of time, presentation of pulp viscosity, increase of the selectivity during the photo bleaching and decrease in the consumption of the bleaching chemicals.

ZnO: Kraft-black liquor is an important effluent from pulp and paper industries. It is easily bleached photocatalytically by ZnO photocatalyst [131]. Photocatalytic degradation of brilliant green and bromopyrogallol over ZnO were investigated by Ameta et al. [132] and Sharma et al. [133], respectively. The effect of the transition metal ions on bleaching of some dyes over ZnO semiconductor has also been studied [134]. Photobleaching of rose Bengal [135] and photodegradation of some textile dyes [136] over ZnO were also reported. Presence of surface charge on the ZnO photocatalyst affects favourably the rate of photocatalytic reduction as established by Zhang et al. [137] and Bhandari et al. [138] in case of photocatalytic degradation of methyl orange and crystal violet. With the decreasing particle size, the photocatalytic activity increases. In this context, Ye et al. [139] reported synthesis of ZnO nanoplatelets as thin as 10 nm, which shows higher efficiency in photodegradation of methylene blue and eosin Y. A reduction in chemical oxygen demand (COD) and removal of color was also achieved in this case.

Ameta et al. [140] used ZnO in presence of surfactant and reported an enhancement in photodegradation of orange–G dye. Photocatalytic degradation of azo dye [141] and photocatalytic bleaching of basic blue [142] in presence of ZnO has also been reported. Amarnath dye [143] bleaches completely in presence of light and ZnO. Photooxidative degradation of reactive orange-60 dye was observed using gas chromatography-mass spectrometry by Bilgi and Demirr [144] while Hashat et al. [145] made a comparative study of photocatalytic degradation of a cationic and an anionic dye. Photocatalytic degradation of malachite green [146], orange-G and rose Bengal [147], crystal violet [148], brilliant red dye and textile waste water [149] has also been reported over ZnO photocatalyst.

CdS and ZnS: Photoreduction of methyl orange dye sensitized by CdS was investigated by Mills and Williams [150] using EDTA as an e^- donor. Photocatalytic degradation of malachite green dye was studied over CdS-TiO_2 [151]. Takizawa et al. [152] compared the photocatalytic behavior of CdS-suspended aerated aqueous solution of methylene blue and rhodamin-B. Photocatalytic reduction of methyl yellow on CdS nanoparticles mediated in reverse micelles was reported by Zang and Shen [153]. Ren et al. [154] reported the effect of surface charge on reduction rate of methyl orange photocatalyzed by zinc sulfide. In efforts to enhance the photocatalytic properties Kobayashi et al. [155] prepared alumina supported ZnS-CdS catalyst. The effect of surface charge on reduction rate of methyl orange photocatalysed by ZnS solutions has also been studied [156].

(b) Chloro Compounds and Phenols:

Although chlorinization is used for the treatment of water, but it is also true that when Cl atom is attached within a moiety of an organic molecule, it also creates a lot of harmful effects on human body as well as on aquatic life. Photocatalysis is the best method to decompose the chloro compounds, as this method does not leave any harmful chemical. It is done either by free radical mechanism (cyclic process in Homogeneous photocatalysis) or by e^-/h^+ redox reactions (reusable photocatalyst in heterogeneous photocatalysis). Both ways provide great benefit in terms of saving time, money, efforts, energy and keeping environment healthy as it is a green chemical pathway, which does not utilize any toxic or hazardous chemical.

The photocatalytic degradation of biphenyl and chlorobiphenyls in the presence of TiO_2 is a last decade issue [157]. In 1999, the photocatalytic degradation of 4- chlorophenol in water over TiO_2 powder and TiO_2 / CdS powder has also been achieved. The degradation of chlorobenzene [158], p-dichlorobenzene [159], monochloro acetic acid [160], trichloroacetic acid[161] and chlorobenzoic acid [161] is not a challenging job now due to photocatalytic degradation. Environmental enemies like polyvinyl chloride and polyurethane are also photodegradable now [162]. Photocatalytic degradation of chloral hydrate in aqueous suspension of ZnO has been reported by Tanaka et al. [163]. Chlorophenols are toxic in nature and since these are resistant to biodegradation, considerable efforts have been made to develop alternative treatment methods for the destruction of chlorophenol wastes. Serpone et al. [164] reported the photocatalysed oxidation of phenol, 2-chlorophenol and pentachlorophenol on coupled semiconductors such as CdS/TiO_2. The use of coupled semiconductor leads to an increase in the rate of photocatalytic reactions as compared to TiO_2 alone.

Photodegradation of dichloromethane, tetrachloroehtylene and 1, 2-dibromo-3-chloropropane in aqueous suspension of TiO_2 under sunlight was observed by Halmann et al. [165] The photocatalytic degradation of $CHCl_3$, $CHBr_3$, CCl_4 and CCl_3COOH was investigated by Choi and Hoffmann [166] in aqueous TiO_2 suspension. A common intermediate, the trichloromethyl radical $^{\bullet}CCl_3$ was involved in the degradation of each substrate expect for CCl_3COOH. In the absence of dissolved oxygen, $CHCl_3$ and $CHBr_3$ are degraded into carbon monoxide and halide ions, while in the presence of oxygen. $CHCl_3$ is degraded into CO_2 and halide ions.

Trichlorobenzene [167] can be photocatalytically decomposed using TiO_2 supported on nickel poly (tetrafluoroethylene) composite plate. In case of dichloro acetic acid [168], it has been observed that Fe-doped TiO_2 colloids show higher yields than pure Q-sized TiO_2 prepared analogously. Along with heterogeneous photocatalysis. Fenton and Photo-Fenton process have been also examined for the degradation of 2-chlorophenol [169]. The photocatalysed transformation of chloroaromatic derivatives such as chlorophenol and dichlorobenzene [170,171] on ZnO which involves the formation of an adduct between $^{\bullet}OH$ radical and aromatic compound.

(c) Insecticides / Pesticides / Herbicides / Agricultural Products

Exponential increase in population produced requirement of larger quantity and better quality food. To increase quantity of the food and to reduce the loss of crop by any calamity i.e., attack of termites, pests, insects, herbs and other enemies. farmers use various chemicals. They also prefer chemical treatment for early ripening of fruits and flowers, long freshness of vegetables, to check dropping of fruits, flowers and vegetables. Use of chemicals such as pesticides, insecticides, herbicides, acetylene has been increased to this extent that they are not consumed completely by crop. Thus, keep their presence in crop as well as in earth is felt. These chemicals are, however, toxic and harmful to human being, animals and our environment also. Most of the pesticides and herbicides are non-biodegradable. But now degradation and decomposition of these harmful and hazardous chemicals is possible due to photocatalytic degradation.

The insecticide chlordane is used for control of termites, ants, wasps and cockroaches. It is of serious mammalian toxicity, both chronic and cumulative, due to accumulation in lipid containing organs. It causes damage to liver and kidneys. It is very toxic to fish and bees [172]. Effective decomposition of chlordane was achieved by photo catalytic degradation in aqueous dispersions of TiO_2 [173].

The fungicide fenarimol is widely used for controlling powdery mildew. It acts by inhibition of ergosterol biosynthesis, under sunlight in natural environmental conditions. Fenarimol in aqueous solution is decomposed slowly over a period of days, because of its low solubility in water. Degradation products due to photolysis were studied in organic solvents. The products were tentatively identified which may be formed due to cleavage of the bond between the quaternary carbon and either the chloroaromatic or the pyrimidine ring [174].

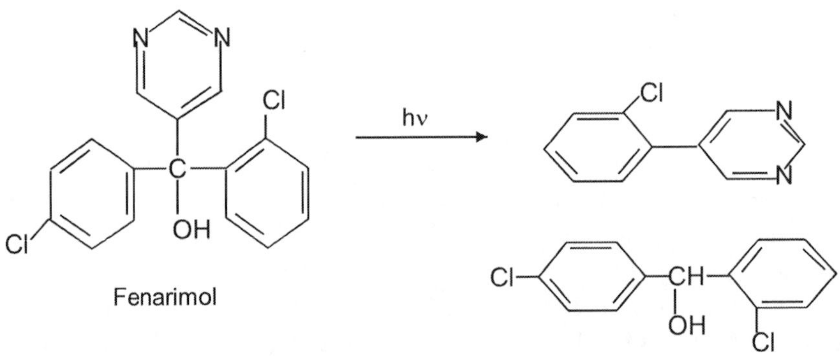

The photodegradation of trimethyl phosphate may be considered as a model reaction for the light-induced decomposition of organophosphorus pesticides and nerve gases. Products of the UV photolysis of trimethyl phosphate were obtained stepwise degradation, leading to dimethyl phosphate, methyl phosphate, and orthophosphate, as well as formaldehyde, formic acid, CO and CO_2. The initial step probably involved hydrogen abstraction, followed by the dimerisation of radicals so produced.

The dimer was detected by mass spectrometry. In the presence of oxygen, photolysis yields were much enhanced, may be possibly due to involvement of OH radicals [175].

$$(CH_3O)_3PO + OH \longrightarrow (CH_3O)_2POCH_2 + H_2O$$

Prapazine is a pre-emergence herbicide of low toxicity in mammals [176]. Its direct photolysis in distilled water is well known under illumination with stimulated sunlight, using a xenon lamp. The rate of photolysis may be enhanced by using proper photocatalyst.

Various researchers have been reported on photocatalytic degradation of pesticides and insecticides by modification of different photocatalysts i.e. TiO_2 in presence of formulating agents for imidacloprid [177], Nayzeolite-supported TiO_2 for omethoate [178], zeolite HY and Hβ supported TiO_2 for 2, 4-dichlorophenoxy acetic acid (DPA) and monocrotophos (MCP) [179], ZnO nanocrystals for insecticide diazinon [180].

Significant detoxification of a commercial pesticide, metasystox can be achieved in only 3h of solar irradiation due to the nearly complete elimination of the active compound, methyloxy-demeton [181]. The degradation of carbendazim pesticide was reported by combination of TiO_2 based photocatalysis and ozonation [182]. Carbendazim was quickly degraded by the combined photocatalytic ozonation process which followed pseudo-first order kinetics. The study also demonstrated that the combination of TiO_2 based photocatalyst and ozonation will overcome the disadvantage of the selectivity of ozonation and the lower efficiency of TiO_2 photocatalysis. Recently, the homogeneous and heterogeneous photocatalytic degradation of imidacloprid, a systematic chloronicotinoid insecticide, has been investigated in aqueous solutions using artificial UV-A or visible illumination [183]. Imidacloprid degradation was also accompanied by the formation of several reaction by-products as confirmed by GC / MS analysis, while ammonium, nitrate and chloride ions have been detected in the liquid phase as mineralization products. By-products were less ecotoxic to marine bacteria than the insecticide itself.

Muszkat et al. [184] observed titanium dioxide photocatalyzed degradation of pesticides in rinse water of agricultural sprayers and in heavily polluted well water under sunlight. The photodegradation of 2-phenylphenol (used as fungicide) can be carried out in presence of TiO_2. The ZnO was found to be more efficient in this case, but it cannot be used in acidic solutions. In both the cases, the main photoproducts identified are – hydroquinone, p-bezoquinone, phenyl hydroquinone, phenyl bezoquinone, 2, 2- and 2, 3- dihydroxy biphenyls [185]. Photocatalysis has also been used for degradation of pesticides like permethrin [186]. Photocatalytic degradation of p-aminoazobenzene (PAAB) and p-hydroxyl azobenzene (PHAB) [187] using various heat treated TiO_2 has also been reported. The study showed that anatase form of TiO_2 annealed at 600-650° C can efficiently photocatalyse the degradation process of PAAB and PHAB as compared to the rutile form, even if annealed above 700°C. The intermediates recognized by UV-visible spectroscopy during the process of degradation of PAAB is o- and p-nitro aniline and for the PHAB, these were o- and p- nitrophenol.

The nitrogen-doped TiO_2 crystalline nanopowder has been used for photocatalytic degradation of the herbicides RS-2-(4-chloro-o-tolyoxy) propionic acid (mecprop) and (4-chloro-2- methylphenoxy) acetic acid (MCPA) [188] using various light sources. When two commercial mixtures of pesticides (Folimat and Ronstar) and 2 fungicides (pyrimethanil and triadimenol) were treated for photocatlytic and biological degradation [189], it was found that the photocatalytic method was able to degrade dicofol, tetradifon, pyrimethanil, triadimenol and the components of Ronstar with the exception of oxadiazon. The photocatalytic method eliminated toxicity of pyrimethanil and reduced the toxicity of triadimenol by 90%, Ronstar by 78% and Folimat by 87%.

Under natural sunlight and at pilot plant scale, photodegradation of eight pesticides belonging to different chemical groups like azoxytrobin, kresoxim-methyl hexaconazole, tebuconazole, triadimenol and pyrimethanil (fungicides), primicarb (insecticide) and propyzamide (herbicide) [190] in leaching water was reported using the tandem ZnO / $Na_2S_2O_8$ as photosensitizer / oxidant and compound parabolic collectors. The use of $Na_2S_2O_8$ implies a significant reduction in treatment time showing a quicker reaction time than ZnO alone. The complete degradation of pesticides occurs from 60 to 120 min. A statistical design of experiment with desirability function was invented for the removal of organophosphorus pesticides from aqueous solution by low cost photocatalyst [191]. Indole is one of the most important component of industrial and agricultural activities. It is very toxic to aquatic organisms and may act as a carcinogen if exposure is chronic. The response surface methodology (RSM) [192] has been developed for the optimization and modeling photocatalytic degradation of indole in UV / TiO_2.

Although there are many methods for water treatment processes and removal of pesticides like (i) Coagulation / Flocculation, (ii) Conventional treatment, (iii) Adsorption process, (iv) Air stripping, (v) Wet oxidation, (vi) Electrochemical oxidation, (vii) Chemical oxidation etc. but all methods have one or more demerits. Photocatalysis is the best alternative for such treatments.

(d) Surfactants

Surfactants are important in daily life in households as well as in industrial cleansing processes. Surfactants in the environment are a prerequisite for the sustainable development of human health and ecosystems. Surfactants (surface active agents) can be anionic, amphoteric, polymeric and non-ionic. They are used in industries such as agro-chemical, mining and oil fields. Surfactants have a hydrophilic head, which attaches to water and a hydrophobic part of the molecule that avoids water. The hydrophobic part of the molecule is also free to attach to grease, fat or oil on the surface and hence, it is used in cleansing processes. As soap is biodegradable but now-a-days, detergents have almost replaced the soaps leaving aside bathing soaps. The detergents are not biodegradable and thus, a now kind of pollution is appearing on the fore fronts i.e. detergent pollution. It is, therefore, important to have a detailed knowledge about their lifetime in the environment, their biodegradability in wastewater treatment plants and in natural waters, and their ecotoxicity.

During decontamination of protective clothing and contaminated materials, in the nuclear and pharma industries and medicinal sector, detergents are employed to bring down the level of radioactive contamination to within safe limits. However, the surfactants present in these wastes interferes in the chemical treatment process, reducing the decontamination [193]. Although surfactants have been studied in complex water soil system [194,195], the effects are not well understood. Studies showed that the surfactant hexadecyltrimethyl-ammonium chloride ($C_{19}H_{42}ClN$), which is a cationic surfactant with the chemical formula of $CH_3(CH_2)_{15}NCl(CH_3)_3$ known as cetrimonium chloride with abbreviations such as HTAB, CTAB and CTACI. The surfactant is photodegradable with the help of TiO_2, H_2O_2 and UV radiations [196].

By using a black light fluorescent tube as irradiation source in photoreactor, two industrial-grade surfactants, sodium lauryl sulphate (SLS) and sodium dodecylbenzenes sulphonate (SDDBS) were degraded photocatalytically and it was found that an 80% removal of these surfactants can achieved in only 60 min of irradiation [197]. The degradation of a cationic surfactant, (dodecylpyridinium chloride (DPC) was carried out[207] in slury reactor as well as in a vessel where the photocatalyst (P_{25}, Degussa) was anchored onto an aluminium surface to avoid the final filtration of the powder at the end of the reaction. The photocatalytic treatment of soil washing wastes, sodium dodecylsulfate (SDS), hexadecyltrimethyl ammonium bromide (HTAB) and polyoxethylene dodecyl ether ($C_{12}E_8$), has been also investigated in presence of TiO_2 dispersions [198]. In TiO_2 based photocatalytic degradation of an anionic detergent (lauryl benzene sulphonate) initial pH of 5.0 has been proved to be most appropriate for higher efficiency of surfactant degradation [199]. Singhal et al. [200] were also successful in degrading cetyl pyridinium chloride photocatalytically. Photocatalytic treatment proved effective for the treatment of various surfactants i.e. 4-nonylphenol polyethoxylate surfactants [201], alkylbenzene sulphonate (LAS)[202], polyethoxylate surfactants [203], surface active and surface inactive naphthalene sulphonate derivatives, sodium butylnaphthalene-1-sulphonate (B-1-NS) and sodium naphthalene-1-sulphonate (1-NS) [204], non-biodegradable p-nitrotoluene-o-sulphonic acid cp-NTS) [205], cationic surfactants cetyltrimethyflammonium bromide (CTMAB) and decabromodiphenyl ether

(BDE 209) [206], nonyphenol ethoxylated [207], dodecyl pyridinium chloride (DPC) [208], an anionic surfactant sodium dodecyl sulfate [209] from wastewater. Many other surfactants have also been removed from waste water photocatalytically.

(ii) Self Cleaning

Just imagine, if you do not need to wash your clothes, there is no feddness to your outside home walls (exterior of your house) due to pollution and dust without cleaning or giving it a new coat of paint. It looked impossible or very difficult for years back, but now onwards, it is not impossible and the whole credit goes to the concept of photocatalysis. It is still in the technology-transfer phase of moving from laboratory to commercial reality. Strong sunlight or UV light decomposes many organic materials in a slow and natural process. Photocatalysts speed up this process and, like other types of catalysts, stimulate a chemical transformation without being consumed or worn out by the reaction.

If photocatalyst used on or in a concrete structure, it will decompose organic materials including dirt (soot, grime, oil and particulates), biological organisms (mold, algae, bacteria and allergens), air borne pollutant (VOCs including formaldehyde and benzene, tobacco smoke, and the nitrous oxides (NO_x) and sulfuric oxides (SO_x) and even the chemicals that cause odours. In the presence of light a photocatalyst breaks down the harmful compounds into oxygen, CO_2, H_2O, sulphate, nitrate and other molecules having relatively benign impact on the environment.

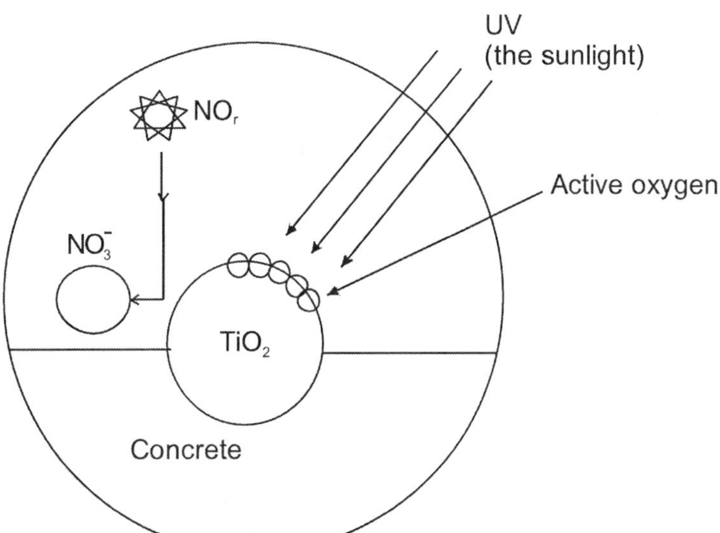

In a study, photocatalytic paving decomposed 15% of the nitrous oxide released by cars traveling the roadway and was found to be more effective, in this regard, than planting trees on both the sides of the roadway. Air-purification capability of photocatalysts are already being used in Japan to make polluted cities like Tokyo more live. Now treated ceramic tiles and pigmented water- based photocatalytic coated hydrotect color coat is also available by which one can make the air in his home or office healthy as well as easier to breath. In addition to decomposing the chemicals that contribute to air pollution, photocatalytic treatments are alleged to have other environmental benefits.

Self cleaning concrete will not require the use of solvents used now to clean buildings because the products of the photocatalytic reaction can be easily removed from the treated surface as the surface becomes hydrophilic. Moisture forms a thin film in this hydrophilic instead of forming beads of water that may cause stains by attracting and holding dirt. Rain or simple rinsing can easily remove this dirt from a hydrophilic surface as the thin moisture film interferes adhesion of dirt. As a result building remained safe, cleaner and more beautiful. On the other hand, the clean concrete will reflect more light, reducing the heat. It may also reduce the formation of smog since the chemical reaction that creates smog increase temperature of air temperature increase. The Jubilee Church (also known as the Dives in Misericordia) in Rome, completed in 2003. It is one of the best example is expected to stay clean for the ages as it has been constructed by photocatalytic additive to the concrete.

In recent years, "self cleaning" coating using photocatalytic titanium dioxide (TiO_2) has gained considerable attention of industries. It is widely used as a white pigment in paint, plastics, cosmetics and a number of other products. New type of nano-scale titanium dioxide particle undergoes a quantum transformation and becomes a semiconductor. By absorbing light energy, the activated TiO_2 creates a charge separation of e^- and h^+ causing chemical reduction, oxidation as well as formation of hydroxyl radicals, which will act as powerful oxidants to decompose organic compounds. This is strong oxidative power can also be used to kill various bacteria attached on the wall and remove foul smell from strains in toilet (e.g. TiO_2 coated tiles and glasses are now available commercially available in the market). Along with this powerful oxidative power, the TiO_2 offers one more unique property i.e., super hydrophilicity, which helps dirt and stains to be easily washed away with water or by rainfall.

However, coating of TiO_2 cannot be applied directly on an organic paint surface because it is highly photoreactive and it may attack the paint surface itself. Along with this we have to think of some more limitations of its usage. The photocatalytic materials do not decompose organic materials until these are exposed to light and secondly, there is no prediction, what is their fate if these compounds drawn out of the earth and reaches humans and other living species through foods and water supply? We can't neglect the fact that these materials may reach into shallow seas and decompose the microorganisms on which the ocean's ecosystem feeds. All such concerns must be addressed well in advance before such materials are commonly used and for this, further researches are needed to know any possible and unknown risk.

Scientists have now found an answer to the dreams of self cleaning clothes that can clean themselves. Almost four decades back in 1951, Alec Guinness played the role of a scientist in "The man in the white suit" where he invented a fabric that never gets dirty or wears out, but it was only a fairy story. Now chemists have made it possible in the real life also by providing coating clothes by small tiny particles of titanium dioxide (20 nm), which will help in break down the carbon-based molecules, food and dirt. The clothes will simply need an exposure to sunlight and this cleaning process begins.

Daoud at Monash University in Victoria, Australia has modified natural fibers such as wool, silk, cotton and hemp that will automatically remove food, grime, and even red wine strains, when exposed to sunlight. He further added that these nanocrystals cannot decompose wool and are almost harmless to skin as these fibers are made of a protein called keratin, which does not have any reactive chemical groups on its surface to bind with TiO_2. Moreover, it is interesting to note that this modification does not change the look and feel of the fabric. TiO_2 has an added property to destroy pathogens such as bacteria in the presence of sunlight as it breaks down the cell wall of the microorganisms. This concept is very useful in making self-cleaning fabric especially to be used in hospital and other medical settings and in reducing the spread of germs.

It has been reported that red wine stains on TiO_2 coated wool disappear within 20 hrs of exposure to sunlight, while the untreated wool remains boldly stained. The stains of coffee fade away in 2 hrs, while blue ink stains disappear in 17 hrs.

(iii) Sterilization Effect:

Almost daily a new disease seems to appear in the headlines. Advances in transportation technology has caused an increase in the movement of people across countries, oceans and continents. It helps infectious diseases to spread rapidly and to long ranges, thus, becoming serious threats for the world. TiO_2 photocatalyst is a revolutionary invention in sterilizing action. It not only kills cells of bacteria, but also decompose the cell itself. The end toxic produces at the death of cell are also expected to be decomposed completely by photocatalytic action. Disinfections by titanium dioxide is 3 times stronger than chlorine, and 1.5 times stronger than O_3. Nano- TiO_2 may be more helpful in preventing SARS, so that it does not strike back. It has been tested with a series of experiments by different centres of research showing very good performance. Photocatalyst nano-TiO_2 can kill Pseudomonas aeruginosa, Ecoli influenza virus, MRSA (Methicillin-resistant Staphylococcus aureus), Tubercle bacillus, etc. including fungi, it also destroys odors and nicotine. Therefore, once applied to the specified surface like room walls, exterior walls, curtains, carpets, fabrics, bedspreads, upholstery or other soft good surfaces that it will actively not only protect the surface from any organic matter even in room light (diffused) will eliminate all offensive odors within 2-3 days of exposure to light. In Asia, during the SARS epidemic, actual usage of photocatalytic TiO_2 in hospitals was exponentially greatly accelerated. This very technology has been adapted by Enviro clean to eliminate complete and continuous malodorous and harmful elements of cigarette and cigar smoke.

Disinfection by nano-TiO_2 photocatalyst may be a promising technology with further research into its practicality for treatment of water. There is no doubt that it may become a possible alternate process for chlorination of water for disinfection.

(iv) Deodorizing Effect

The human sense of smell, senses CH_3SH (methyl mercaptan) at a concentration as low as 0.00012 ppm. Its concentration lower than 0.002 – 0.01 ppm is only allowed as it not harmful directly. Along with concentration of CH_3SH, concentration of other melodorous pollutant i.e., ammonia (NH_3), hydrogen sulphide (H_2S) and acetaldehyde (CH_3CHO) are also regulated by the Japanese Government with the help of TiO_2 photocatalyst [220].

In photocatalytic reactions of TiO_2 $^{\bullet}OH$ hydroxyl radicals are formed as intermediate. These OH radicals accelerate the breakdown of any volatile organic compounds (VOCs) by destroying the molecular bonds. This helps in combining the organic gases to form a molecule that is not harmful to humans. Some of the examples of the molecules with odour are – Tobacco, formaldehyde, NO_2, urine and fecal odor, gasoline and many other hydro carbon molecule in the atmosphere. Titanium dioxide on silica is being developed as a form of odor control in cat litter.

(v) Anti-fouling / Anti-Fogging

The anti-fouling effect of photocatalyst TiO_2 has also been observed [210,211] only qualitatively. A self cleaning function can also be expected with TiO_2 – coated substrate, when small amount of organic contaminants gradually accumulate on it. In outdoor spaces, major cause for stains may be sand, which is an inorganic contaminant. In such cases, photocatalytic oxidative power may not be considered useful for self cleaning process. However, TiO_2-coated substances show very effective anti-fouling effect even against inorganic contaminants [212]. It is assumed that inorganic ones adhere on the substrate surface by oil spots coming from automotive exhaust fumes, which serves as a binder. So ultimately, the hydrocarbon from automotive exhaust is oxidized and the dirt on the walls washes away with rainfall, keeping the building exterior clean at all times.

REFERENCES

[1] Photocatalysis, C. N. Serpone and E. Pelizzetti (Eds.), John Wiley and Sons, New York, 1989.
[2] M. Anpo, Catal., Surveys Jpn., 1997, 2, 167.
[3] M. Anpo, Proc. 1st *Int. Conf. Protect. the Environ.*, 1998, Rome, 75.
[4] M. Anpo, in: Green Chemistry (P. Tundo and P. Anastas eds.), 1, Oxford University Press, 2000.
[5] IUPAC glossary of terms used in photochemistry, *Pure Appl. Chem.*, 1988, 60, 1055.
[6] E. P. Gratzel, Energy Resources through Photochemistry and Catalysis, Academic, New York, 1983.
[7] E. Pelizzetti and N. Serpone, Homogeneous and Heterogeneous Photocatalysis, D. Reidel, Holland, 1986.
[8] G. Somorjai, *J. Photochem.*, 1987, 11, 197.
[9] "Photocatalytic and Environment : Trends and Applications", ed. M. Schiavello, Kulwer Acad. Pub., Dorddrecht., 1987, 237.
[10] S. C. Ameta, M. Bala, M. Dak, N. K. Jain and S. Sahasi, *Asian J. Chem. Rev.*, 1991, 2, 90.
[11] S. C. Ameta, R. Ameta, J. Vardia, R. Ameta and Zehra Ali, *J. Indian Chem. Soc.*, 1999, 76, 281.
[12] K. A. Alexander and D. M. Roundhill, *J. Mol. Cat.*, 1983, 19, 85.
[13] V. Balzani and V. Carassiti, *Photochemistry of Coordination Compounds* (Academic Press, New York, 1970) Ch.5.

[14] C. Katul and A. W. Adamson, in Comprehensive Coordination Chemistry, edited by G. Wilkinson, R. D. Gillard and J. A. McCleverty, Vol.1 (Pergamon Press, Elmsford, New York, 1987, 385-414.

[15] M. Schiavello, V. Augugliaro and L. Palmisan, *J. Catal.,* 1991, 127, 332.

[16] V. Augugliaro, L. Palmisano and M. Schiavelo, *AI. Ch. EJ.*, 1991, 37, 1096.

[17] M. Voinov and J. Augustynski, Heterogeneous Photocatalysis, M. Schiavello (Ed.), John Willey and Sons, 1997.

[18] M. Lannoo, in Materials Science and Technology, Vol. 4, Electronic Structure and Properties of Semiconductors, W. Schroter, VCH (Ed.), Weinheim 1991.

[19] U. Ozgur, *J. Appl. Phys*. 2005, 98, 41301.

[20] D. C. Look, J. W. Hemisky and J. R. Sizelove, J. R., *Phys. Rev. Lett.*, 1999, 82, 2552.

[21] R. W. Matthews, *J. Catal,* 1998, 111, 264.

[22] Y. Cui, Z. Yiren and H. Dongbao, *J. Photogra. Sci. Photochem.*, 2001, 19, 131.

[23] L. Lecheng and W. Dacui, Chem. Indus. Press, 2001, 252.

[24] H. Chun, L. Xingjuan and L. Shuang, *Acta Scientiae Circumstantiae*, 18, 81. M. A. Hema, V. Ramakrishna and J. C. Kuriacose, *Indian J. Chem.*, 1977, 15B, 947.

[25] Z. Yiren, Z. Zhenchao and X. Hengcan, *J. Environ. Prot. Chem. Indus.*, 2005, 25, 290.

[26] Y. Yongli, Z. Xiujing and W. Naizhi, *J. Environ. Prot. Chem. Indus.,* 2005, 25, 182.

[27] E. Shanhu, S. Yong and C. Liuping, *Chin. J. Catal.*, 2002, 23, 100.

[28] Z. Changwei, H. Hong and W. Jinbao, *J. Guizhou Environ. Prot. Sci. Technol.*, 2003, 4, 44.

[29] F. I. J. Wha C and C. Yue Chin, *Sci. Sin.*, 1976, 19, 716.

[30] H. Liang Lui and C. K. Thomas Yang, *Proc. Biochem.*, 2003, 39, 475.

[31] C. A. Gouvea, F. Wypych, S. G. Moraes, N. Duran, N. Nagata, and P. Peralta-Zamora, Chemosphere, 2000, 40, 433.

[32] J. Domenech and M. Andres, *New J. Chem.*, 1987, 11, 443.

[33] J. Domenech, M. Andres and J. Munoz, *Electrochim. Acta*, 1987, 32, 773.

[34] F. Steinbach and R. Harborth, *Faraday Disc. Chem. Soc.*, 58 (1947).

[35] T. Inoue, A. Fujishima, S. Konishi and K. Honda, *Nature*, 1979, 277, 637.

[36] J. C. Kuriacose and M. C. Markham, *J. Catal.*, 1962, 1, 498.

[37] M. C. Markham, M. C. Hannan and S. W. Evans, *J. Am. Chem. Soc.*, 1954, 76, 820.

[38] M. C. Markham, J. C. Kuriacose, J. A. D. Marco and C. Giaquint, *J. Phys. Chem.*, 1962, 66, 932.

[39] M. C. Yeber, Rodriquez, J. Freer, N. Duran and H. D. Mansilla, *Chemosphere*, 2000, 41, 1193.

[40] S. Sakthivel, B. Neppolian, M. V. Shankan, B. Arbindo, M. Palanichamy and V. Murugesan, *Solar Energy Material Solar Cells*, 2003, 77, 65.

[41] E. L. Gorsey. A, Chen. M, Dubrovinsky. L, Gillet. P, Graup. G, *Scienc*e, 2001, 293, 5534.

[42] E. L. Gorsey, Ahmed, *Earth Planetary Sci. Lett.*, 2001, 192, 485.

[43] N. Greenwood N., A. Earnshaw, Chemistry of the elements, Oxford: Pergamon, 1984, pp. 117-19.

[44] Fujishima, *Nature*, 1972, 238, 37.

[45] Japan Nanonet Bulletine, 44th Issue, May 12, 2005.

[46] Jenny Hogan, Newscientist.com, 4 February 2004.

[47] TIME's Best Inventions of 2008, October 31, 2008.
[48] Carbon – doped titanium dioxide is an effective photocatalyst.
[49] B. Sun, P. G. Smirniotis and P. Boolchand, *Langmuir*, 2005, 21, 11397.
[50] G. Colon, J. N. S. Espana, M. C. Hidalgo and J. A. Navio. *J. Photochem Photobiol*, 2006, 179A, 20.
[51] Nakajima, S. Koizumi, T. Watanabe and K. Hashimoto, *Langmuir*, 2008, 16, 7048.
[52] R. D. Sun, A. Nakajima, T. Watanabe, A. Fujishima and K. Hashimoto, *J. Phys. Chem.*, 2001, 105, 1984.
[53] N. Sakai, A. Fujishima, T. Watanabe and K. Hashimoto, *J. Phys. Chem.*, 2001, 105, 3023.
[54] L. Muszkat, L. Feigelson L, L. Bir and K. A. Muszkat. *J. Photochem Photobiol.* B, 2001, 60, 32.
[55] J. B. De Heredia, J. Torregrosa, J. R. Dominguez, J. A. Peres, *J. Hazard. Mater.*, 2001, 83, 255.
[56] Hager S, Baver R, *Kudielba* G, 2000, 41, 1219.
[57] K. Charles, Introduction to Solid State Physics, 7th Edition 1995, Wiley – India.
[58] M. Schiavello, Photoelectrochemistry, Photocatalysis and Photoreactors, Fundamentals and Developments Springer, 1985.
[59] F. Borgarillo, K. Kalyansundaram, M. Gratzel and E. Pelizzetti, *Helv. Chim. Acta.*, 1982, 65, 243.
[60] L. Chen Sha, T. Ya Ping, K. Bo Nan, W. Bin Song, Z. Feng, M. Qiang, X. Ji, W. Da Zhi and L. Ji, Sci. China Ser. E : *Technol. Sci.,* 2007, 50, 279.
[61] Doong Ruey. An, Chen Chih – Hua, Maithrupala R. A., Chang Sue Min, *Water Research*, 2001, 35, 2873.
[62] Henglin, *J. Phys. Chem.*, 1982, 86, 2291.
[63] H. Yin, Y. Wada, T. Kitamura, T. Sabata, H. Mori and S. Yanagida, *Chem. Letts.*, 2001, 30, 34.
[64] K. Kobayakawa, T. Miura, A. Suzuki, Y. Sato and A. Fujishima, *Sol Energy Mater.* Solar Colls, 1993, 30, 201.
[65] K. Kobayakawa, T. Inoue, Y. Sato and A. Fujishima, *Denki Kagaku*, 1993, 61, 874.
[66] E. Borgarello, R. Terzian, N. Serpone, E. Pellizetti and M. Barbeni, *Inorg. Chem.*, 1986, 25, 2135.
[67] S. S. Wang, Z. H. Wang and Q. X. Zhuang, *Appl. Catal.*, 1992, 1, 257.
[68] V. N. Pak and S. Kharangi, Z. *Fiz. Khim.*, 1997, 70, 1696.
[69] M. Parakh, Z. Ali, J. Vardia and S. C. Ameta, Res. J. Chem. Environ., 1998, 2, 49.
[70] Kumar and S. Kumari, *J. Phys. Org. Chem.*, 1998, 11, 277.
[71] K. J. Gu, Y. Cao and B. W. Zheng, *Acta Cnimica Sinica*, 1989, 47, 668.
[72] Mills and G. Williams, *J. Chem. Soc. Frarday Trans.* I., 1987, 83, 2647.
[73] Q. L. Li, Z. S. Chen, X. H. Zheng and J. S. Jin, *J. Phys. Chem.*, 1992, 96, 5962.
[74] M. M. Taqui Khan, R. C. Bhardwaj and C. Bhardwaj, Indian J. Chem., 1986, 25A, 1.
[75] N. Rappon and R. T. Syvitski, *J. Photochem. Photobiol.*, 94A, 243 (1996).
[76] M. A. Hasnat, L. A. Siidiquey and A. Nuruddin, Dyes and Pigments, 2005, 66, 185.
[77] W. Hetterich and H. *Kisch, Chim. Ber.*, 1991, 121, 15.
[78] A. F. Wells, Structural Inorganic Chemistry, 5th Ed., Oxford: Clarendon Press, 1984.
[79] S. Yanagida, T. Azuma, H. Kawakami, H. Kizumoto and H. Sakurai, *J. Chem. Soc. Chem. Commun.*, 1984, 21.

[80] H. Schuchmann and S. Sonntag. J. Photochem. Photobiol., 1981, 16A, 289.
[81] M. Oyama, *J. Org. Chem.*, 1965 30, 2429.
[82] S. Yanagida, H. Kawakami, Y. Midori, H. Kizumoto, C. Pac and Y. Wada, *Bull. Chem. Soc.* Japan, 1995, 68, 1811.
[83] K. Anil, and J. Anshuman, *J. Colloid Interface Sci.,* 2006, 297, 607.
[84] W. Helterich and H. *Kisch, Chem. Ber.*, 1988, 121, 15.
[85] X. M. Ren, L. Zhang and C. Y. Liu, *J. Photochem. Photobiol.*, 1995, 85A, 239.
[86] Clavdia L., Torres-Martinez, R. Kho, O. I. Mian and R. K. Mehra. *J. Coll. Inter. Sci.,* 2001, 240, 525.
[87] H. Zhang, X. Chen, Z. Li, J. Kou, T. Yu and Z. Zou. *J. Phys. D: Appl. Phys.*, 40 (2007) 6846 – 6849.
[88] Y. Xiaodan, W. Qingyin, J. Shicheng and G. Yihang, *Mater. Charact.*, 2006, 57, 333.
[89] J. Lui, B. Geng and S. *Wang. Cryst. Growth*, 2009, 9, 4384.
[90] R. Dangi, J. Gandhi and S. Bhardwaj, R. J. C Rasayan, *J. Chem.*, 2008, 1, 355.
[91] S. S. Priya, M. Premalatha and N. Anantharaman, *ARPN J. Eng. Appl. Sci.*, 2008, 3, 36.
[92] J. Li, Y. Xu, Y. Lui, D. Wu and Y. *Sun China Particuology*, 2004, 2, 266.
[93] Monaci and A. Ginestra, *J. Photochem. Photobiol.*, 1996, 93A, 65.
[94] X. Zhao and Y. Zhu. *Environ. Sci. Technol.*, 2006, 40, 3367.
[95] J. Killasenor, N. Duran and H.D. Mansilla, *J. Environ. Technol.*, 2002, 23, 955.
[96] R. R. Sharma, A. Venkatachalam and S.J. Garg, *Samyak J. Chem.,* 1998, 2, 16.
[97] J. Jose, Thesis "Synthesis, Characterization and use of some ternary oxides of Lanthanum in treatment of polluted water," M.L. Sukhadia University, 2008.
[98] J.H. Sin, X.L. Wang and J. Y. Sun, *J. Mol. Catal.*, 2006, 260, 235.
[99] J. Jose, J. Ameta, P.B. Punjabi, V.K. Sharma and S.C. Ameta, Bulletin of the Catalysis Society of India, 2007, 6, 110.
[100] M. Sun, D. Li, Y. Zheng, W. Zhang, Y. Shao, Y. Chen, W. Li and X. Fu. *Environ. Sci. Technol.* 2009, 43, 7877.
[101] K. Y. Jung, S. B. Park, M. Matsuoka, M. Anpo, *Int. J. Photoenerg.*, 2003, 5, 31.
[102] L. Z. Zhang, I. Djerdj, M. Cao, M. Antonietti, M. Niederberger, *Advanced Materials*, 2007, 19, 2083.
[103] S. Sahoo, S. Sumithra, R.J.N. Guru and G. *Madras Bull. Mater. Sci.,* 2009, 32, 1-6.
[104] D. Saha, S. Mahapatra, M. Giridhar and R.T.N. Guru, Ind. Eng. Chem. Res. submitted Apr 15, 2009, accepted DOI : 10, 1021/i.e. 9006134.
[105] S. Mahapatra, R. Vinu, G. Madras, G. Row, *Appl. Catal. A.,* 351 (2008) 45-53.
[106] Mahata, P., Madras, G. and Natarajan, S. *J. Phys. Chem. C.*, 2007, 111, 1665.
[107] Baidya, T., Marimuthu, A., Hedge, M.S. and Madras G., *J. Phys. Chem.* B., 2007, 111, 830.
[108] Muktha, B., Madras G., Guru R.T.N., *J. Photochem. Photobiol.* A., 2007, 187, 177.
[109] Mukhta, B., Madras G., Guru R.T.N., *J. Solid. Struct. Chem.*, 2006, 179, 3919.
[110] Mukhtha B., Priya, M.H., Madras G. and Guru, R.T.N. *J. Phys. Chem.* B., 2005, 109, 41442.
[111] Nagaveni, K., Hedge, M.S., Madras G., J. Phys. Chem. B., 2004, 108, 20204.
[112] W. Wang, J. Bi, L. Wu, Z. Li, X. Wang and X. Fu. *Nanotechnol.,* 2008, 19, 505705.
[113] L. Ren, L. Ma, L. Jin, J.B. Wang, M. Qui and Y. Yu *Nanotechnol.,* 2009, 20, 405602.
[114] Alton and J.L. Ferry, *J. Photochem. Photobiol.*, 2003, 152A, 175.

[115] S. C. Ameta, R. Choudhary, R. Ameta and J. Vardia, *Indian J. Chem. Soc.,* 2003, 80, 257.
[116] G. T. Brown and J.R. Darwent, *J. Chem. Soc. Faraday Trans.* I., 1984, 80, 1631.
[117] Y. Mu, H. Q. Yu, J. C. Zheng and S. J. Zheng, *J. Photochem. Photobiol.*, 2004, 163A, 311.
[118] S. Papic, N. Koprivanac and AL. Bozic, *Water Environ. Res.*, 2006, 78, 572.
[119] C. C. Chen, C. S. Lu, and Y. C. Chung, *J. Photochem. Photobiol.*, 2006, 181A, 120.
[120] R. K. Wahi, W. W. Yu and Y. P. Liu, *J. Catal.,* 2005, 242A, 4.
[121] H. Kominami, H. Kumamoto, Y. Kera and B. Ohtani, *J. Photochem. Photobiol.*, 2003, 160A, 99.
[122] K. Vinodgopal, D. Wynkoop and P. V. *Kamat, Chem. Mater.*, 1996, 8, 2180.
[123] G. M. Liu, X. Z. Li and I. C. Zhao, *J. Mol. Catal.*, 2000, 153A, 221.
[124] M. Mrowtez and E. Selli, *J. Photochem. Photobiol.*, 2004, 162A, 81.
[125] K. V. S. Rao, B. Lavedrine and P. Boule, *J. Photochem. Photobiol.*, 2003, 154A, 189.
[126] Y. Zhiyong, E. Mielezarski, J. Mielezarski, D. Laub, P. Buffat, U. Klehm, P. Albers, K. Lee, A. Kulik, A. Renken, J. Kiwi and L. Kiwi Minsker, *Water Res.*, 2003, 41, 862.
[127] A. Poulios, E. Micropoulou, R. Panou and E. Kostopolou, Appl. Catal., 2003, 41, 345.
[128] R. W. Mathew, *J. Chem. Soc. Faraday Trans.*, 1989, 1, 86, 1291.
[129] R. W. Mathew's, *Water Res.*, 1991, 25, 1169.
[130] H. D. Mansilla and J. Villasnov, *J. Photochem. Photobiol.*, 1994, 78A, 267.
[131] R. Ameta, J. Vardia, C. V. Bhatt and S. C. Ameta, *Sam. J. Chem.*, 1997, 1, 29.
A. Sharma, J. Vardia, R. P. Mathur and S. C. Ameta, *Res. J. Chem. Environ.*, 1998, 2, 17.
[132] S. Bhandari, Ph.D. Thesis, M. L. Sukhadia University, Udaipur, India, 2004.
[133] S. Gulati, F. Heetawala, V. K. Sharma and S. C. Ameta, *Poll. Res.*, 2003, 22, 213.
[134] B. Neppolian, S. Sakthivel, B. Arbindo, M. Palanicham and Y. Y. Murugesan, *Bull. Cat. Soc. India*, 1999, 9, 164.
[135] L. Zhang, C. Y. Liu and X. M. Kin, *J. Photochem. Photobiol.*, 1995, 85A, 239.
[136] S. Bhandari, J. Vardia, R. K. Malkani and S. C. Ameta, *Toxicological and Environmental Chemistry*, 2006, 88, 35.
[137] C. Ye, Y. Bando, G. Shen and D. Golberg, *J. Phys. Chem.*, 2006, 110, 15146.
[138] R. Ameta, S. Jain, C. V. Bhatt and S. C. Ameta, *Rev. Rowen Chim.*, 2000, 45, 49.
[139] N. Daheshvar, D. Salari and R. Khataee, *J. Chem. Soc. Faraday Trans.*, 2003,152A, 111.
[140] R. Ameta, C. Kumari, C. V. Bhatt and S. C. Ameta, *Indust. Quim.*, 1998, 33, 36.
[141] S. Kothari, N. Jain and R. Ameta, *Indian J. Chem. Tech.,* 2004, 11, 423.
[142] S. Bilgi and C. Demir, *Dyes and Pigments*, 2005, 66, 69.
[143] M. A. Hashat, L. A. Siddiqui and A. Nasuddin, *Dyes and Pigments*, 2005, 66, 185.
[144] N. Modirshala and M. A. Behnajudy, *Dyes and Pigments*, 2006, 76, 675.
[145] A. Sharma, R. Sharma, S. Singh and S. C. Ameta, *J. Indian Chem. Soc.*, 2002, 79, 929.
[146] P. Rao, G. Patel, S. L. Sharma and S. C. Ameta, *J. Toxicol. Environ. Chem.*, 1997, 60, 155.

[147] A. F. Martins, M. C. Wilde and C. Dasilveria, *Ind. J. Chem. Tech.*, 2006, 13, 114.
[148] I. Mills and G. Williams, *J. Chem. Soc. Faraday Trans. I.,* 1987, 83, 2647.
[149] Anju Jain, Ph.D. Thesis, Sukhadia University, Udaipur, India, 2004.
[150] T. Takizawa, T. Watanabe and K. Honda, *J. Phys. Chem.*, 1978, 82, 1391.
[151] L. Zhang and T. Shen, *J. Chem. Soc. Chem. Commun.*, 1996, 473.
[152] X. M. Ren, L. Zhang and C. Y. Liu, *J. Photochem. Photobiol.*, 1995, 85A, 239.
[153] K. Kobayashi, K. Kitaguchi, H. Tanaka, H. Tsniki and A. Venu, *J. Chem. Soc. Faraday Trans.* I, 1987, 63, 1395.
[154] Ling Zhang, Chun-Xan Liu and Xin-Min Ren, *J. Photochem. Photobiol.*, 1995, 85A, 239.
[155] J. H. Carey, J. Lawrence, H. M. Tosine, *Bull. Environ. Contam. Toxicol.*, 1976, 16, 697.
[156] D. F. Ollis, C. Hasiao, L. Budirman and C. Lee, *J. Catal.*, 1984, 88, 89.
[157] Jatinder Kaur, Ph.D. Thesis, Sukhadia University, Udaipur, India, 1993.
[158] P. Kopf, E. Gilbert and S. H. Eberie, *J. Photochem. Photobiol.*, 2000, 136A, 163.
[159] H. Tahiri, Y. A. Icnou and J. Herrmann, *J. Photochem. Photobiol.*, 1998, 114A, 219.
[160] Z. Osawa and T. Sunakani, Polymer Degrad. *Stability*, 1994, 150, 61.
[161] K. Tanaka, T. Hisanaga and K. Harada, *J. Photochem. Photobiol.*, 1989, 48A, 155.
[162] N. Serpone, P. Muruthamuthu, P. Pichat, E. Pelizzetti and H. Hidaka, *J. Photochem. Photobiol.*, 1995, 85A, 247.
[163] M. Halmann, A. J. Hunt and D. Spath, *Solar Energy Mater. Solar Cells*, 1992, 26, 1.
[164] W. Choi and M. R. Hoffmann, *Environ. Sci. Technol.*, 1997, 31, 99.
[165] H. Uchida, S. Katoh and M. Watarloe, *Chem. Lett.*, 1995, 4, 261.
[166] D. W. Bahnemann, *Isr. J. Chem.*, 1993, 33, 115.
[167] V. Kavitha and K. Palanivelu, *J. Environ. Sci.* Health A Tox. Hazard. Subst. Environ. Eng., 2003, 38, 1215.
[168] T. Sehili, P. Boule and J. Lemaire, *J. Photochem. Photobiol.*, 1989, 50A, 103.
[169] T. Sehili, P. Boule and J. Lemaire, *J. Photochem. Photobiol.*, 1989, 50A, 117.
[170] D. Hartley and H. Kidd (Eds.), "The Agrochemicals Handbook, Royal society of chemistry", Nottingham, England (1987).
[171] K. Hustert, D. Kotzioas and F. Karte, *Chemosphere,* 12, 55 (1983).
[172] M. C. Amateus, A. M. Silva and H. D. Burrows, *J. Photochem. Photobiol.*, 1994, 80A, 409.
[173] H. Benschop and M. Halmann, *J. Chem. Soc.*, 1974, 1175.
[174] D. Barcelo, G. Durand, N. De Bertrand and *J. Albaiges, Sci. Tot. Environ.*, 1993, 132, 283.
[175] A. Guera, E. Almansa, S. Malato, M. I. Maldonado and A. R. Fernandez Alba, *Analusis.*, 1998, 26, 245.
[176] D. Zhao, J. Wang, Z. Zhang and J. Zhang, Frontiers of Chemical Engineering in China, 2009, 3, 206.
[177] M. V. Shankar, S. Anandan, N. Venkatachalam, B. Arabindoo, V. Murugesan, *J. Chem. Tech. and Biotech.,* 2004, 79, 1279.
[178] N. Daneshvar, S. Aber, M. S. S. Dorraji, A. R. Khataee, M. H. Rasoulifard, *Separ. Purific. Tech.,* 2007, 58, 91.
[179] Arques, A. Garcia – Ripoll, R. Sanchis, L. Santos-Juanes, A. M. Amat, M. F. Lopez and M. A. Miranda, *J. Sol. Energy Eng.*, 2007, 129, 74.

[180] R. Rajeshwari and S. Kanmani, *Iran J. Environ. Health Sci. Eng.*, 2009 6, 61.
[181] V. Kitsiou, N. Filippids, D. Mantzavinos and I. *Poulis, App. Catal. B: Environ.*, 86, 2009, 27.
[182] L. Muszkat, L. Bir and L. Feigelson, *J. Photochem. Photobiol.*, 1995, 87A, 85.
[183] A. A. Khodja, T. Sehili, J. F. Pilichouski and P. Boule, *J. Photochem. Photobiol.,* 2001, 741A, 231.
[184] K. Nehara, J. Zhao, W. Serpone and E. Pelizzetti, *J. Photochem. Photobiol.*, 1992, 64A, 2.
[185] L. G. Devi and G. M. Krishnaiah, *J. Photochem. Photobiol.*, 1999, 121A, 141.
[186] I. F. Abramovic, D. V. Sojic, V. B. Anderluh, N. D. Abazovic and M. I. *Comor Desalination,* 2009, 244, 293.
[187] J. Arana, C. Cabo, C. F. Rodriguez, J. A. H. Melian, J. A. O. Mendez, J. M. D. Rodriguez, and P. Pena, *Chemosphere*, 2008, 71, 788.
[188] Navarro S., Fenoll J., Vela N, Ruiz E,, Navarro G., *J. Hazard. Mater.*, 2009 Aug 8 [Epub ahead of print].
[189] A. M. Azharul, S. Vasilios, A. Triantafyllos, *J. Hazard. Mater.*, 2009, 170, 230 - 238.
[190] S. Merabet, D. Robert, J. V. Weber, M. Bouhelassa and S. M. Benkhanouche, *Environ. Chem. Lett.*, 2009, 7, 45.
[191] Chitra, S., Chandran, S., Sasidhar, P., Lal, K.B., Amalraj, R.V.," Biodegradation of surfactant bearing wastes", *Indian Environ Prot.* 11(9)(1991), 689-692.
[192] I. N. Aronstein, Y. M. Calvillo and M. Alexander, *Environ. Sci. Technol.*, 1991, 25, 1728.
[193] S. Laha and R. G. Luthy, *Environ. Sci. Technol.*, 1991, 25, 1920.
[194] Mukhtar-al-Hassan, Jamil Anwar and Muhammad Jaurwad Saif, *J. Sci. Res.,* 2008, 38, 29.
[195] I. Lizama, C. Bravo, C. Caneo, M. *Ollino Environ. Technol.*, 2005, 26, 909.
[196] Clandia Bianchi, "Photocatalysis for the degradation of ionic surfactants in water : The Case of DPC." Paper : 1171-507-19, DO1 = 10.1557 PROC – 1171-507-19.
[197] I. Fabbri, A. Crime, M. Davezza, C. Medana, C. Baiocchi, A. Bianco Prevot and E. Pramauro, *Appl. Catal. B : Environ.*, 2009, 92, 318.
[198] Book – Colloids for Nano and Biotechnology, Publisher – Springer Berlin/Heidelberg, Vol. 135/2008 ISBN – 978-3-540-85133-2. Erzselet Szabo-Bardos, Zoltan Zsilak rotto Horvath.
[199] H. Sathoshi, W. Natasuki, H. Hisao. *J. Japan Oil Chem. Soc.*, 49, 631.
[200] M. Mehrab and V.S. Haffield. J. Environ. Sci. Health, Part A, *Toxic hazardous substances and environmental engineering*, 2005, 40, 1003.
[201] K. B. Sherrard, P. J. Marriott, M. J. Mc Cormick, R. Cotton and G. Smith. *Anal. Chem.*, 1994, 66, 3394.
[202] H. Hidaka, T. Koike and N. Serpone, *J. Oleo Sci.,* 2003, 52, 245.
[203] S. Parra, S. Malato, J. Blanco, P. Peringer and C. Pulgarin, *Water Science and Technology*, 2001, 44, 219.
[204] Taicheng, A., Jiaxin, C., Guiying, L., Xuejun, Guoying, S., Jiamo F., Bixian, M. and Kevin, E.S. *Catalysis Today*, 2008, 139, 69.
[205] I. Pelizzetti, C. Minero, V. Maurino, A. Sclafani, H. Hidaka and N. Serpone. *Environ. Sci. Technol.*, 1989, 23, 1380.

[206] C. L. Bianchi, S. Ardizzone, G. Cappelletti, A. Naldoni, C. Pirola and V. Ragaini http://inpact. inp-toulouse. fr/GPE-EPIC 2009/abstracts/abstract 66. doc.
[207] S.B. Mortazani, A. Khavanin, G. Moussani and A. Azhdarpoor, *Pak. J. Biol. Sci.*, 2008, 11, 290.
[208] P. Sawunyama, A. Fujishima and K. Hashimoto, *Langmuir*, 1999, 15, 3551.
[209] T. Minabe, D. A. Tryk, P. Sawunyama, Y. Kikuchi, K. Hashimoto and A. Fujishima, *J. Photochem. Photobiol. A. Chem.*, 2000, 137, 53.
[210] J. Arana, J. A. Herrera Melian, J. M. Dona Rodriguez, O. Gonzalez Diaz, A. Viera, J. Perez Pena, P. M. Marrero Sosa and V. Espino Jimenez, *Catalysis Today*, 2002, 76, 279.

In: Recent Progress in Chemistry…
Editor: A. K. Haghi, pp. 205-221
ISBN: 978-1-61668-501-0

Chapter 12

MODIFIED ATMOSPHERE PACKAGING OF FISHERY PRODUCTS

T. K. Srinivasa Gopal*[*]*, C. O. Mohan and C. N. Ravishankar
Fish Processing Division, Central Institute of Fisheries Technology,
Cochin 682 029, India

ABSTRACT

The demand for fresh or mildly preserved fish products is on rise globally. Chilling is the best method to preserve the fish in fresh condition. However the limited shelf life of chilled products has prompted the industry to search for new and more advanced preservation methods. Modified atmosphere packaging is one such advanced preservation technique which considerably extends the shelf life of fish products. This involves the replacement of the air surrounding the product with desired gases mainly carbon dioxide, oxygen and nitrogen depending on the type of fish products. By applying this technique shelf life of fish products can be doubled which helps in improved distribution of this nutrient rich perishable commodity to distant places.

INTRODUCTION

In recent times, human life style has changed in a variety of ways and presently food is not only a basic need but also became a measure of living standards. Consumers have great expectations regarding the food what they eat. It should have unadulterated taste, longer shelf life and attractive appearance of packed perishable goods. To cope up with these ever increasing demands, food industry has is always looking for an advanced and cost effective method which extends the shelf life of perishable food commodities. Food preservation and distribution has undergone two major revolutions in the last century, canning and freezing. These methods gave consumers easy availability to most type of produce. However, the cost

[*] Corresponding author: Dr. T. K. Srinivasa Gopal, Head and Principal Scientist, CIFT Matsyapuri, Cochin 682 029, India E-mail: tksgopal@gmail.com, Phone : +91-484-2666845 Fax: +91-484-2668212

involved in these methods, high energy consumption, ecological awareness and demand for healthy and mildly preserved fresh food has created a need for an innovative technology that allows distribution of fresh products around the year. The shelf-life of perishable foods, particularly fish products is limited in the presence of normal air by two principal factors-the chemical effect of atmospheric oxygen and the growth of aerobic spoilage micro organisms. These factors either individually or in association with one another bring about changes in odour, flavour, colour and texture leading to an overall deterioration in quality. Chilled storage will slow down these undesirable changes but will not necessarily extend the shelf - life sufficiently for retail distribution and display purposes. Preservation of food products in modified atmosphere packaging (MAP) could be a reliable technique to extend the shelf life. This concept is rapidly growing in the food packaging market. It improves the product quality, freshness and increases the shelf-life of the product as well as provides convenience to the consumer and adds value to the product. It is mainly used to extend the shelf life of fresh perishable food products.

What Is MAP?

The normal composition of air is 21% oxygen, 78% nitrogen and less than 1% is constituted by argon, carbon dioxide, hydrogen etc. Modification of the atmosphere within the package by reducing the oxygen content while increasing the levels of carbon dioxide and/or nitrogen has been shown to significantly extend the shelf-life of perishable foods at chill temperatures. Thus MAP is the enclosure of food in a package in which the atmosphere inside the package is modified so that its composition is other than that of air. Modification can be achieved by removing air and replacing it with a controlled mixture of gases or by creating vacuum. This helps in controlling the biochemical, enzymatic and microbial actions so as to avoid or decrease the main degradations that might occur. This allows the preservation of the fresh state of the fish product without the temperature or chemical treatments used by competitive preservation techniques, such as canning, freezing, dehydration and other processes. MAP is the replacement of air in a pack with a single gas or mixture of gases; the proportion of each component is fixed when the mixture is introduced. No further control is exerted over the initial composition, and the gas composition is likely to change with time owing to the diffusion of gases into and out of the product, the permeation of gases in to and out of the pack, and the effects of product and microbial metabolism (Church, 1994).

History

The first recorded use of MAP was in 1927 to extend the shelf life of apples by storing them in reduced O_2 and increased CO_2 atmosphere. Then onwards its application has extended to most of the perishable food products mainly for bulk transportation and for extension of shelf life. The technology became commercially applicable only in the early 1970s in Europe for retail packs. Spenser and Marks in the UK introduced MAP meat in

1979, the success of this led to the introduction of MAP for fish products. MAP technique is now used on a wide range of fresh or chilled, cured fishery products.

PRINCIPLE GASES USED IN MAP

The basic concept of the MAP of fresh foods is the replacement of the air surrounding the food in the package with a mixture of atmospheric gases different in proportion from that of air. The normal composition of air is 78% nitrogen and 21% oxygen, with the balance made up of carbon dioxide (0.035%), other gases and water vapour. An increase in the proportion of carbon dioxide and/or a reduction in the proportion of oxygen within specified limits maintain the original product quality and extend the product shelf life. Various gases both alone and in combination are being used for MAP products to maintain the quality. Three principal gases used in Modified atmosphere packaging are Carbon dioxide, Nitrogen and Oxygen. The proportion of each component gas is fixed when the mixture is introduced into the package. However, no control is exercised during storage. The composition of the gas mixture changes from its initial composition as a result of chemical, enzymatic and microbial activity of the product during storage.

Carbon Dioxide (CO_2)

Carbon dioxide is the major gas used in modified atmosphere. It is soluble in both water and lipid and although it is not a bactericide or fungicide, it has bacteriostatic and fungistatic properties. The overall effect on micro organisms is an extension of the lag phase of growth and a decrease in the growth rate during the logarithmic growth phase. However, the former effect is greater and therefore as bacteria move from the lag to log phase of growth, the inhibitory effects are reduced. Thus, the earlier the product is gas packaged the more effective CO_2 will be (Brody, 1989). This bacteriostatic effect is influenced by the concentration of CO_2, the partial pressure of CO_2, volume of headspace gas, the type of micro organism, the age and load of the initial bacterial population, the microbial growth phase, the growth medium used, the storage temperature, acidity, water activity, and the type of the product being packaged (Church, 1994; Farber, 1991; Phillips, 1996; Church and Parsons, 1995). Storage temperature has major role to play in the antimicrobial effect of CO_2. Temperature should be kept as low as possible, because the solubility of CO_2 decreases dramatically with increasing temperature. Thus keeping low temperature will helps in dissolution of CO_2 into the product to excert its antibacterial effect. The solubility of CO_2 leads to dissolved CO_2 in the food. A drop in surface pH is observed in MAP products because of the acidic effect of dissolved CO_2, but this could not entirely explain all of CO_2's bacteriostatic effect. Although the bacteriostatic effect of CO_2 has been known for many years, the precise mechanism if its action is still a subject of considerable interest. There have been many theories regarding the way in which CO_2 exerts its influence on a bacterial cell. The effect of CO_2 on bacterial growth is complex and four activity mechanisms of CO_2 on microorganisms have been identified (Daniels et al, 1985; Dixon and Kell, 1989):

a. Alteration of cell membrane function including effects on nutrient uptake and absorption
b. Direct inhibition of enzymes or decreases in the rate of enzyme reactions
c. Penetration of bacterial membranes, leading to intracellular pH changes
d. Direct changes to the physio-chemical properties of proteins.

Probably a combination of all these activities account for the bacteriostatic effect. The absorption of CO_2 is highly dependent on the moisture and fat content of the product. If product absorbs excess CO_2, the total volume inside the package will be reduced, giving a vacuum package look commonly known as package collapse. Dissolved CO_2 takes up much less volume compared with CO_2 gas and after packaging a product in CO_2 atmosphere, a drop in pressure is developed within the package and package collapse may occur. The ratio between the volume of gas and volume of food product (G/P ratio) should usually be 2:1 or 3:1 (The volume of gas is two to three times the volume of food). This high G/P ratio is also necessary to prevent package collapse because of the CO_2 solubility in wet products. Increase in dripping is caused by the dissolution of gases on the muscle surface in atmospheres containing high CO_2 levels (>60%), reduced pH and, consequently, low protein water retention ability. Excess CO_2 absorption in combination with package collaps can also reduce water holding capacity of the fish products, resulting in very high drip.

For some fishery products which are associated with pathogens like *Clostridium perfringens* and *Clostridium botulinum,* their growth will be encouraged by the use of CO2 as it creates anaerobic conditions. In such cases a certain amount of 02 is introduced in the package. In general carbon dioxide is most effective in foods where the normal spoilage organisms consist of aerobic, gram-negative psychrotropic bacteria (Hotchkiss, 1998; Phillips, 1996).

Oxygen (O_2)

Food deteriorates due to physical, chemical and microbiological factors. Oxygen is probably the most important gas in this context being used metabolically by aerobic spoilage micro-organisms and taking part in some enzymic reactions in food including the compounds such as vitamins and flavours. The presence of oxygen may cause oxidative rancidity problems in fish with high lipid, promoting the formation of low molecular weight aldehydes, ketones, alcohols and carboxylic acids. For these reasons, in modified atmosphere packaging, oxygen is either excluded or the levels set as low as possible. The exception is where O_2 is needed to maintain the colour in red meat fishes and to avoid anaerobic conditions (Parry, 1993). O_2 levels are normally kept as low as possible to reduce the oxidative changes of fish products. Oxygen generally stimulates the growth of aerobic bacteria and may inhibit the growth of exclusively anaerobic bacteria, although anaerobic microorganisms show different sensitivity levels to oxygen. One of the major functions of O_2 in MAP meats is to maintain myoglobin in its oxygenated form, oxymyoglobin. This is the form responsible for the bright red colour, which most consumers associate with fresh red meat (Farber, 1991). Oxygen in MAP packages of fresh fish will also inhibit reduction of TMAO to TMA.

Nitrogen (N_2)

Nitrogen is an insipid and inert gas which has little or no antimicrobial activity on its own. It shows very low solubility in water and lipids and due to this, the presence of N_2 in a MAP food can prevent pack collapse that can occur when high concentrations of CO_2 are used. In addition, N_2, by displacing O_2 in the pack, can delay oxidative rancidity and also inhibit the growth of aerobic micro organisms.

Carbon monoxide (CO)

Carbon monoxide has little inhibitory effect on micro organisms. It is normally applied as pigment fixative for maintaining the fresh red colour of red meat in MAP. It is very effective in maintaining the red colour in fresh meat due to the formation of carboxymyoglobin. It has not been used commercially for this purpose however since carbon monoxide, a highly toxic gas is not approved by the regulatory authorities owing to the possible health hazard. Although the US Food and Drug Administration (FDA) approved the practice as safe for use in 2004 in response to requests from two food companies, the EU prohibits food companies from using carbon monoxide. However, FDA is facing critics for its approval from many sectors including consumers. Several countries including Japan, Canada and Singapore also ban the use of carbon monoxide in tuna.

Other Gases

The other potential gases like argon, chlorine, etyhlene oxide, nitrogen dioxide, ozone, propylene oxide and sulphur dioxide for modified atmosphere packaging have been investigated experimentally but their commercial use for packaging foods is unlikely to meet with approval from the regulatory authorities.

Gas Mixtures

Commonly CO_2, O2 and N_2 gases are commonly used for MAP of fishery products. Using these three gases, there are three types of gas mixtures used in modified atmosphere packaging (Goodburn and Halligan, 1988) namely,

1. Inert blanketing (N_2)
2. Semi-reactive blanketing (CO_2/N_2 or $O_2/CO_2/N_2$)
3. Fully-reactive blanketing (CO_2 or CO_2/O_2)

The recommended gas mixtures for typical fish products (Parry, 1993) are listed below:

Product	% Oxygen	%Carbon dioxide	%Nitrogen
Red meat fish	60-85	15-40	-

Lean Fish	30	40	30
Fatty Fish	-	60	40

MACHINERY

Appropriate equipment is required for optimum gas packaging and there are two different techniques to replace the air. They are Gas flushing and compensated vacuum. The gas flush technique is normally accomplished on a form fill-seal machine. The replacement of air inside a package is performed by a continuous gas stream. This gas stream dilutes the air in the atmosphere surrounding the food product. The package is then sealed. Since the replacement of air inside the package is accomplished by dilution, there is a limit on the efficiency of this unit. Typical residual oxygen levels in gas flushed packs are 2-5% O2. Therefore, if the food item to be packaged is very oxygen sensitive, the gas flush technique is normally not suitable. So when considering a packaging system it is important to consider the oxygen sensitivity of the food product. The great advantage of the gas flush technique is the speed of the machine. Since the action is continuous, the product rate can be very high.

The compensated vacuum technique removes the air inside by pulling a vacuum on the atmosphere inside the package and then breaking the vacuum with the desired gas mixtures. Since the replacement of the air is accomplished in a two-step process, the speed of operation of the equipment is slower than the gas flush technique. However, since the air is removed by vacuum and not simply diluted, the efficiency of the unit with respect to residual air levels is better. Therefore, if the food product is extremely sensitive to oxygen like fatty fishes, a compensated vacuum machine must be used.

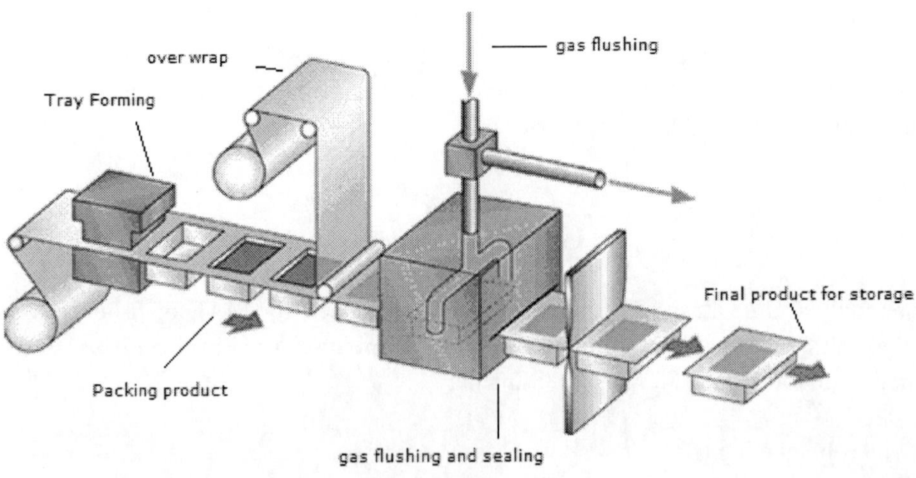

Figure 1. Form fill seal machine for gas flushing system.

PACKAGING MATERIALS

The choice of packaging material is an extremely important part of the MAP operation. The materials must be cost effective, have low water vapour transmission rate, high gas barrier, mechanical strength to withstand machine handling and subsequent storage and distribution of the finished pack as well as have the capability of giving high integrity seals to ensure retention of gas within the pack until opened by consumer.

Also, once a gas atmosphere is applied, the level and proportion of headspace gas/gases is controlled only by judicious selection of packaging material with specified permeability characteristics.

The permeability of the packaging material determines the atmospheric conditions in the headspace and ultimately the shelf-life of the product. If an atmosphere higher in carbon dioxide and / or lower in oxygen is required, the material should be impermeable to the gases. There are six main characteristics to consider when selecting packaging material for MAP foods:

1. Carbon dioxide permeability
2. Oxygen permeability
3. Sealing reliability
4. Resistance to puncture
5. Water transmission rate
6. Antifogging properties

No single polymer offers all the properties required for MAP. In addition to barrier properties, properties like machinability / sealability must also be taken into account. One inherent requirement for all MAP packs is the ability to retain the desired atmosphere as long as possible.

This is achieved first by choosing a film or films to provide the required gas and moisture vapour permeability characteristics and second by ensuring seal integrity of the packs. To achieve the above film characteristics the different plastic films are either laminated or co-extruded.

Although an increasing choice of packaging materials is available to the MAP industry, most packs are still constructed from single or coextruded films or laminates of ethylene vinyl alcohol (EVOH), polyvinyl dichloride (PVDC), polyvinyl chloride (PVC), polyethylene terephthalate (PET), polyproylene (PP) and polyethylene (PE), polyester, amorphous nylon (polyamide resin) and nylon. Films are usually coated on the inside of the pack with an antifogging agent, typically a silicone or stearate material, to disperse droplets of condensed moisture and permit the food to be visible.

Factors Influencing Shelf Life of Fish Products Packaged in Modified Atmospheres

Shelf life of fish products packed in MAP varies with fish species and other parameters (Table. 1). Several interrelated factors influence the shelf life and keeping quality of fish packaged under CO_2 enriched atmospheres. These include (a) concentration of CO_2 in gas mixture, (b) packaging film permeability, (c) storage temperature and (d) types and numbers of microbial contaminants.

Table. 1. Shelf life of fresh fishery products packaged under MAP compared to Vacuum and air

fish products	Storage temperature (∘C)	Atmosphere CO_2:N_2:O_2	Shelf life (days)	Reference
Cod (*Gadus morhua*) whole	2	100:0:0	10	Jensen et al (1980)
	2	60:40:0	10	
	2	40:60:0	9-10	
	2	Vacuum	8-9	
	2	Air	7	
Rockfish (*Sebastes* spp.)fillets	1.7	CA 80:20 air	13	Parkin et al (1981)
		Air	6	
Salmon,king (*Oncorhynchus tshawytsche*)fillets	2	Air	8	Stier et al (1981)
	4.4	60:15:25	12	
	4.4	Air	6	
	22.2	60:15:25	2	
	22.2	Air	1	
Cod fillets	0	40:30:30	12.5	Cann et al (1983)
	5	40:30:30	<7	
	10	40:30:30	3	
	0	Vacuum	9	
	5	Vacuum	<4	
	10	Vacuum	2	
Crayfish (*Pacifastacus leniusxulus*) whole cooked	4	80:20air	21	Wang and Brown (1983)
	4	air	14	
Swordfish(*Xiphias gladius*) steaks	2	Air	6*	Oberlender et.al (1983) *Shelf life based on microbial data only
	2	CA 100:0:0	>22*	
	2	CA 70:0:30	>22*	
	2	CA 40:0:60	14*	
	2	CA 70:30:0	>22*	
	2	CA 40:60:0	20*	
Trout whole	0	60:40:0	8	Cann et al (1984)
	5	60:40:0	8	
	10	60:40:0	3.8	
	0	Vacuum	9	
	5	Vacuum	6.5	
	10	Vacuum	3.7	
Snapper (*Chrysophrys auratus*) fillets	3	100:0:0	6-8	Scott et al (1984)
	3	Vacuum	3*	*No/medium 0_2 barrier mat
	3	Vacuum	6**	**High oxygen barrier material
	3	Air	3	
Cod (*Gadus morhua*) fillets	1	CA 60:40:0	12	Woyewoda et.al (1984)
	1	60:40:0	12	
	1	Air	9	
Salmon steaks	0	60:40:0	12.9	Cann et.al (1984)
	5	60:40:0	7.1	
	10	60:40:0	3.4	
	0	Vacuum	11.8	
	5	Vacuum	8	

fish products	Storage temperature (◦C)	Atmosphere CO_2:N_2:O_2	Shelf life (days)	Reference
Flounder (*Limanda*	26	Air	2	Post et al (1985)
ferruginal) fillets	26	Vacuum	2	
	26	0:100:0	4	
	26	100:0:0	1	
	12	Air	5	
	12	Vacuum	8	
	12	0:100:0	7	
	12	100:0:0	8	
	8	Air	5	
	8	Vacuum	7	
	8	0:100:0	4	
	8	100:0:0	10	
Shrimp,spotted	0	CA100:0:0	>14	Matches and
(*Pandalus*	0	Air	7	Layrisse (1985)
platyceros) whole	0	Air	7	
head on/off				
Cod (*G.morhua*)	26	100:0:0	2-3	Post et al. (1985)
fillets	26	Other	2	*All packaging
		atmosphere*		types at 26°C
		Air		(air,vacuum,
	12		6	100%Nitrogen,
		Vacuum		90:8:2 and
	12	0:100:0	10	65:31:4) except
	12	100:0:0	13	100%Co_2 had a
	12	Air	11	shelf life of 2 days
	8	Vacuum	06	
	8	0:100:0	16	
	8	100:0:0	17	
	8	90:8:2	23	
	8	65:31:4	17	
	8	100:0:0	16	
	4		40-53	
Cod fillets	0-2	Air	11	Gopal (1985)
		CAP(50%	19	
		O_2+50%CO_2)		
		Air	5	
		CAP(50%	11	
	5-7	O_2+50%CO_2)		
Whiting	26	All atmosph*	2*	Post et al (1985)
(*Merlucclus*	12	Air	5	*All packaging
bilinearis) fillets	12	Vacuum	9	types at 26°C (air,
	12	0:100:0	9	vacuum, 100%N_2,
	12	100:0:0	12	100%$C0_2$,90:8:2
	3	Air	4	and 65:31:4) had a
	8	Vacuum	10	shelf life of 2 days
	8	0:100:0	10	
	8	100:0:0	15	
	8	90:8:2	13	
	8	65:31:4	7	
	4	100:0:0	15	
Snapper	-1	40:60:0	9	Scott et al. (1986)
(*C.auratus*) fillets	-1	Air	9	
	-1	100:0:0	18	
Cod (*G.morhua*)	0	Air	12-13	Villemure et al
whole/fillets	0	25:75:0	20	(1986)

Table 1. (Continued)

fish products	Storage temperature (◦C)	Atmosphere CO_2:N_2:O_2	Shelf life (days)	Reference
Trout (*Salmo gairdneri*)fillets	1.7 1.7 1.7	80:20:0 80:20:0 Air	20 20* 10	Barnett et al (1987) *Treated with 2% potassium sorbate dip
Sardines (*Sardinops melanostictus*)	5 5 5	80:20:0 20:80:0 Air	4 4 2	Fuji et al (1989)
Catla (*Catla catla*), fillets	0-4	Air 50% O_2+50%CO_2 20% O_2+80%CO_2	12 20 28	Gopal et al. (1990)
Cod (*G.morhua*)fillets	0 0 0 0 0	2:98:0 3:97:0 29:71:0 48:52:0 97:3:0	14 13 16 20 15-16	Dalgaard et al (1993)
Haddock (*Melanogrammus aeglefinus*)whole	0 0 5 5 10 10	40:30:30 Air 40:30:30 Air 40:30:30 Air	10 8 7 7 4 4	Dhananjaya and Stroud (1994)
Catfish, channel (*Catalurus punctatus*) fillet strips	2	80:20 air	28	Silva and White (1994)
Cod,blue (*Arapercis colias*) fillets commercially smoked	3 3 3 -1.5 -1.5 -1.5	100:0:0 Vacuum Air 100:0:0 Vacuum Air	49 14 14 113 35 28	Penney et al (1994)
Herring (*Clupea harengus*) fillets	0 0	60:40:0 Air	14 12	Dhananjaya and Stroud (1994)
Herring (*C.harengus*) whole	0 0	60:40:0 Air	14 12	Dhananjaya and Stroud (1994)
Hybrid striped bass (*Marane saxatilis, M.chrysops*) strips	2 2	60:34:6 Air	13 7	Handumrongkul and Silva (1994)
Tilapia fillets	4 8 16 4 8 16	75:25:0 75:25:0 75:25:0 Air Air Air	>25 13-16 9-13 9-13 6-9 3-6	Reddy et al (1995)
Trout , rainbow (*Oncorhynchus mykisss*) fillets	2 2 2 2 2	20:80:0 20:80:0 40:60:0 40:60:0 Vacuum	6* 9** 6* 9** 6	Randell et al (1995) *G/P ratio 0.4 **G/P ratio 1.0

fish products	Storage temperature (◦C)	Atmosphere CO_2:N_2:O_2	Shelf life (days)	Reference
Herring, Baltic fillets	2 2 2 2 2	20:80:0 20:80:0 40:60:0 40:60:0 Vacuum	3* 3** 6* 8** 3	Randell et al(1995) *volume of gas to product ratio 0.4 **volume of gas to product ratio 1
Mackerel (*Scombrus scombrus L*) fillets	-2	100:0:0	>21	Hong et al (1996)
Sea bream (*Argyrozona argyrozona*)	0-4	Air 20% O_2+80%CO_2	22 40	Gopal et al (1996)
Salmon, atlantic (*Salmo salar*)slices	2	100:0:0	18	Pastoriza et al (1996)
Catfish fillets	16 16 16 8 8 8 4 4 4	Air 75:25:0 Vacuum Air 75:25:0 Vacuum Air 75:25:0 Vacuum	3 4 3 6 13 6 13 38-40 20-24	Reddy et al (1997a)
Salmon fillets	16 16 16 8 8 8 4 4 4	Air 75:25:0 Vacuum Air 75:25:0 Vacuum Air 75:25:0 Vacuum	4 5-6 3 13-17 20-24 >6<10 24-27 55-62 34-38	Reddy et al(1997b)
Hake (*Merluccius merluccius*)slices	2 2 2	50:45:5 50:45:5 Air	14 16* 7-8	Pastoriza et al (1998) *5 min 5%Nacl dip
Cod fillets	16 16 16 8 8 8 4 4 4	Air 75:25:0 Vacuum Air 75:25:0 Vacuum Air 75:25:0 Vacuum	3-4 6 3-4 13-17 24-27 13 20-24 55-60 24-27	Reddy et.al (1999)
Salmon (*S.salar*) fillets	10 2 2 2 2	Vacuum 60:40:0 40:60:0 Vacuum Air	3 17 17 17 11	Randell et.al (1999)

Note: Initial atmosphere in percent $C0_2$:N_2:0_2. If mixtures of carbon dioxide and air is used this is reported as % $C0_2$: % air (e.g. 80:20 air), if initial atmosphere is maintained during storage either in controlled atmosphere tank or by reflushing, this is indicated with CA.

CONCENTRATION OF GASES

Concentration of gases is a very important factor for shelf life of fish packaged under MAP conditions. The bacteriostatic effect of MAP storage increases with increasing CO_2 concentrations (Gill and Tan, 1980). With cod fillets stored in different concentrations of CO_2 and O_2/N_2, Stenstrom (1985) showed that the shelf life increased as the CO_2 concentration increased. Haines (1932) reported that concentrations as low as 10 to 20% CO_2 were sufficient to inhibit the growth of *Pseudomonas* and *Achromobacter* species. High CO_2 concentrations might be used for many types of seafood with little risk of heme protein discoloration because of the low pigment concentration (Parkin and Brown, 1982). Several studies have shown that carbon dioxide inhibits some types of microorganisms but has no direct effect on others. Most non-sterile food products contain a variety of organisms, and those organisms favored by environmental conditions may grow. In MAP, the inhibition of gram negative, rod shaped spoilage organisms such as *Pseudomonas* coincides with the growth of Gram positive, rod shaped spoilage organisms such as lactic acid producing organisms such as Lactobacillus. Carbon dioxide in sufficient concentrations can inhibit the growth of spoilage molds (Hintlian and Hotchkiss, 1986). Shewan (1949) recommended concentrations of between 30 and 40% CO_2 atmospheres and reported superior quality for fish stored in the higher concentration of CO_2. Tarr (1954) recommended a minimum of 40 to 50% CO_2 in the package headspace to derive maximum benefit of CO_2 storage of fresh fish. Coyne (1933), in his studies on use of carbon dioxide atmospheres to prolong fish quality, recommended concentrations of 40 and 60% CO_2. Based on these studies, approximately 50-60% CO_2 in the package headspace is commonly used, with the remainder comprising of a mixture of O_2 and N_2. The pioneering work of Coyne (1932, 1933) demonstrated that high levels of CO_2 (20-100%) would inhibit microbial growth and extend the shelf life of fish.

Packaging Film Permeability

MAP can only be successful if used in conjunction with packaging materials of correct O_2/CO_2 permeability characteristics. It is no use having the correct atmosphere if the film allows the atmosphere to change too rapidly. The properties required for a suitable packaging film are seldom formed in one polymer and individual polymers are laminated to produce films of superior barrier properties. Detailed note on the packaging materials requirement for MAP is given is already discussed above.

STORAGE TEMPERATURE

MAP is not a substitute for proper storage temperature (Ogrydziak and Brown, 1982). The effectiveness of MAP is decreased as the storage temperature increases due to the fact that the solubility of CO_2 also decreases at higher temperatures. For respiring products, increasing the temperature also increases the rate of respiration, resulting in a decrease in shelf life. The effects of temperature abuse are particularly important from the standpoint of safety. Temperature abuse of MAP muscle foods may result in the rapid growth of both spoilage and pathogenic bacteria. The minimum reported temperature for *Salmonella* and *Escherichia coli* inoculated in ground meat grew equally well at 12.5 °C when the meat was packaged in low and high permeability film (Goepfert and Kim, 1975). *Staphylococcus aureus* can grow and produce enterotoxin under anaerobic conditions at 10°C or less. The minimum-recorded temperature for enterotoxin production under anaerobic conditions is 10°C, while Lysteria monocygoens has been reported to grow at temperatures as low as -1.5°C and *Yersinia enterocolitica* even at –2°C. Of major concern, with respect to safety of MAP fish, is the growth and toxin production by *Clostridium botulinum* type E that is capable of growth at 3.3 °C (Palumbo, 1986). Proper refrigeration is therefore essential in order to assure the effectiveness of CO_2 as an anti microbial agent and to prevent potential growth of pathogenic organisms.

Levels of Microbial Contamination

The shelf life extension of fish under modified atmospheres is dependent on the initial microbial load and the types of organisms. Higher initial counts will accelerate spoilage and shorten the shelf life of fish products.

Furthermore, high numbers of aerobic bacteria will also consume rapidly headspace O_2 and may change electronegative potential of product and enhance the growth of anaerobic organism such as *Clostridium botulinum* type E.

Advantages and Disadvantages of MAP

Advantages	Disadvantages
Increased shelf life of 50–400%. Extended storage results in reduced economic losses and wider distribution	Temperature control required
Reduction in retail waste	Different gas compositions for each type of product
Fewer distribution deliveries leads to lower Costs	Requirement for special gas packing equipment, gases and packaging material and operator training adds to increased cost
Little or no need for chemical preservatives	Cost of analytical equipment to ensure that correct gas mixtures are being used
Easier separation of sliced foods (except vacuum packing)	Increased pack volume has impact on transport and retail display costs
Improved presentation-clear view of product and all round visibility	Benefits of modified gas packaging are lost once the pack is opened or leaks
Hygienic stackable pack, sealed and free from product drip and odour	Potential growth of food-borne pathogens due to temperature abuse by retailers and consumers

Adapted from Davies (1995), Farber (1991) and Blakistone (1998).

SAFETY OF MAP PRODUCTS

Control of microbial spoilage and growth of cold tolerant pathogens in MAP products is a matter of public health concern. For fishery products, modified atmospheres should be as low as possible in oxygen and as high as possible in CO_2 without causing the pack to collapse or produce changes to the flavour or appearance of the product. Reducing the concentration of oxygen inhibits the development of normal spoilage micro-organisms, especially *Pseudomonas* sp. (Walker, 1992). Other spoilage bacteria that can grow in low oxygen concentrations grow more slowly and so extend the time taken for spoilage to occur, for example lactic acid bacteria or *Brochothrix thermosphacta,* which cause spoilage by souring (Nychas and Arkoudelos, 1990). Concern has been expressed over potential risks to consumer safety from modified atmospheres packaging because they inhibit normal spoilage microorganisms and thus allow food to appear fresh, while permitting the growth of anaerobic pathogens. Several pathogens including *Clostridium botulinum*, *Listeria monocytogenes*, *Yersinia enterocolitica*, *Salmonella* sp., and *Aeromonas hydrophila* are anaerobes or facultative anaerobes (Blakistone, 1998). Although they are present in very low numbers in fish products, they pose a great threat to the consumer's health. Hence to overcome the potential safety hazard, it is recommended that one or more of the following criteria are met:

- Water activity is maintained below 0.92
- pH should be below 4.5
- Use of sodium nitrite or other preservative
- The temperature is maintained below +3°C

Different fish have different composition, moisture content and microbial flora. Hence, optimum gas composition should be established for each fish products. Care should be taken to prevent temperature abuse during processing and distribution. Use of very fresh fish, hygienic handling of fish products and the adoption of Hazard Analysis and Critical control points (HACCP) system plays a major role in ensuring the safety of all MAP fishery products.

REFERENCES

Barnett, H. J., Conrad, J. W and Nelson, R. W. (1987). Use of laminated high-and low density polythene flexible packaging to store trout (*Salmo gairdneri*) in modified atmosphere, *J. Food Prot.* 50 (8), 645-651.

Blakistone, B. A. (1998) Introduction. In: *Principles and Applications of Modified Atmosphere Packaging of Foods*, 2nd edn (Blakistone, B. A. ed.), pp. 1–13, Blackie Academic and Professional, London.

Brody, A. L. (1989). Modified atmosphere packaging of seafoods. In; Controlled/ Modified Atmosphere/Vacuum packaging of foods (Brody, A.L. ed), pp. 59-65, Food and Nutrition Press. Trumbull, CT.

Cann, D. C., Smith, G. L. and Houston, N. C. (1983) *Further studies on Marine fish Stored under Modified Atmosphere Packaging*. Ministry of Agriculture, Fisheries and Food, Torry Research Station, Aberdeen, Scotland.

Cann, D. C., Houston, N. C., Taylor, L. Y., Smith, G. L., Thomson, A. B. and Craig, A. (1984). *Studies of Salmonids Packed and Stored under a Modified Atmosphere. Aberdeen*: Torry Research Station.

Church, I. J. and Parsons, A. L. (1995). Modified Atmosphere Packaging Technology: A Review, *J. Sci. Food Agric.* 67, 143-152.

Church, N. (1994). Developments in Modified-Atmosphere Packaging and Related Technologies, *Trends in Food Science and Tech.* 5, 345-352.

Coyne, F. P. (1932). The effect of carbon dioxide on bacterial growth with special reference to the preservation of fish. Part I, *J. Soc. Chem. Ind.* 51, 119T-121T.

Coyne, F. P. (1933). The effect of carbon dioxide on bacterial growth with special reference to the preservation of fish. Part II- Gas storage of fresh fish. *J. Soc. Chem. Ind.* 52: 19T-24T.

Dalgaard, P., Gram, L. and Huss, H. H. (1993). Spoilage and shelf life of cod fillets packed in vacuum or modified atmospheres. *Int. J. Food Microbiol.* 19, 283-294.

Daniels, J. A., Krishnamurthy, R., and Rizvi, S. S. H. (1985). A Review of Effects of CO_2 on Microbial Growth and Food Quality, *J. Food Prot.* 48, 32-537.

Davies, A. R. (1995). Advances in Modified-Atmosphere Packaging, New Methods of Food Preservation. (Gould, G.W. ed.), pp.304-320, Blackie, Glasgow, UK.

Dhananjaya, S., and Stroud, J. (1994). Chemical and sensory changes in haddock and herring stored under modified atmosphere. *Int. J. Food Sci. and Technol.* 29, 575-583.

Dixon, N. M., and Kell, D. B. (1989). The Inhibition by CO_2 of the Growth and Metabolism of Microorganisms, *J.Appl.Bacteriol.,* 67, 109-136.

Farber, J. M. (1991). Microbiological aspects of modified atmosphere-packaging technology – a review. *J. Food Prot.*, 54, 58-70.

Fuji, T., Hirayama, M., Okuzumi, M., Yasuda, M., Nishito, H. and Yokotyama, M. (1989). Shelf life studies on fresh sardine packaged with carbon dioxide-nitrogen gas mixture. *Nippon Sulsan gaikkashi,* 55: 1971-75

Gill, C.O. and Tan, K.H. (1980). Effect of carbon dioxide on growth of meat spoilage bacteria. *Appl. Environ. Microbiol.* 39, 317-321.

Goepfert, J. M. and Kim, H. V. (1975). Behaviour of selected foodborne pathogens in raw ground beef, *J. Milk Food Technol.* 38, 443-448.

Goodburn, K. E., and Halligan, A. C. (1988). Modified Atmosphere Packaging: A Technology Guide, Leatherhead Food RA.

Gopal, T. K. S, Nambiar, V. N, Battacharya, S. K, Joseph, J and Prabhu, P. V.(1990). Modified Atmosphere storage of freshwater fish fillets of Catla catla. Proc. International Institute of Refrigeration on Chilling and Freezing of new fish proucts. pp. 143-146, Aberdeen, U.K.

Gopal, T. K. S. (1985). Studies on the storage life of cod fillets using controlled atmosphere packaging. Proc. Symposium on Recent Developments in Food Packaging, pp. 86-89, AFSTI, India.

Gopal, T. K. S., Nambiar, V. N, Joseph, J, Lalitha, K. V, Prabhu, P. V. (1996). Modified atmosphere storage of sea bream fillets from sea bream (*Argyroxona argyrozona*). Proc. Workshop on Scientific results of FORV Sagar Sampada, pp. 513-519, DOD, New Delhi.

Haines, R. B. (1932). The influence of carbon dioxide on the rate of multiplication of certain bacteria as judged by viable counts. *J. Soc. Chem Ind.* 52, 13-18.

Handumrongkul, C. and Silva, J. L. (1994). Aerobic counts, colour, and adenine nucleotide changes in CO_2 packed refrigerated striped bass strips. *J. Food Sci.*, 59, 67-69.

Hintlian, C. B. and Hotchkiss, J. H. (1986). The safety of modified atmosphere packaging: A review. *Food Technol.*, 50, 70-74.

Hong, L., Leblanc, E. L., Hawrysh, Z. J. and Hardin, R. T. (1996). Quality of Atlantic mackerel (*Scomber scombrus L.*) fillets during modified atmosphere storage. *J. Food Sci.* 61, 646-651.

Hotchkiss, J. H. (1989). Microbiological Hazards of Controlled / Modified Atmosphere Food Packaging, *J. Assoc. Food Drugs Offic.,* 53 (3), 41-49.

Jensen, M. H., Petersen, A. R. and Jepsen, A. (1980). Storage of chilled cod under vacuum at various concentrations of carbon dioxide. In: Advances in Fish Science and Technology (Connell. J. J. ed.), p.294, Fishing News, Torry Research Station Jubilee conference. Farnham.

Matches, J. R. and Layrisse, M. E. (1985). Controlled atmosphere storage of spotted shrimp (*Pandalus platyceros*). *J. Food Prot.* 48, 709-711.

Nychas G. J. and Arkoudelos, J. S. (1990). Microbiological and physiochemical changes in minced meats under carbon dioxide, nitrogen or air at 3ºC. *Int. J. Food Sci. Technol.* 25, 389–398.

Oberlender, V., Hanna, M. O., Miget, R., Vanderzant, C. and Finne, G. (1983). Storage characteristics of fresh swordfish steaks stored in carbon dioxide enriched controlled (flow through) atmospheres. *J. Food Prot.* 46, 434-440.

Ogrydziak, D. M. and Brown, W. D. (1982). Temperature effects in modified atmosphere storage of seafoods. *Food Technol.* 36, 86-96.

Palumbo, S. (1986). Is refrigeration enough to restore foodborne pathogens? *J. Food Prot.* 49, 1003-1005.

Parkin, K. L. and Brown, W. D. (1982). Preservation of seafoods with modified atmospheres. In: *Chemistry and biochemistry of marine food products*. AVI Publishing Co., Westport, CT.

Parkin, K. L., Wells, M. J. and Brown, W. D. (1981). Modified atmosphere storage of rockfish fillets. *J. Food Sci.* 47, 181-185.

Parry, R. T. (1993). Introduction. In: *Principles and applications of modified atmosphere packaging of food,* (Parry, R. T. ed.), pp. 1-18, Blackie, London.

Pastoriza, L., Sampedro, G., Herrera, J. J. and Cabo, L. M. (1996). Effect of modified atmosphere on shelf life of iced fresh hake slices. *J. Sci. Food Agr.* 71, 541-547.

Pastoriza, L., Sampedro, G., Herrera, J. J. and Cabo, L. M. (1998). Influence of sodium chloride and modified atmosphere packaging on microbiological, chemical and sensorial properties in ice storage of slices of hake (*Merluccius merluccius*). *Food Chem.* 61, 23-28.

Penney, N., Bell, R. G. and Cummings, T. I. (1994). Extension of the chilled storage life of smoked blue cod (*Parapercis colias*) by carbon dioxide packaging. *Int. J. Food.Sci. Technol*, 31, 463-479.

Phillips, C. A. (1996). Review: Modified Atmosphere Packaging and Its Effects on the Microbiological Quality and Safety of Produce, *Int. J. Food Sci. Tech.*, 31, 463-479.

Post, L. S., Lee, D. A., Solberg, M., Furgang, D., Specchio, J. and Graham, C. (1985). Development of botulinal toxin and sensory deterioration during storage of vacuum and modified atmosphere packaged fish fillets. *J. Food Sci.* 50, 990-996.

Randell, K., Abvenainen, R. and Hattula, T. (1995). Effect of gas/product ratio and CO_2 concentration on the shelf life of MA packed fish. *Packaging Technol. and Sci.* 8, 205-218.

Randell, K., Hattula, T., Skytta, E., Sivertsvik, M., Bergslien, H. and Ahvenainen, R. (1999). Quality of filleted salmon in various retail packages. *J. Food Quality.* 22, 485-497.

Reddy, N. R., Roman, M. G., Solomon, H. M., Kauter, D. A. and Rhodehamel, E. J. (1997a). Shelf life and *Clostridium botulinum* toxin development during storage of modified atmosphere packaged fresh catfish fillets. *J. Food Sci.* 62, 878-884.

Reddy, N. R., Solomon, H. M., Yep, H., Roman, M. G. and Rhodehamel, E. J. (1997b) Shelf life and toxin development by *Clostridium botulinum* during storage of modified atmosphere packaged fresh aqua cultured salmon fillets. *J.Food Prot.* 60,1055-1063.

Reddy, N. R., Villanueva, M. and Kautter, D. A. (1995) Shelf life of modified atmosphere packaged tilapia fillets stored under refrigeration and temperature abuse conditions. *J. Food Prot.* 58, 908-914.

Reddy, N. R., Solomon, H. M. and Rhodehamel, E. J. (1999). Comparison of margin of safety between sensory spoilage and onset of *Clostridium botulinum* toxin development during storage of modified atmosphere- packed fresh marine cod fillets with MA packaged aqua cultured fish fillets. *J. Food Saf.* 19, 171-183.

Scott, D. N., Fletcher, G. C. and Hogg, M. G. (1986). Storage of snapper fillets in modified atmosphere s at -1°C. *Food Technol. Australia*, 38, 234-238.

Scott, D. N., Fletcher, G. C. and Summers, G. (1984). Modified atmosphere and vacuum packaging of snapper fillets. *Food Technol. Australia,* 36, 330-334.

Shewan, J. M. (1949). Some bacteriological aspects of handling, processing, and distribution of fish. *J. Roy. Sanit. Inst.* 69, 394-398.

Silva, J. L. and White, T. D. (1994). Bacteriological and color changes in modified atmosphere packaged refrigerated channel catfish. *J. Food Prot.* 57, 715-719.

Stenstrom, I. M. (1985). Microbial flora of cod fillets (*Gadus morhua*) stored at 2°C in different mixtures of carbon dioxide and nitrogen/ oxygen. *J. Food Prot.* 48 (7), 585-589.

Stier, R. F., Bell, L., Ito, K. A., Shafer, B.D., Brown, L. A. Seeger, M. L., Allen, B. H., Porcuna, M. N. and Lerke, P. A. (1981) .Effect of modified atmosphere storage on *Clostridium botulinum* toxigenesis and the spoilage microflora of salmon fillets. *J. Food Sci.* 46, 1638-1642.

Tarr, H. L. A. (1954). Microbiological deterioration of fish postmortem: its detection and control. *Bacteriol. Rev.* 18, 1-6.

Villemure, G., Simard, R. F. and Picard, G. (1986). Bulk storage of cod fillets and gutted cod (*Gadus morhua*) under carbon dioxide atmosphere. *J. Food Sci.* 51, 317-320.

Walker, S. J. (1992). Chilled foods microbiology. In: *Chilled foods – a comprehensive guide.* (Dennis, C. and Stringer, M. eds.), pp. 165–195, Ellis Horwood, London.

Wang, M. Y. and Brown, W. O. (1983). Effect of elevated CO_2 atmosphere on storage of freshwater crayfish (*Pacifastacus leniusculus*). *J. Food Sci.* 48, 158-159.

Woyewoda, A. D., Bligh, E. G. and Shaw, S. J. (1984). Controlled and modified atmosphere storage of cod fillets. *Canadian Institute Food Science and Technology,* 17, 24-27.

In: Recent Progress in Chemistry...
Editor: A. K. Haghi, pp. 223-239

ISBN: 978-1-61668-501-0

Chapter 13

ADVANCES AND APPLICATION OF GALACTOSIDASES IN FOOD PRODUCTS

Michele Rigon Spier**, Adenise Lorenci Woiciechowski, Ashok Pandey and Carlos Ricardo Socco
Universidade Federal do Paraná, Setor de Tecnologia,
Engenharia de Bioprocessos e Biotecnologia, Brazil

ABSTRACT

Some vegetables and dairy products have nutritional properties, but its consumption is somewhat limited in part of the population due to some digestion problems or intolerances it provokes. This chapter focuses on foods containing high level of some oligosaccharides and disaccharides, and the importance of α-galactosidases and β-galactosidases to improve health and quality of these products. Several studies have shown the production and the application of pure α-galactosidases to reduce undigested components naturally present in foods to minimize digestion problems. β-galactosidases are another group of galactosidases also important in food processing mainly in dairy-products processing to achieved lower lactose content to people who has lactose intolerance. Researches of β-galactosidases production and purification are being done to apply in new dairy-products. The development of new food products containing healthy benefits and which minimized digestive problems are world-wide consumption trends.

INTRODUCTION

Many vegetables and dairy products important to the human dietary have nutritional properties, but contain considerable levels of undesirable components such as tannins, lectins, protease and amylase inhibitors and the presence of oligosaccharides due to some digestion problems it provokes. Raffinose, stachyose, mellibiose and verbascose are examples which are not hydrolyzed in the small intestine due to a lack of the specific hydrolase enzyme; Once

* Corresponding author Email: Michele Rigon Spier : spier@ufpr.br

being in the low intestinal tract, bacterial colonies are able to ferment these sugars producing CO_2, H_2 and CH_4, often causing flatulence (Iyer et al, 1980; Vidal-Valverde et al, 1993). Other symptoms that can appear are nausea, cramps, abdominal pain (Cristofaro et al, 1974), headache, mental and concentration disturbance (Sannie et al, 1997).

The raffinose family of oligosaccharides (RFOs) is present in soybean and other legumes in relatively small quantities as components of the carbohydrate fraction (Bach Knudsen and Li, 1991). In general, the predominant RFO is stachyose, followed by raffinose and verbascose, depending on the type of grain (Petterson and Mackintosh, 1994). On the oligosaccharide structure, sucrose may contain one or more α-D-galactopyranosyl (α-D-Gal) groups, in which the α-galactosyl units are always to the left of sucrose, connected to the molecule of glucose (Silva et al., 1992). The structural relationship of these sugars is showed in Figure 1.

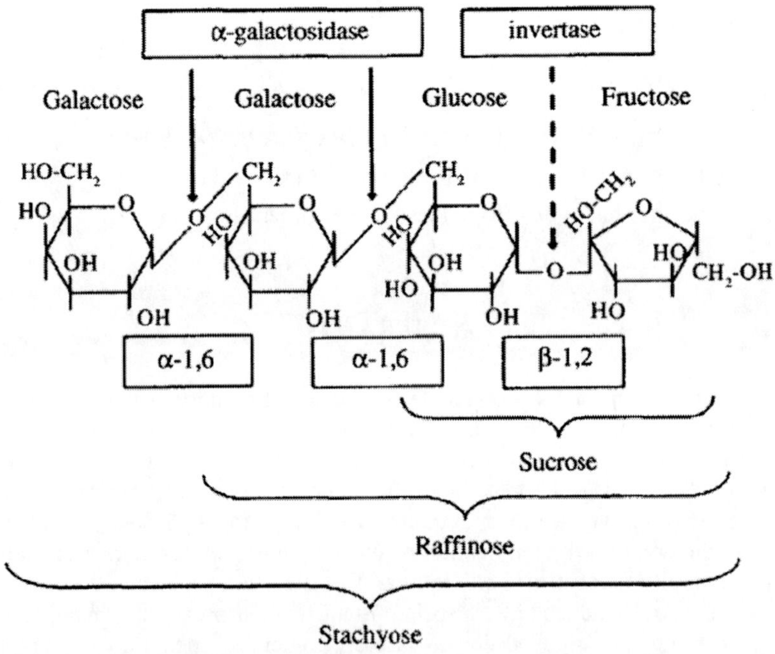

Figure 1. Structures of the α-galactosides in soybean, stachyose and raffinose, and the enzymes that break the glucosidic linkage (LeBlanc et al., 2004; Sanada et al, 2009).

The raffinose family of oligosaccharides (RFOs) is considered the cause of flatulence (Dey and Pridham, 1972). The absence of the α-galactosidase enzyme in the low gastrointestinal tract (responsible for the hydrolysis of α-1,6 galactosides linkages) and its accumulation in the large intestine, result in the fermentation by anaerobic bacteria (Fleming, 1981). The ingestion of galactooligosaccharides (GOS) showed an increase in the incidence of diarrhoea in mice. This fact was confirmed by Liying et al. (2003), when a diet supplemented with 2% purified stachyose increased the occurrence of diarrhoea in pigs when compared to control.

Oligosaccharides

The potential of carbohydrates in the maintenance of health and as therapeutic agents has been realized (McAuliffe and Hindsgaul 1997; Sharon and Lis, 1993). Oligosaccharides consist of two to ten monosaccharide residues linked via glycosidic linkages and can be liberated by depolymerisation (Khadem 1988; Pazur 1970).

Oligosaccharides in Vegetables

Beans (*Phaseolus vulgaris* L.) are important sources of proteins, carbohydrates, fibers, vitamins, unsaturated fatty acids, and micronutrients (Deshpande, 1992). Brazil is the first world producer of the species *Phaseolus vulgaris* (Hungria et al, 2000). In 2007, the Brazilian cultivation of beans achieved 3.2 million of tons, being a basic food to the population. The consumption of cooked beans is 12.7 Kg/person/year in Brazil (Embrapa 2008).

Bean grains contain 20–35% protein and 50–60% carbohydrates, but present antinutritional factors such galactooligosaccharides (GO), mainly raffinose, stachyose and verbascose (Lyimo et al 1992). The GO are heat resistant and remain in the end product (Jangchud and Bunnang, 2001). Due to the absence of α-galactosidase (EC 3.2.1.22) in the intestinal tract of human, the GO are accumulated and fermented, producing a large amount of gases. The GO are considered inducers of gastrointestinal disorders, diarrhoea and flatulence, inhibiting nutrient uptake (De Lumen, 1992).

Raffinose negatively impacts sugarbeet (*Beta vulgaris* L.) industrial processing by decreasing extractable sucrose yield and altering sucrose crystal morphology which reduces filtration rates and slows processing. Although increased raffinose concentrations have been observed during cold storage, the physiological and biochemical mechanisms associated with **raffinose accumulation** in sugarbeet are poorly understood. (Haagenson et al, 2008).

α-Galactosidases

The enzyme α-D-galactoside galactohydrolase (EC 3.2.1.22), also named α-galactosidase or α-gal, catalyzes the hydrolysis of α-1,6- galactosidic bonds, releasing α-D-galactose. These bonds are found in oligosaccharides such as melibiose, raffinose and stachyose (Figure 1), which are associated to flatulence in monogastric animals and humans (Suarez et al., 1999). Human consumption of soy-derived products has been limited by the presence of these nondigestible oligosaccharides (NDOs), since most mammals, including man, lack pancreatic α-galactosidase. However, such NDOs can be fermented by gas producing microorganisms present in the cecum and large intestine, thus inducing flatulence and other gastrointestinal disorders in sensitive individuals (LeBlanc et al., 2004). α-galactosidases are applied in biocatalytic processes to improve quality to products destined to human nutrition and also as component in animal diets to increase digestibility and reduce the fermentation of NDOs. Nutritional studies using commercial preparations of α-galactosidase as supplement in the feed of monogastric animals demonstrated to significantly increase weight profit simultaneously with digestibility in swine and chickens. The food industry, especially the soy derivatives sector, has great interest in the reduction of galacto-oligosaccharides, which is resistant to heat and, therefore, are not eliminated during conventional processing (Said and Pietro, 2004). Mulimani, et al (1997) reported that crude α-galactosidase treatment on soybean flour reduced the raffinose and stachyose contents by 90.4% and 91.9%, respectively. In the beet-sugar production process, this enzyme hydrolyses raffinose present in the beetroot boiled must, contributing to product crystallization.

α-Galactosidases present in bean grains can be potentially used for several biotechnological applications, mainly in the food industry to hydrolyze the GO in soybean products (Guimarães, 2001; Viana et al 2005).

Sources of α-Galactosidases

Several sources of α-galactosidases have been studied (Shivanna et al, 1989). According to Manzanares et al (1998), the bacterial enzyme shows some advantages comparing to fungal α-gal, since its higher production. However, fungal galactosidases are obtained easily due its extracellular location and its wide stability profile, being therefore viable its production in large scale. (Annunziato et al, 1986; Somiari and Balochi, 1992). This enzyme presents different forms, monomeric or tetrameric, and presents thermal stability (Alani et al, 1989; Porter et al, 1992). The great majority of the enzymes are glycoproteins, with glucose and manose residues in soya enzyme (Porter et al 1992; Zeilinger et al 1993). In soybean, the galactose protects partially the catalytic site in tetrameric form against heat denaturation, but this protection is pH-dependent. Studding the enzyme of vegetable source, Mulimani and Ramalingam (1997) performed experiments with soy flour and concluded that galactosidases from guar (*Cyamopsis tetragonolobus*) better reduced oligosaccharides content than microbial enzymes.

Microbial Enzyme Source

Sakai et al. (1987) studied the performance of Bifidobacterium breve in soy milk and observed raffinose and staquiose reduction. The maximal enzyme activity was detected at 50ºC at pH 5.0. These optimal conditions are similar to those observed to enzyme produced by Bifidobacterium longum and to Lactobacillus fermentum (Garro et al., 1994; Garro et al., 1996). However, differences on enzyme behavior are observed according to inhibitors action: the addition of p-chloromercuribenzoato (PCMB) affected the enzymatic activities of B. breve and L. fermentum suggesting the presence of sulfidrile groups in catalytic site of enzyme, while α-galactosidase produced by B. longum do not suffer with the addition of PCMB, revealing different structures to enzymes produced by distinct microorganisms. The addition of Hg^{2+} íons inhibits the enzyme activity produced by the three microorganisms due the presence of tiol groups.

Porter et al. (1992) showed the use of α-galactosidase produced by Bacillus stearothermophilus presented thermo stability at 65-70ºC. This thermal resistance allows the application of this enzyme in food and feed industries, where thermal processing is required.

Similar studies were done by Eveleigh (2000) using the bactéria Thermotoga neopolitana, for α-galactosidase to industrial application.

Fialho et al (2006) demonstrated the specific α-galactosidase activity in the bean grain. The values varied from 0.032 to 0.049 U/mg and suggested that the varied 'Mexico 222' is a promising source of α-galactosidase for industrial purposes.

Sanada et al (2009) studied the production of α-galactosidase by submerged fermentation using soybean vinasse as substrate, a residue of the alcoholic fermentation of soybean molasses. Soybean molasses is a by-product of the protein-concentrate soybean meal production. Lactobacillus agilis was selected for presenting the highest enzymatic activity, equivalent to 11.07 U/mL after 144 h of fermentation. The effects of soluble solids concentration in the fermentation medium, C:N relation and size of inoculum were investigated. Results demonstrated that the medium concentration of 30% soluble solids, with a C:N relation of 9, and size of inoculum of 25% (v/v) were the best conditions for α-galactosidase production.

β-Galactosidases

β-Galactosidases catalyze the hydrolysis of β(1-3) and β(1-4) galactosyl bonds in oligosaccharides. Also catalyse the inverse reaction of enzymatic condensation and trans-glycosylation. β-galactosidase facilitates the hydrolysis of lactose in the two monosaccharides (glucose and galactose) which are then absorbed.

A great part of world population presents some food intolerance. Food intolerance is an adverse food reaction resulting from enzymatic defects of the gastrointestinal tract. The symptoms are manifested in the gastrointestinal tract after ingestion of certain foods.

The most common food intolerance is caused by the consumption of lactose. This comes from a genetically predetermined reduction in β-galactosidase activity, a hydrolase also known as lactase which hydrolyzes the carbohydrate lactose from dairy products in the small intestine (Ortolani and Pastorello 2006; (Moriwaki and Matioli, 2000). During the neonatal period, the enzyme is essential in human nutrition. The decline of the intestinal β-galactosidase (β-gal or commonly known as lactase) activity is a biological characteristic of the maturing intestine in the majority of the world's population. This declination, termed hypolactasia, causes insufficient lactose digestion in the small intestine. Hypolactasia and lactose malabsorption accompanied with clinical symptoms, such as bloating, flatulence, nausea, abdominal pain and diarrhoea, are termed lactose intolerance. Numerous studies have shown that individuals with hypolactasia could tolerate fermented dairy products better than an equivalent quantity in milk (Vasiljevic, 2009).

Later, the specific activity of **lactase** decreases to low adult levels (Moriwaki and Matioli, 2000). It can affect 6–12% of caucasians, but in some ethnic groups the prevalence can exceed 60%. Due to β-galactosidase deficiency lactose cannot be completely hydrolysed and the molecules reach the colon where they are degraded by bacteria in H_2O, CO_2 and H_2.

The lactose that is not easily absorbed in the small intestine passes on to the colon where it produces fluid accumulation by osmotic action and undergoes microbial fermentation causing intestinal distress (Potter and Hotchkiss, 1995). The fermentation in the colon causes disturbances such as bloating, abdominal pain and sometimes diarrhoea (Ortolani and Pastorello 2006). This frequent gastrointestinal symptoms (are abdominal pain, flatulence, or diarrhea), may result from the fermentation of undigested lactose by colonic bacteria. The reduction or elimination of milk and dairy products from the diet of lactose-intolerant children may compromise their uptaken of protein, riboflavin and calcium. Accordingly, alternatives to milk elimination from the diet have been developed. Microbially produced β-galactosidase, is added to milk to prehydrolyse lactose, or can be consumed when foods containing lactose are ingested. Powdered fermented milk that retains β-galactosidase activity also reduced lactose maldigestion in children. It has been proposed that improved fermented milk products could be produced using strains of bacteria that possess high β-galactosidases activity at acidic pH. Another approach to the management of lactose digestion is the consumption of fermented (unfermented) milk containing live lactic bacteria like *Lactobacillus acidophilus* and *Bifidobacterium longum* (Moriwaki and Matioli, 2000).

It is now well established that population who are **lactase** non-persistent digest lactose more efficiently from yogurt and fermented milks than from milk. An in vitro assay was used to study the **lactase** activity of *Bifidobacterium bifidum*. Frozen cells and fermented milk supplemented with such living organisms were tested. **Lactase** activity was measured by lactose disappearance. Glucose and galactose by-products from lactose hydrolysis were used differently by the bacterial cells (Passerat and Desmaison, 1995).

Rojas et al (2004) reported the crystallographic structure of *Penicillium* sp. β-galactosidase, which is a glycoprotein containing seven N-linked oligosaccharide chains.

The lactase hydrolyses the lactose in glucose and galactose, which edulcorant capacity is 0.8 times of sacarose; the hydrolysed is also 3 to 4 times more soluble than lactose. Besides, monosaccharides are easily absorbed in the intestine (Wiseman, 1991).

β-Galactosidases Production by Microorganisms

Commercial sources of β-galactosidases are *Saccharomyces lactis*, *Aspergillus niger*, *Aspergillus oryzae*, *Escherichia coli*, *Saccharomyces*, *Kluyveromyces*. The lactase from milk yeast *S. lactis* is the most used commercial enzyme. The optimal activity conditions are 35-40ºC and pH 6.6-6.8, are similar to the milk yeasts, but it is most used in the milk and non-acid whey treatments.

The second known lactase preparation is derived from A. niger, which optimum activity conditions are 50ºC and pH 3.5-4.5 to application in acid-whey. The enzymes from *A. niger* and *S. lactis* are inhibited by the product, the galactose, and it is too difficult to get a complete lactose hydrolysis (Wiseman, 1991).

β-galactosidase from *Kluyveromyces marxianus* was produced under submerged fermentation using a culture medium based on cheese whey. The fermentations were carried out in a shaker at 30°C and initial pH 5.5 under agitation, starting with an initial cellular concentration of 10^7 cells per mL, where the initial concentrations of lactose and yeast extract were varied.

Enzymes were extracted from the cells by lysis with chloroform in potassium phosphate buffer. In the medium with a initial lactose concentration of 50 g/L, supplemented with salts, yeast extract 12 g/L, the enzymatic activity and cellular concentration were 28 U/mL and 5.3g/L respectively (Santiago et al, 2004).

Table 1. β-galactosidases produced by microorganisms

Microorganism	Conditions	Reference
Alicyclobacillus acidocaldarius	Submerged fermentation, 100 rpm, 65°C, 24h; 0.69 U/mg; It belong to the glycoside hydrolyse family 42	Gul-Guven et al, 2007
Bacillus subtilis	Appeared to be independent of starch or lactose availability. Medium containing tryptona. Probable inducing effect of oat and chestnut flour. Yeast extract improve enzyme yield	Konsoula et al, 2007
Kluyveromyces marxianus	Submerged fermentation, whey medium with 50 g/L lactose, 12 g/L yeast extract, (NH4)2SO4, KH2PO4 e MgSO4.7H2O, 150rpm, 30°C, pH 5.5; 28 U/mL	Santiago et al, 2004
Bifidobacterium bifidum	Submerged fermentation using milk, initial pH 6.5. In 240 h, utilized 5.94 mmol/ 100 mL of lactose, final pH 6.13	Passerat and Desmaison, 1995
K. lactis	Submerged fermentation 200 rpm, 30°C, 48 h, Medium composition (in g L^{-1}): lactose, 30; yeast extract, 1; K_2HPO_4, 2; $NH_4H_2PO_4$, 1; $(NH_4)_2HPO_4$, 1; $MgSO_4$. $7H_2O$, 0.1 and pH was adjusted to 7.0, intracellular enzyme	Dagbagli and Goksungur (2008)
Bififobacteria longum	Submerged fermentation, 4 % lactose, 3.5 % yeast extract, 0.3% K2HPO4, 0.1 KH2PO4, 0.05 MgSO4.7H2O; 0.03% L-systeine, 16 h of fermentation, 37°C; around 18.6 U/mL	Hsu et al, 2005
Bifidobacteria infantis	Submerged fermentation,	Hsu et al, 2005
Kluyveromyces lactis	Solid state fermentation, corn grits and wheat bran, 250 rpm, 30°C, 20 h	Becerra and Siso, 1996

α-Galactosidase Production by Microrganisms

α-Galactosidases are widely distributed in microorganisms, plants and animals (Said and Pietro, 2004). Microorganisms are able to produce α-galactosidases with high yield. The production of these enzymes by microorganisms as fungi, yeasts and bacteria consists of microorganism growth in a selective and generally liquid culture medium, containing an enzymatic activity inductive carbon source.

Sugars as galactose, melibiose, raffinose and stachyose have already been described as inductors of α-galactosidase activity (Rezende, Guimarães, Rodrigues, and Felix, 2005; Said and Pietro, 2004).

Table 2. α-galactosidases produced by microorganisms

Microorganism	Conditions	Reference
Aspergillus awamori	Solid state fermentation using raw pinto bean flour contained 0.38% raffinose and 3.34% stachyose on a dry weight basis treated with α-galactosidase (60 U/mL). Complete hydrolysis after 2 h	Song et al, 2009
Lactobacillus agilis	Submerged fermentation, medium at 30% (w/v) soluble solids containing soybean vinasse, yeast extract, C:N 9, 144 h,	Sanada et al, 2009,
Aspergillus oryzae	Solid state fermentation in tray, red gram plant waste + wheat bran, 81% moisture content, pH 5.5, 4 days; 30ºC; 4.37 U/g	Shankar and Mulimani, 2007
Rhizopus sp.	Enzyme-induced medium containing 3% soybean meal, bran, lactose, galactose, xylose, glucose, cellulose. Submerged fermentation at 250 rpm, 30ºC, 120 h, 0.4 U/mL, 78 KDa	Cao et al, 2007
Bacillus stearothermophilus	Submerged fermentation, 0.2% (w/v) galactose, phosphate salts and yeast extract, pH=7.0, 200 rpm, 60ºC, 24h	Gote et al, 2006
Penicillium sp.	Submerged fermentation, 250 rpm, 30ºC, 7 days	Mi et al, 2007
Bifidobacterium lactis	Submerged fermentation in sterilized commercial soymilk for 48h at 42ºC reduced raffinose in 77.4%,	Donkor et al, 2007
Humicola sp.	Solid state fermentation using soya flour at 45ºC, 10 days, initial moisture of 95%, final pH 8.5, 44.6 U/g	Kotwal et al, 1997
Lactobacillus reuteri	Raffinose as substrate, highest yield at the beginning of the stationary growth phase; 64 KDa enzyme	Tzortzis et al, 2003
Lactobacillus fermentum *Lactobacillus plantarum*	Solid state fermentation using pinto bean and fiend pea flours, moisture adjusted to 20%, 72h, stachyose reduction in 27%	Duszkiewicz-Reinhard et al, 2006
Lactobacillus fermentum *Lactobacillus plantarum*	Solid state fermentation using field pea flours, moisture adjusted to 20%, 72 h, stachyose reduction in 43%	Duszkiewicz-Reinhard et al, 2006
Gibberella fujikuroi	Submerged fermentation using Czapek-Dox media with 20 g/L of galactose as carbon source, pH=5.5, 160 rpm, 27ºC	Thippeswamy and Mulimani, 2002
Streptomyces griseoloalbus	Solid state fermentation composed by soybean flour supplemented with 1.5% galactose, salts solution, incubated at 30ºC, 40% initial moisture, inoculum $1.9x10^6$ CFU/g initial dry substrate, 5 days, 117 U g^{-1} of dry fermented substrate.	Anisha and Prema, 2008
Streptomyces griseoloalbus	Submerged fermentation, 20.4 U/mL, initial pH 6.0, 1% locust bean gum, 30ºC, 10% (v/v) inoculum	Anisha and Prema, 2007
Pichia guilliermondii	Submerged fermentation; maximal α-galactosidase detected in media containing lactose and the soytone product at pH 7. Enzyme production was stimulated by addition of metal salts to the culture broth.	Church and Meyers, 1980
Thermomyces lanuginosus	Sucrose substrate is promising for development of an enzyme fermentation technology. Submerged fermentation 3% of sucrose, 0.9% ammonium acetate, 9 days of fermentation, 90-100 U/mL	*Rezessy-Szabó* et al, 2003

Crude cell-free extracts from *Lactobacillus reuteri* grown on cellobiose, maltose, lactose and raffinose were assayed for glycosidic activities. When raffinose was used as the carbon source, α-galactosidase showed the highest yield at the beginning of the stationary growth phase. A 64 kDa enzyme was purified by ultra- and gel filtration, and characterized for its hydrolytic and synthetic activity. Highest hydrolytic activity was found at pH 5.0 at 50 °C (K_M 0.55 mM, V_{max} 0.80 μmol min^{-1} mg^{-1} of protein). At a substrate concentration of 23% (w/v) oligosaccharide mixtures were formed with main products being a trisaccharide at 26% (w/w) yield from melibiose after 8 h and a tetrasaccharide at 18% (w/w) yield from raffinose after 7 h. Methylation analysis revealed the trisaccharide to be 6′ α-galactosyl melibiose and the tetrasaccharide to be stachyose. In both cases synthesis ceased when hydrolysis of the substrate reached 50% (Tzortzis et al, 2003).

Lactic acid bacteria such as *Lactobacillus plantarum*, *Lactobacillus fermentum*, *Lactobacillus brevis*, *Lactobacillus buchneri* and *Lactobacillus reuteri* are also able to hydrolyse α-galacto-oligosaccharides into digestible carbohydrates during vegetable fermentations. Recently the characterization of genes involved in α-galacto-oligosaccharides hydrolysis by *Lactococcus raffinolactis* was described (LeBlanc et al, 2005). Bacterial enzyme-producing species also include *L. fermentum* (Garro et al, 1996), *Bacillus stearothermophilus* (Gote et al, 2004). Fungal enzymes can be produced by *Aspergillus fumigatus* (Rezende et al., 2005), Humicola sp. (Kotwal et al, 1998) and *Aspergillus niger* (Manzanares et al, 1998).

Legumes constitute one of the richest and least expensive sources of protein in the human diet in many parts of the world. Proteins are essential components of the diet, needed for survival of humans and animals. The basic function of proteins in nutrition is to supply adequate amounts of required amino acids. Soybean (Glycine max) is one such important plant in human and animal nutrition. About 60% of all processed food products include ingredients derived from soy. Soy products have superior nutritional characteristics in terms of high protein, amino acids contents and it is suitable as a lowcost substitute to dairymilk in developing countries and as a nutritive supplement for the lactose intolerant population (Biranjan et al, 1983). Furthermore it contains a large amount of galactooligosaccharides such as raffinose and stachyose that are indigestible by mammals and are flatulence-causing factors (Reddy et al, 1980). Flatulence is the common symptom associated with consumption these oligosaccharides but more serious accompanying consequences such as abdominal pain and diarrhoea overshadow its social implications (Salunkhe and Kadam, 1989). α-galactosidase is abundant in plants, microorganisms and animals. It hydrolyzes a variety of simple a-D galactoside as well as more complex molecules such as oligosaccharides and polysaccharides (Cruz and Park, 1982; Dey and Pridham, 1972). Several groups have developed enzymatic processes using α-galactosidase from plant, bacterial and fungal source for the hydrolysis of galacto-oligosaccharides of raffinose family. Khare et al. (1994) have successfully demonstrated the removal of significant amount of oligosaccharides by *Aspergillus oryzae* cells entrapped in agarose. Their result indicates a possibility of using whole cells for pre-treatment of soymilk to overcome the physiological problems of flatulence. They suggest to constructing a fluidized bed reactor for continuous hydrolysis of RFOs in soymilk. The aim of that work is to investigate the applicability of the immobilized cells for the reduction of RFOs in soymilk in continuous mode using a fluidized bed reactor to solve the physiological problem of flatus in a systematic manner.

Raffinose oligosaccharides (ROs) make up a substantial part (40%) of the soluble sugars found in soybean seeds and are responsible for flatulence after the ingestion of soybean and other legumes. Consequently, soy-based foods would find a broader approval if the ROs were removed from soybean products or hydrolysed by α-galactosidases. During soybean seed germination, the content of ROs decrease substantially, while the α-galactosidase activity increased. α-Galactosidase was partially purified from germinating seeds by partition in an aqueous two-phase system and ion-exchange chromatography. The enzyme preparation presented maximal activities against q-nitrophenyl-a-D-galactopyranoside (qNPGal) at 60°C and a pH of 5.0 and the K_M values for qNPGal, melibiose, and raffinose of the enzyme preparation were 0.33, 0.42, and 6.01 mM, respectively. The enzyme was highly inhibited by SDS, copper, and galactose. Hydrolysis of soybean flour ROs by enzyme preparation reduced the stachyose and raffinose contents by 72.3% and 89.2%, respectively, after incubation for 6 h at 40°C.

A-Galactosidase in Foods

α-galactosidase has been used in some countries for the hydrolysis of raffinose into sucrose and galactose in concentrated sugar syrups and molasses from sugar beet processing. This process provides for improved sugar crystallization and a higher sucrose yield. For example, Hokkaido Sugar Company has treated 300 tons of molasses with 1-5% raffinose per day in 1990 (Buchholz et al, 2005).

Soy-Based Products

Soy-based foods may provide a range of health benefits to consumers. Hypolipidemic, anticholesterolemic, antiatherogenic, allergenicity reducer, reducer risk of most hormone-associated health disorders are some properties described in literature.

However, consumption of soymilk is hindered, due to the presence of unpleasant off-flavours carried over from soy beans. These characteristic flavours are caused by n-hexanal and -pentanal, which occur in beans as a product of breakdown of unsaturated fatty acids (Arai, Suzuky, Fujimake, and Sakurai, 1996; Scalabrini et al, 1998). In addition to these aldehydes, soymilk contains various oligosaccharides, including raffinose and stachyose, which may cause a gastrointestinal discomfort to consumers (Tsangalis and Shah, 2004).

Raffinose and stachyose are α-galactosides of sucrose, comprising three and four monomeric units, respectively. They are non-digestible in the gut due to the absence of α -galactosidase in the human intestinal mucosa. Consequently, intact oligosaccharides pass directly into the lower intestine, where they are metabolized by bacteria that synthesize this enzyme, resulting in the production of gases (Tsangalis and Shah, 2004). This problem could be alleviated by using α-galactosidases, or an organism that possesses high α-galactosidase activity, to minimize the content of flatulence-causing oligosaccharides in the product (Scalabrini et al., 1998). Several Bifidobacterium strains have been reported to produce varying levels of α-galactosidase, which metabolize α-galactosyl oligosaccharides in soymilk (Scalabrini et al., 1998).

Soymilk is a good medium for growing *Bifidobacterium* because it contains oligosaccharides that are fermented by most of the strains belonging to this genus (Liu, 1997; Scalabrini et al., 1998).

Soy Sauce

In soy sauce production, high levels of protein and oligosaccharides (raffinose, stachyose, melibiose and sucrose) presents in soybeans are reduced by koji fermentation. Fermentation of koji is performed with a mixture of soybeans or a mixture of beans and wheat which are inoculated with *Aspergillus oryzae* or *Aspergillus soyae*, allowed to stand for 3 days at 25-35°C and moisture around 30%. So, significant level of exogenous enzymes, including α-galactosidases acted in saccharifying sugars during koji fermentation. The sugars are required to the second step of soy sauce fermentation by lactic acid bacteria and yeast.

In the mash stage, called moromi, brine id added to koji until a salt concentration of 17-19% and fermented at room temperature for 2-4 months (commercial preparation). Bacteria and yeast involved in the fermentation include *P. halophilus*, *Lb. delbrueckii*, *Z. rouxii* and *Torulopsis* species. The souce is removed from the mash and pasteurised at 70-80°C before bottling (Caplicec e Fitgerald, 1999).

Tempeh

Tempeh is a Indonesian soy-based food obtained by solid state fermentation by *Rhizopus oligosporus*. First, soybeans are soaked in water overnight at ambient temperature. Then the grains are dehulled, partially cooked and inoculated with the spores solution of *R. oligosporus*.

Tempeh contains 40% of protein and is an important source of vitamin B12, may be meat substitute and commonly used in soups or sliced, salted and deep fat fried in coconut oil. Lactic bacteria strains such as *L. casei* and *Lactococcus* species dominate the fermentation, which may be initiated by the addition of a commercial culture or a small amount of a previous batch (Caplice and Fitgerald, 1999). The fermented beans are bound together with the fungal mycelia forming a compact block similar to a cake. The oligosaccharides content in Tempeh is reduced by fermentation by α-galactosidases produced by *R. oligosporus*.

The tempeh fermentation of the germinated soybeans caused a further decrease in oligosaccharides (raffinose and stachyose) (Markakis 2006).

Tofu

Although many high-protein foods such as meats, dairy products and beans are quite difficult for some people to digest, tofu is soft and highly digestible food. It presents a digestion rate of 95%, it is certainly the most digestible of all natural soybean foods, and is much more digestible than cooked whole soybeans (68%) due the reduced oligosaccharides content (Shurtleff and Aoyagi, 1999).

Yambean Based Foods

Yambean (*Sphenostylis stenocarpa* L.) is a legume commonly found in Africa, rich in protein but underutilized due to the flatulence and diarrhoea caused by raffinose family oligosaccharides (raffinose, stachyose and verbascose) consumption present in these seeds. In order to reduce the limitation of yambean uses as food, non-traditional processing methods are required. The yambean seeds were fermented by tempeh fungal (*Rhizopus oligosporus*). In fermented yambea, instead of water for soaking and cooking, 1% citric acid solution was used (Azeke et al, 2007). The fermentation reduced 98% of α-galactosides in three varieties of yambea seeds tested, resulting in a bacteria-free tempeh made with yambea seeds instead soybeans. Instead the fermentation, boiling seeds for 4h resulted in 8-30% reduction of total α-galactosides. The authors concluded that both tempeh production processes were clearly more effective than was traditional cooking in reducing the flatulence potential of the yambean seeds.

The reduction of stachyose in legume flours was studied (Duszkiewicz-Reinhard et al, 2002). Different populations of *Lactobacillus fermentum* or *Lactobacillus plantarum* were inoculated in pinto bean and field pea flours, adjusted with water to 20% moisture and incubated. After 72 h of incubation stachyose was reduced in pinto bean (bv 27%) and field pea (43%) flours (Duszkiewicz-Reinhard et al, 2002).

B-Galactosidases in Food

Relief of the symptoms of lactose intolerance is probably the most widely accepted health benefit of probiotic organisms. The traditional cultures used in making yoghurt (i.e., *L. delbrueckii* ssp. *bulgaricus* and *Str. thermophilus*) contain substantial quantities of β-D-galactosidase (Shah, 2007), and so both yoghurt and probiotic yoghurt are tolerated well by lactose intolerant organism. However, reduced levels of lactose in fermented products due to partial hydrolysis of lactose during fermentation are only partly responsible for this tolerance to yoghurt (Rodgers, 2008; Bernardeau, et al., 2008; Donkor, 2007).

Foods that contain pro and prebiotics are called symbiotic. An example is the water soluble extract of soybeans containing the oligosaccharides raffinose and stachyose, as prebiotics to stimulate the growth of probiotics, and an alternative used by people who have allergies and/or disturbances caused by the ingestion of milk (Spier et al, in press).

The high lactose content in milk products such as ice-cream, frozen milks, whey spreads and condensed milk, can lead to excessive lactose crystallisation resulting in products with a mealy, sandy or gritty texture. The application of β-galactosidase to these dairy products could reduce lactose concentration to acceptable values, and so improve some technological and sensorial quality of dairy foods; e.g. increasing the digestibility, softness, creaminess, etc (Zadow, 1993).

In the cheese industry whey is a waste containing lactose, which causes many environmental problems. Whey containing high level of lactose trends to crystallize easily. Besides lactose is a relatively insoluble sugar which provokes several problems during storage and transportation by its crystallization (Wiseman, 1991). Lactose is associated with the high biochemical and chemical oxygen demand and uncertain solubility (Guimarães et al, 1992).

Lactose may be hydrolysed by β-galactosidases into very useful sweet syrup, which can be used in the dairy, confectionary, baking and soft drinks industries (Pivarnik et al 1995).

Some important applications of β-galactosidases include milk destined to population whom has lactose intolerance, especially children; milk to dairy products processing such as cheeses, yoghurts, ice-cream and frozen dairy products; whey or lactose to produce edulcorants and soluble hydrolyzed syrups from whey; concentrated products such as condensed milk and to avoid sandy texture in frozen milk and ice cream prepared with lactose (Wiseman, 1991).

The milk may be treated with β-galactosidases (lactases) in the course of its processing. An alternative is add the enzyme to regular milk and hold the milk in the refrigerator for a prescribed time to hydrolysis process. Commercial β-galactosidases are available for this purpose.

TRENDS ON GALACTOSIDASES APPLICATION

The food industries focus on the production of new products by the evolution of technological processes for extending the shelf-life and increasing human health. Besides the reduction of collateral and adverse effects to their digestive systems are also population demand and necessity. Some components in vegetables such as legumes and grains bring anti-nutritional or indigestible effects to human consumption.

In the last decades, bean processing was basically made by grain processing companies which used basic unit operations such as washing, drying, selection, classification, cooking and packing to commercialization.

Another way of beans, grains and legumes industrialization is pre-cooking the grains, drying in a drum dryer, getting flakes to instantaneous preparation. This product is used in soups and composition of culinary preparation. The demand increase in bean derived foods, the industries should apply enzymes that degradate oligosaccharides non-digestible to minimize gastrointestinal problems described above. After some pre-treatment of these vegetables, α-galactosidases are the main group of enzymes that hydrolysis these components in digestible molecules that may permit the use of bean product in baby food after research studies about digestibility of the components in the hydrolyzed product.

The increasing demand by products based in grains such as soy and black beans turns an alternative to added value to these raw materials, promote economy growth, employee generation and offer to the consumer new alternatives of consumption of healthy foods.

New alternatives for children products and baby foods could be developed using legumes and grains not recommended for children before a year old due to the presence of indigestible compounds. So, vegetables submitted to enzymatic treatment in order to eliminate these compounds such as non-digestible oligosaccharides could be an alternative to the children food industries and new perspectives and consumption options to this important and promissory sector.

Dairy products with low lactose content are been developed to lactose-intolerant population. Products are obtained by lactic acid bacteria or enzymatic hydrolysis of lactose in milk, where glucose and galactose are release during processing, and such sugars are known as secure and avoid digestion problems related before.

REFERENCES

Alani, S.R.; Smith, D.M.; Markakis, P. (1989), α-galactosidases of Vigna unguiculata. *Phytochemistry,* 28 (8), 2047-2051.

Anisha G.S.; Prema, P. (2008), Reduction of non-digestible oligosaccharides in horse gram and green gram flours using crude α-galactosidase from Streptomyces griseoloalbus *Food Chemistry* 106, 1175–1179.

Annunziato, M.E.; Mahoney, R.R.; Mudgett, R.E. (1986), Production of α-galactosidase from Aspergillus oryzae grown in solid state culture. *Journal of Food Science*, 51 (5), 1370-1371.

Arai, S.; Suzuky, H.; Fujimake, M.; Sakurai, H. (1996), Studies on flavour components in soybean. Part 2. Phenolic acids in defatted soybean flour, *Agricultural and Biological Chemistry* 30 (1996), 364–369.

Azeke, M.A.; Fretzdorff B, Buening-Pfaue H.; Betsche T. (2007), Comparative effect of boiling and solid substrate fermentation using the tempeh fungus (Rhizopus oligosporus) on the flatulence potential of African yambean (Sphenostylis stenocarpa L.) seeds. *Food Chemistry* 103 (4), 1420-1425.

Becerra, M.; González Siso, M.I. (1996), Yeast β-galactosidase in solid-state fermentations *Enzyme and Microbial Technology* 19, 39-44.

Biranjan, J. R., Ingole, D. R., and Sanyal, S. R. (1983). Utility of soymilk in dietary therapy of undernourished children. *Indian Medical Gazette*, 9, 313–316.

Buchholz, K.; Kasche, V.; Bornscheuer, U.T. (2005), *Biocatalysts and enzyme technology*, Weinheim: Wiley.

Cao Y.; Yang P.; Shi, P.; Wang, Y.; Luo, H.; Meng, K.; Zhang, Z.; Wu, N.; Yao B. Fan, Y. (2007), Purification and characterization of a novel protease-resistant -galactosidase from Rhizopus sp. F78 ACCC 30795. *Enzyme and Microbial Technology* 41 (2007) 835–841.

Caplicec, E.; Fitzgerald, G.F. (1999), Food fermentations: role of microorganisms in food production and preservation. *International Journal of Food Microbiology* 50:131–149.

Church Jr., F.C.; Meyers, S.P. (1980), Alpha-Galactosidase from Pichia guilliermondii. *Mycologia*, 72, (2), 279-287.

Cruz, R.; Park, Y.K. (1982), Production of fungal a-galactosidase and its application to the hydrolysis of galacto-oligosaccharides in soybean milk. *Journal of Food Science*, 47, 1973–1975.

Dagbagli, S.; Goksungur, Y. (2008), Optimization of α-galactosidase production using Kluyveromyces lactis NRRL Y-8279 by response surface methodology. *Electronic Journal of Biotechnology* 11 (4).

Darrin M. Haagenson, Klotz, K.L.; Campbell, L. (2008), Impact of storage temperature, storage duration, and harvest date on sugarbeet raffinose metabolism. *Postharvest Biology and Technology* 49, 221–228.

De Lumen OB (1992) Molecular strategies to improve protein quality and reduce flatulence in legumes: a review. *Food Structure* 11: 33–46.

Deshpande SS (1992), Food legumes in human nutrition: a personal perspective. *Crit. Rev. Food Sci. Nutr.* 32: 333–363.

Dey, P.M.; Pridham, J.B. (1972), Biochemistry of a-galactosidase. *Advance Enzymology*, 36, 911–930.

Donkor, O.N.; Henriksson A.; Vasiljevic, T.; Shah, N.P. (2007), a-Galactosidase and proteolytic activities of selected probiotic and dairy cultures in fermented soymilk. *Food Chemistry* 104, 10–20.

Duszkiewicz-Reinhard, W.; Gujska E., Khan K. (2002), Reduction of Stachyose in Legume Flours by *Lactic Acid Bacteria Journal of Food Science*, 59 (1), 115 – 117.

Fialho, L.S.; Guimaraes, V.M.; Barros, E.G.; Moreira, M.A.; Dias, L.A.S.; Oliveira, M.G.A.; Jose, I.C.; Rezende, S.T. (2006), *Biochemical Composition and Indigestible.*

Garro, M., Valdez, G.F., Oliver, G., Giori, G.S. (1996). Purification of α-galactosidase from Lactobacillus fermentum. *Journal of Biotechnology*, 45, 103–109.

Giacomini C., Irazoqu G., Batista-Viera F., Brena B.M. (2001): Influence of immobilization chemistry on the properties of immobilized β-galactosidase. Journal of Molecular Catalysis B: Enzymatic, 11, 597–606.

Gote, M., Umalkar, H., Khan, I., Khire, J. (2004). Thermostable α-galactosidase from Bacillus stearothermophilus (NCIM 5146) and its application in the removal of flatulence causing factors from soymilk. *Process Biochemistry*, 39, 1723–1729.

Gote, M.M.; Khan, M.I.; Gokhale, D.V.; Bastawde, K.B.; Khire, J.M.(2006), Purification, characterization and substrate specificity of thermostable α-galactosidase from Bacillus stearothermophilus (NCIM-5146), *Process Biochemistry* 41, 1311–1317.

Guimarães VM, de Rezende ST, Moreira MA, de Barros EG, Felix CR (2001) Characterization of α-galactosidases from germinating soybean seed and their use for hydrolysis of oligosaccharides. *Phytochemistry* 58, 67–73.

Guimarães W.V., Dudey G.L., Ingram L.O. (1992), Fermentation of sweet whey by ethanologenic Escherichia coli. *Biotechnology and Bioengineering*, 40: 41–45.

Gul-Guven R.; Guven,K.; Poli, A.; Nicolaus, B. (2007), Purification and some properties of a β-galactosidase from the thermoacidophilic Alicyclobacillus acidocaldarius subsp. rittmannii isolated from Antarctica. *Enzyme and Microbial Technology* 40, 1570–1577.

Hsu, C.A.; Yu, R.C.; Chou, C.C. (2005), Production of β-galactosidase by Bifidobacteria as influenced by various culture conditions, *International Journal of Food Microbiology* 104, 197– 206.

Hungria M, Andrade DS, Chueire LMO, Probanza A, Guttierrez- Mañero FJ, Megıas M (2000), Isolation and characterization of new efficient and competitive bean (Phaseolus vulgaris L.) rhizobia from Brazil. *Soil. Biol. Biochem.* 32, 1515–1528.

Jangchud K, Bunnang N (2001), Effect of soaking time and cooking time on qualities of Red Kidney bean flour. *Kasetsart J.* 35, 409–415.

Kestwal, R.M.; Bhide, S.V. (2007), Purification of β-galactosidase from Erythrina indica: Involvement of tryptophan in active site. *Biochimica et Biophysica Acta* 1770, 1506–1512.

Khare, S.K.; Jha, K.; Gandhi, A.P. (1994), Hydrolysis of flatulence-causing oligosaccharides by agarose-entrapped Aspergillus oryzae cells. *Food Chemistry*, 51, 29–31.

Konsoula, Z.; Liakopoulou-Kyriakides, M. (2007), Co-production of a-amylase and b-galactosidase by Bacillus subtilis in complex organic substrates. *Bioresource Technology* 98, 150-157.

Kotwal, S.M., Gote, M.M., Sainkar, S.R., Khan, M.I., Khire, J.M. (1997). Production of α-galactosidase by thermophilic fungus Humicola sp. in solid-state fermentation and its application in soyamilk hydrolisis. *Process Biochemistry*, 33, 337–343.

Kumar, S.K.P.; Mulimani, V.H. (2010) Continuous hydrolysis of raffinose family oligosaccharides in soymilk by fluidized bed reactor. *LWT - Food Science and Technology* 43, 220–225.

LeBlanc, J.G., Piard, J.C., Sesma, F., Giori, G.S. (2005), Lactobacillus fermentum CRL 722 is able to deliver active alpha-galactosidase activity in the small intestine of rats. *FEMS Microbiology Letters*, 248, 177–182.

Lyimo M, Mugula J, Elias T (1992), Nutritive composition of broth from selected bean varieties cooked for various periods. *J. Sci. Food Agric.* 58, 535–539.

Manzanares, P., Graaff, L.H., Visser, J. (1998). Characterization of galactosidases from Aspergillus niger: Purification of a novel α-galactosidase activity. *Enzyme and Microbial Technology,* 22, 383–390.

Markakis, S.P. (2006), Tempeh Prepared from Germinated Soybeans. *Journal of Food Science* 52 (6), 1736 – 1737.

Moriwaki, C.; Matioli, G. (2000), Influence of β-galactosidase in milk technology and lactose metabolism. *Arq. Ciên. Saúde Unipar* 4(3): 283-290.

Nakayama, T.; Goto, Y.; Hasegawa, S.; Inohara-Ochiai, M.; Shibano, Y.; Toshihiko, Ashikari, T.; Nishino, T. (2000), Mutational Analysis of the Role of His452 of Saccharopolyspora rectivirgula P-Galactosidase, *Journal of Bioscience and Bioengineering*, 90 (5), 535-539.

Oligosaccharides in Phaseolus vulgaris L. *Seeds Plant Foods for Human Nutrition* 61: 87–89.

Ortolani C.; Pastorello, E.A. (2006). Food allergies and food intolerances. *Best Practice and Research Clinical Gastroenterology* 20 (3), 467–483.

Osaana N. Donkor, A. Henriksson, T. Vasiljevic, N.P. Shah (2007), α-galactosidase and proteolytic activities of selected probiotic and dairy cultures in fermented soymilk. *Food Chemistry* 104, 10–20.

Papayannakos n., Markas g. (1993): Studies on modelling and simulation of lactose hydrolysis by free and immobilized β-galactosidase from *Aspergillus niger. The Chemical Engineering Journal*, 52: B1–B12.

Passerat, B.; Desmaison, A.M. (1995), Lactase activity of Bifidobacterium bifidum. *Nutrition Research* 15 (9), 1287-1295.

Pivarnik L.F., Senacal A.G., Rand A.G. (1995), Hydrolytic and transgalactosylic activities of commercial β-galactosidase (lactase) in food processing. *Advances in Food and Nutrition Research*, 38, 1–102.

Porter, J.E. (1992), Effect of pH on subunit association and heat protection of soybean α-galactosidase . *Enzyme Microbial Technology*, New York, 14, 609-613, 1992.

Porzucek H, Duszkiewicz-Reinhard, W.; Piecyk, M.; Klepacka, M., Gniewosz, M. Changes of Flatulence-Causing Sugars in Legume Protein Samples by High Hydrostatic Pressure. *Food Science and Technology*.

Potter, N.N.; Hotchkiss (1995), Food science, 5th ed. New York: Chapman and Hall. 608 p.

Reddy, N.R., Salunkhe, D.K., Sharma, R.P. (1980), Flatulence in rats following ingestion of cooked and germinated black gram and a fermented product of black gram and rice blend. *Journal of Food Science*, 45, 1161–1164.

Rezende, S.T., Guimarães, V.M., Rodrigues, M.C., Felix, C.R. (2005). Purification and characterization of an alpha-galactosidase from Aspergillus fumigatus. *Brazilian Archives of Biology and Technology*, 48(2), 195–202.

Rezessy-Szabó, J.M.; Nguyen, Q.D.; Bujna, E.; Takács, K.; Kovács, M.; Hoschke, A.

Thermomyces lanuginosus CBS 395.62/b Strain as Rich Source of α-Galactosidase Enzyme. *Food Technol. Biotechnol.* 41 (1) 55–59 (2003).

Rojas, A.L.; Nagem, R.A.P; Neustroev, K.N; Arand, M; Adamska, M; Eneyskaya, E.V; Kulminskaya, A.A; Garratt, R.C; Golubev, A.M; Polikarpov, I; Rojas, A.L. (2004), Crystal structures of β-Galactosidases from *Penicillium* sp. and its complex with galactose. *Journal of Molecular Biology*, 343 (5), 1281-1292.

Roy I.; Gupta M.N. (2003), Lactose hydrolysis by LactozymTM immobilized on cellulose beads in batch and fluidized bed modes. *Process Biochemistry*, 39: 325–332.

Salunkhe, D.K.; Kadam, S.S. (1989). Hand book of world legumes: Nutritional chemistry processing technology and utilization. Boca Raton, Florida: CRC Press Inc.

Sanada, C.T.N; Karp S.G.; Spier, M.R.; Portella, A.C.; Gouvêa P.M., Caroline T. Yamaguishi, C.T.; Vandenberghe, L.P.S.; Pandey, A.; Soccol, C.R. (2009) Utilization of soybean vinasse for α-galactosidase production. *Food Research International* 42, 476–483.Santiago, P.A.; Marquez, L.D.S.; Cardoso, L.D.S.; Ribeiro, E.J. (2004) Estudo da Produção de β-galactosidase por fermentação de soro de queijo com Kluyveromyces marxianus. *Ciênc. Tecnol. Aliment.*, 24(4): 567-572.

Scalabrini P.; Rossi, M.; Spettoli P.; Matteuzzi, D. (1998), Characterization of bifidobacterium strains for use in soymilk fermentation. *Int. J. Food Microbiol.* 39: 213-219.

Shankar, S.K.; Mulimani, V.H. (2007), a-Galactosidase production by Aspergillus oryzae in solid-state fermentation. *Bioresource Technology* 98 (2007) 958–961.

Shivanna, B.D., Ramakrishan, M. (1985), α-Galactosidase from germinating guar (Cyamopsis tetragonolobus). *Journal of Biosciences*, 9, 109–116.

Shurtleff, W.; Aoyagi, A., (1999), The book of tofu, Tokyo.

Soh, C.P.; Ali, Z.M.; Lazan, H. (2006), Characterisation of an α-galactosidase with potential relevance to ripening related texture changes, *Phytochemistry* 67 (2006), 242–254

Song, D., Chang, S.K.C.; Ibrahim, S.A. (2009), Descriptive sensory characteristics of no flatulence pinto bean. Journal of Food Quality. DOI 10.1111/j.1745-4557.2009.00278. *Aspergillus awamori.*

Spier, M.R.; Habu, S.; Scheidt, G.N.; Woiciechowski, A.L., Pandey, A.; Soccol, C.R. (in press), Fermented functional foods, In: Comprehensive Food Fermentation Biotechnology, chapter n. 29.

Thippeswamy, S.; Mulimani, V.H. (2002), Enzymic degradation of raffinose family oligosaccharides in soymilk by immobilized a-galactosidase from Gibberella fujikuroi. *Process Biochemistry* 38, 635-640.

Tsangalis D.; Shah NP. (2004), Development of an isoflavone aglycone-enriched soymilk using soy germ, soy protein isolate and bifidobacteria. *Food Res. Int.* 37(4):301–12.

Tzortzis, G.; Jay, A.J.; Baillon, M.L.A.; Gibson, G.R.; Rastall, R.A. (2003), Synthesis of a-galactooligosaccharides with α-galactosidase from Lactobacillus reuteri of canine origin. *Applied Microbiol and Biotechnol* 63 (3), 1432-0614.

Vasiljevic, T.; Jelen, P. Innovative Food Science and Emerging Technologies Production of β-galactosidase for lactose hydrolysis in milk and dairy products using thermophilic lactic acid bacteria.

Viana SF, Guimarães VM, Jose IC, Oliveira MGA, Costa NMB, de Barros EG, Moreira MA, de Rezende ST (2005) Hydrolysis of oligosaccharides in soybean flour by soybean α-galactosidase. *Food Chem* 93: 665–670.

Wiseman, A. Manual de biotecnología de los enzimas. Zaragoza, Ed. Acribia, 1991.

Yamaguishi C.T., Sanada, C.T.; Gouvêa, P.M.; Pandey, A.; Woiciechowski, A.L.; Parada, J.L.; Soccol, C.R. (2009), Biotechnological process for producing black bean slurry without stachyose. *Food Research International*, 42 (4), 425-429.

Zadow J.G. (1993), Economic considerations related to the production of lactose and lactose by-products. *IDF Bulletin*, 289: 10.

INDEX

B

C

D

E

I

J

K

L

M

T

U

V

W

X

Y

Z